THE SKILLED HELPER

EIGHTH EDITION

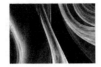

THE SKILLED HELPER

A PROBLEM-MANAGEMENT
AND OPPORTUNITY-DEVELOPMENT
APPROACH TO HELPING

Gerard Egan

Professor Emeritus
Loyola University of Chicago

THOMSON
™
BROOKS/COLE

Australia • Brazil • Canada • Mexico • Singapore • Spain
United Kingdom • United States

THOMSON
―――★―――™
BROOKS/COLE

Senior Acquisitions Editor:
  Marquita Flemming
Assistant Editor: *Alma Dea Michelena*
Technology Project Manager:
  Inna Fedoseyeva
Marketing Manager:
  Meghan McCullough
Senior Marketing Communications
  Manager: *Tami Strang*
Project Manager, Editorial Production:
  Christine Sosa
Creative Director: *Rob Hugel*
Art Director: *Vernon Boes*

Print Buyer: *Rebecca Cross*
Permissions Editor: *Bob Kauser*
Production Service: *Anne Draus,*
  Scratchgravel Publishing Services
Copy Editor: *Mary Anne Shahidi*
Illustrator: *Scratchgravel Publishing*
  Services
Cover Designer: *Irene Morris*
Cover Image: © *Brand X Pictures*
Cover Printer: *Phoenix Color Corp*
Compositor: *International Typesetting*
  and Composition
Printer: *R.R. Donnelley/Crawfordsville*

Printed in the United States of America

1 2 3 4 5 6 7 10 09 08 07 06

Library of Congress Control Number:
2006904374

Student Edition: ISBN 0-495-09203-7

Thomson Higher Education
10 Davis Drive
Belmont, CA 94002-3098
USA

For more information about our products,
contact us at:
Thomson Learning Academic Resource
Center
1-800-423-0563

For permission to use material from this
text or product, submit a request online at
http://www.thomsonrights.com.

Any additional questions about permissions can be submitted by e-mail to
thomsonrights@thomson.com.

CHAPTER 3

THE HELPING RELATIONSHIP: VALUES IN ACTION   47

Part Two

THE THERAPEUTIC DIALOGUE   69

CHAPTER 4

COMMUNICATION: THE SKILLS OF TUNING IN AND ACTIVELY LISTENING TO CLIENTS   71

CHAPTER 5

COMMUNICATING EMPATHY: WORKING HARD AT UNDERSTANDING CLIENTS   99

Chapter 6

The Art of Probing and Summarizing   121

Chapter 7

Helping Clients Challenge Themselves   138

Chapter 8

Challenging Skills and the Wisdom to Use Them Well   159

Chapter 9

Helping Difficult Clients Move Forward: Reluctance, Resistance, and Resilience   183

Gerard Egan, Ph.D., Professor Emeritus of Organization Development and Psychology at Loyola University of Chicago, has written over 15 books. His books in the field of counseling and communication include *The Skilled Helper, Interpersonal Living, People in Systems* (with Michael Cowan), *TalkWorks: How to Get More Out of Life Through Better Conversations* (with Andrew Bailey), *TalkWorks at Work: How to Become a Better Communicator and Make the Most of Your Career* (with Andrew Bailey), and *Essentials of Skilled Helping. The Skilled Helper*, translated into both European and Asian languages, is currently the most widely used counseling text in the world. He has written *Exercises in Helping Skills*, a manual to accompany *The Skilled Helper*.

His other books, dealing with business and management, include *Change Agent Skills in Helping and Human Service Settings; Change Agent Skills A: Designing and Assessing Excellence; Change Agent Skills B: Managing Innovation and Change; Adding Value: A Systematic Guide to Business-Based Management and Leadership*, and *Working the Shadow Side: A Guide to Positive Behind-the-Scenes Management*. Through these writings, complemented by extensive consulting, he has created a comprehensive and integrated business-based management system focusing on strategy, operations, structure, human resource management, the managerial role itself, and leadership. The management system also includes a framework for initiating and managing change and a model of effective interpersonal communication. *Working the Shadow Side* provides a framework for managing such shadow-side complexities as organizational messiness, personalities, social systems, politics, and culture. It also focuses on the messiness and politics of change management. He has consulted, lectured, and given workshops on these topics in Africa, Asia, Australia, Europe, and North and South America. He consults to a variety of companies and institutions worldwide.

He works with chief executives and senior managers and their teams as consultant, coach, and counselor, often on a long-term basis. He consults in strategy, business and organization effectiveness, management and management development, performance management, leadership, the design and management of change, and challenging and redesigning corporate culture. He has been involved in Executive Leadership Programs with BP, Merck, Medco Health Solutions, and Bank of America. He sees interpersonal communication as one of the major enabling skills of life and works with BT in promoting more effective interpersonal communication throughout the world.

The eighth edition of *The Skilled Helper* is designed primarily for those undergraduates or graduates who are preparing themselves for a career in professional helping. Because people should enter the profession with their eyes wide open, both the problems and potential of the helping professions are discussed. It is a splintered profession with widely different views on science itself, the philosophical underpinnings of science, what useful research looks like, the nature of scientific evidence, what professionals should look like, what professional education and training should be, and what makes helping work. Pride, prejudice, and posturing are not uncommon. Professional discussions are often rough and tumble and sometimes downright mean. Like other professions it has its fads, its fakes, and its felons.

But it is also a profession with a lot of heart. It has a lot of intelligent helpers and researchers—street-smart, wise, and principled helpers. It is filled with people who care deeply about other people and want to become better and better at helping them. It is filled with people who care about the world and would like to do something about the injustices that abound in it.

This is a book about three things—the values needed to establish authentic helping relationships with others, the communication skills needed to engage others in a collaborative helping dialogue, and a universal problem-management and opportunity-development framework that outlines the fundamental stages, tasks, and processes of helping. The model is client-centered in the sense that the starting point is the troubled person's humanity and concerns. The model is not about schools of psychology, interesting theories, or the latest fads. It is about people with problem situations and unused opportunities. At the heart of the model is listening and understanding, but the model also recognizes that people often need to be invited to challenge themselves in a host of ways if they are to improve their lot in life. Caring, tough love is also client-centered.

The helping model here is cognitive, affective, and behavioral—that is, it deals with the way people think, the way they come to feel and express emotions, and the way they act. Although past and present thinking, emotional expression, and acting are often part of the problem situation or unused opportunities, different ways of thinking, managing emotions, and acting are pathways to a better future. Helpers who use the model best recognize that in many ways people create their own reality, but helpers also understand that reality itself has a way of biting back.

Although the model recognizes the pervasiveness and importance of feelings and emotions, it also stands by the importance of emotional intelligence. Competent helpers espouse individual freedom, but they do not confuse liberty with license. They also appreciate the wisdom of the French saying, "Nos actes nous suivent," our actions have a way of pursuing us. Clients often act their way into problem situations; counselors help them act their way out of them. Although wise helpers appreciate individuality, they also recognize everyone's need for some form of community. The model as spelled out in these pages recognizes and delights in human diversity, but it remains rooted in the commonalities of our humanity.

Although the model does not discount the importance of the past, it is not deterministic and refuses to subscribe to the tyranny of the past. Instead of focusing on problems and speculations about their causes, it unapologetically focuses more on possibilities for a better future. That said, the past can be both a source of learning for a better future and a warehouse of possible solutions. "When your marriage was at its best, what was it like?"

The model appreciates the power of both goal setting and planning even when the troubled person—or the world at large—does not. Talking about goal setting and planning may evoke a yawn in some quarters, but the challenge of the helper is to make these processes living, vibrant, useful realities in the eyes of those seeking or needing help. Helping is about decision making. Effective helpers understand both the bright and dark sides of decision making and become guides as troubled people muse about, make, glide toward, flirt with, or fall into decisions—or attempt to avoid them altogether.

The model is frank about the obstacles to change, even life-enhancing change. Discretionary change—that is, change that would be helpful but is not demanded or imposed—has a poor track record in human affairs. Consider New Year's resolutions even among the resolute. Helping is about real change—internal, external, or both. Therefore, the model encourages small actions—small, problem-managing changes—right from the beginning.

Effective helpers know that grappling with problems in living is hard work and don't hesitate to caringly invite those seeking solutions to buckle down and engage in that work—but not work for the sake of work. In the end, helping is about work that produces outcomes that favorably impact help seekers' lives.

Acknowledgments

I would like to thank the following reviewers for their helpful insights and suggestions: Daisy B. Ellington, Wayne State University; Shirley A. Hess, Shippensburg University; Mary Kaplan, University of South Florida; Jeanmarie Keim, University of Arizona South; Rich McGourty; Kanoa Meriwether, University of Hawaii, West Oahu; Suzanne Skinner, Ivy Tech State College Region 6; Shawn L. Spurgeon, The University of North Carolina at Greensboro; and Fernelle L. Warren, Troy State University.

THE SKILLED HELPER

LAYING THE GROUNDWORK

Although the centerpiece of this book is a problem-management and opportunity-development helping model and the methods and communication skills that make it work, we will need to lay some groundwork. This includes the nature and goals of helping and some challenges facing the helping professions (Chapter 1), an overview of the helping process itself illustrated through a case (Chapter 2), and the helping relationship and the values that should drive it (Chapter 3).

CHAPTER

1

INTRODUCTION TO HELPING

Formal and Informal Helpers: A Very Brief History

Throughout history there has been a deeply embedded conviction that, under the proper conditions, some people are capable of helping others come to grips with problems in living. This conviction, of course, plays itself out differently in different cultures, but it is still a cross-cultural phenomenon. Today this conviction is often institutionalized in a variety of formal helping professions. In Western cultures, counselors, psychiatrists, psychologists, social workers, and ministers of religion are counted among those whose formal role is to help people manage the distressing problems of life.

There is also a second set of professionals who, although they are not helpers in the formal sense, often deal with people in times of crisis and distress. Included here are organizational consultants, dentists, doctors, lawyers, nurses, probation officers, teachers, managers, supervisors, police officers, and practitioners in other service industries. Although these people are specialists in their own professions, there is still some expectation that they will help those they serve manage a variety of problem situations. For instance, teachers teach English, history, and science to students who are growing physically, intellectually, socially, and emotionally and struggling with developmental tasks and crises. Teachers are, therefore, in a position to help their students, in direct and indirect ways, explore, understand, and deal with the problems of growing up. Managers and supervisors help workers cope with problems related to work performance, career development, interpersonal relationships in the workplace, and a variety of personal problems that affect their ability to do their jobs. This book is addressed directly to the first set of professionals and indirectly to the second.

To these professional helpers can be added any and all who try to help relatives, friends, acquaintances, strangers (on buses and planes), and themselves come to grips with problems in living. In fact, only a small fraction of the help provided on any given day comes from helping professionals. Informal helpers—bartenders and hairdressers are often mentioned—abound in the social settings of life. Friends help one another through troubled times. Parents need to manage their own marital problems while helping their children grow and develop. Indeed, most people grappling with problems in living seek help, if they seek it at all, from informal sources (Swindle, Heller, Pescosolido, & Kikuzawa, 2000). In the end, of course, all of us must learn how to help ourselves cope with the problems and crises of life. This book is about the fundamental models, methods, processes, and skills of helping. It is designed to help you become a better helper no matter which category you fall into.

WHAT HELPING IS ABOUT

Let's call the person seeking or needing help with problems in living a *client* and consider two questions: Why do people seek help in the first place? And what is the principal goal of helping?

To determine what helping is about, it is useful to consider (1) why people seek—or are sent to get—help in the first place and (2) what the principal goals of the helping process are.

Clients With Problem Situations and Unused Opportunities

Many people become clients because, either in their own eyes or in the eyes of others, they are involved in problem situations that they are not handling well. Others seek help because they feel they are not living as fully as they might. Many come because of a mixture of both. Therefore, clients' problem situations and unused opportunities constitute the starting point of the helping process.

Problem situations. Clients come for help because they have crises, troubles, doubts, difficulties, frustrations, or concerns. Often called "problems," generically, they are not problems in a mathematical sense because these problems often cause emotional turmoil and have no clear-cut solutions. It is probably better to say that clients come not with problems but with *problem situations*—that is, with complex and messy *problems in living* that they are not handling well. These problem situations are often poorly defined. Or, if they are well defined, clients still don't know how to handle them. Or clients feel that they do not have the resources needed to cope with them adequately. If they have tried solutions, they have not worked.

Problem situations arise in our interactions with ourselves, with others, and with the social settings, organizations, and institutions of life. Clients—whether they are hounded by self-doubt, tortured by unreasonable fears, grappling with the stress that accompanies serious illness, addicted to alcohol or drugs, involved in failing marriages, fired from jobs because of personal behavior, office politics, or disruptions in the economy, confused in their efforts to adapt to a new culture, suffering from a catastrophic loss, jailed because of child abuse, wallowing in a midlife crisis, lonely and out of community with no family or friends, battered by their spouses, or victimized by racism—all face problem situations that move them to seek help or, in some cases, move others to refer them or even send them for help.

With some help, even people with devastating problem situations can often handle them more effectively. Consider the following example.

Martha S., 58, suffered three devastating losses within 6 months. One of her four sons, who lived in a different city, died suddenly of a stroke. He was only 32. Shortly after, she lost her job in a downsizing move stemming from the merger of her employer with another company. Finally, her husband, who had been ill for about 2 years, died of cancer. Though she was not destitute, her financial condition could not be called comfortable, at least not by middle-class North American standards. Two of her other three sons were married with families of their own. One lived in a distant suburb. The other in a different city. The unmarried son, a sales rep for an international company, traveled abroad extensively.

After her husband's death, she became agitated, confused, angry, and depressed. She also felt guilty. First, because she believed that she should have done "more" for her husband. Second, because she also felt strangely responsible for her son's early death. Finally, she was deathly afraid of becoming a burden to her children. At first, retreating into herself, she refused help from anyone. But eventually she responded to the gentle persistence of her church minister. She began attending a support group at the church. A psychologist who worked at a local university provided some direction for the group. Helped by her interactions within the group, she slowly began to accept help from her sons. She began to realize that she was not the only one who was experiencing a sense of loss. Rather she was part of a "grieving family" the members of which needed to help one another to cope with the turmoil they were experiencing. She began relating with some of the members of the group outside the group sessions. This filled the social void she experienced when her company laid her off. She had an occasional informal chat with the psychologist who provided services for the group. Through contacts within the group she got another job.

Note that help came from many quarters. Her newfound solidarity with her family, the church support group, the active concern of the minister, the informal chats with the psychologist, and upbeat interactions with her new friends helped Martha enormously. Furthermore, because she had always been a resourceful person, the help she received enabled her to tap into her own ingenuity.

It is important to note that none of this "solved" the losses she had experienced. Indeed, the goal of helping is not to "solve" problems but to help the troubled person *manage* them more effectively or even to transcend them by taking advantage of new possibilities in life. Problems have an upside. They are opportunities for learning. What clients learn in the give-and-take of helping sessions is to be applied first of all to managing the presenting problem. But such learnings often have wider applications. When a client learns how to sort out a difficult relationship, her learning can be applied both to sorting out other problems in relationships and to preventing relationship problems from arising in the first place.

Missed opportunities and unused potential. Some clients come for help not because they are dogged by problems like those listed above but because they are not as effective as they would like to be. And so clients' *missed opportunities* and *unused potential* constitute a second starting point for helping. Most clients have resources they are not using or opportunities they are not developing. People who feel locked in dead-end jobs or bland marriages, who are frustrated because they lack challenging goals, who feel guilty because they are failing to live up to their own values and ideals, who want to do something more constructive with their lives, or who are disappointed with their uneventful interpersonal lives—such clients come to helpers not to manage their problems better but to live more fully.

In this case, it is a question not of what is going wrong but of what could be better. It has often been suggested that most of us use only a small fraction of our potential. Most of us are capable of dealing much more creatively with ourselves, with our relationships with others, with our work life, and, generally, with the ways in which we involve ourselves with the social settings of our lives. Consider the following case.

After 10 years as a helper in several mental health centers, Carol was experiencing burnout. In the opening interview with a counselor, she berated herself for not being dedicated enough. Asked when she felt best about herself, she said that it was on those relatively infrequent occasions when she was asked to help provide help for other mental health centers that were experiencing problems, having growing pains, or reorganizing. The counselor helped her explore her potential as a consultant to

ce organizations and make a career adjustment. She enrolled in an organ-
lopment program at a local university. Carol stayed in the helping field,
w focus and a new set of skills.

In this case, the counselor helped the client manage her problems (burnout, guilt) by
helping her identify, explore, and develop an opportunity (a new career).

The Two Principal Goals of Helping

Helpers have two basic goals—one relating to clients' managing specific problems in
living more effectively or developing unused opportunities and the other relating to
their clients' general ability to manage problems and develop opportunities in everyday life.

> *Goal One: Help clients manage their problems in living more effectively
> and developing unused or underused resources and opportunities more
> fully.*

Helpers are successful to the degree to which their clients—through client–helper
interactions—see the need to manage specific problem situations and develop specific unused resources and opportunities more effectively and are in a better position
to do so. Notice that I stop short of saying that clients actually end up managing
problems and developing opportunities better. Although counselors help clients
achieve valued outcomes, they do not control those outcomes directly. In the end,
clients can choose to live more effectively or not.

Clients are successful to the degree that they commit themselves to the helping
process and capitalize on what they learn from the helping sessions to manage problem situations more effectively and develop opportunities more fully "out there" in
their day-to-day lives.

The importance of results. A corollary to Goal One is that *helping is about constructive change that leads to life-enhancing results, outcomes, accomplishments, and
impact.* Helping is an "-ing" word: It includes a series of activities in which helpers
and clients engage. These activities, however, have value only to the degree that
they lead to valued outcomes in clients' lives. Ultimately, statements such as "We
had a good session," whether spoken by the helper or by the client, must translate
into more effective living on the part of the client. If a helper and a client engage in
a series of counseling sessions productively, something of value will emerge that
makes the sessions worthwhile. Unreasonable fears will disappear or diminish to
manageable levels, self-confidence will replace self-doubt, addictions will be con-
quered, an operation will be faced with a degree of equanimity, a better job will be
found, a woman and man will breathe new life into their marriage, a battered wife
will find the courage to leave her husband, a man embittered by institutional racism
will regain his self-respect and take his rightful place in the community. The president of the American Counseling Association put it this way:

> If counseling is constructive, clients grow and open up. They make purpose-
> ful decisions. They stop behaviors that are destructive and nonproductive
> such as fighting, being passive, or just blabbering. They quit trying to exter-
> nally control others and begin taking charge of themselves. They assess and
> utilize their strengths. Outside of sessions, they practice new behaviors through
> role-plays and simulations. They confront injustices and abuse. They give

themselves permission to seek wellness. The outcome is substantial, for old habits become history and new skills, realities, and lifestyles emerge. (Gladding, 2005, p. 5)

Helping is about constructive change that makes a substantive difference in the life of the client. Today there is another reason for focusing on outcomes. In the United States many psychological services are offered in managed-care settings. As such, more and more third-party payments depend on meaningful treatment plans and the delivery of problem-managing outcomes (Meier & Letsch, 2000). But economics should not force helpers to do what they should be doing anyway in the service of their clients.

The kind of results discussed in this book can be perceived by clients and people who interact with clients such as family, friends, peers, and co-workers. Kazdin (2006), who works with families and children, emphasizes the importance of moving beyond change based on statistical significance and even clinically significant change to change that has palpable impact on clients' daily lives. Here Kazdin provides some examples of outcomes that make a difference in the everyday life of a child with "conduct disorder" and the lives of those who interact with him.

> For example, one can see that the child no longer beats up a parent, teacher, or siblings; has stopped running away from home; does homework for the first time; no longer steals from neighbors; no longer brandishes a knife with younger siblings or peers; interacts appropriately with an infant sibling (e.g., talk, gentle play) rather than physically abusing him or her; and becomes only mildly upset during a tantrum at home without any of the usual property destruction. (p. 47)

The need for "palpable" results is seen clearly in the case of a battered woman, Andrea N., outlined by Driscoll.

> The mistreatment had caused her to feel that she was worthless even as she developed a secret superiority to those who mistreated her. These attitudes contributed in turn to her continuing passivity and had to be challenged if she was to become assertive about her own rights. Through the helping interactions, she developed a sense of worth and self-confidence. This was the first outcome of the helping process. As she gained confidence, she became more assertive; she realized that she had the right to take stands, and she chose to challenge those who took advantage of her. She stopped merely resenting them and did something about it. The second outcome was a pattern of assertiveness, however tentative in the beginning, that took the place of a pattern of passivity. When her assertive stands were successful, her rights became established, her social relationships improved, and her confidence in herself increased, thus further altering the original self-defeating pattern. This was a third set of outcomes. As she saw herself becoming more and more an "agent" rather than a "patient" in her everyday life, she found it easier to put aside her resentment and the self-limiting satisfactions of the passive-victim role and to continue asserting herself. This constituted a fourth set of outcomes. The activities in which she engaged, either within the helping sessions or in her day-to-day life, were valuable because they led to these valued outcomes. (1984, p. 64)

Andrea needed much more than "good sessions" with a helper. She needed to work for outcomes that made a difference in her life.

Goal Two: Help clients become better at helping themselves in their every-day lives.

Clients often are poor problem solvers, or whatever problem-solving ability they have tends to disappear in times of crisis. What G. A. Miller, Galanter, and Pribram said many years ago is, unfortunately, probably just as true today.

> In ordinary affairs we usually muddle about, doing what is habitual and customary, being slightly puzzled when it sometimes fails to give the intended outcome, but not stopping to worry much about the failures because there are still too many other things still to do. Then circumstances conspire against us and we find ourselves caught failing where we must succeed—where we cannot withdraw from the field, or lower our self-imposed standards, or ask for help, or throw a tantrum. Then we may begin to suspect that we face a problem. . . . An ordinary person almost never approaches a problem systematically and exhaustively unless he or she has been specifically educated to do so. (1960, pp. 171, 174)

Most people in our society are not "educated to do so." And if many clients are poor at managing problems in living, they are equally poor in identifying and developing opportunities and unused resources. We have yet to find ways of making sure that our children develop what most consider to be essential "life skills" such as problem management, opportunity identification and development, sensible decision making, and the skills of interpersonal relating.

It is no wonder, then, that clients—often poor problem solvers to begin with—often struggle when crises arise. If the second goal of the helping process is to be achieved—that is, if clients are to go away better able to manage their problems in living more effectively and develop opportunities on their own—then helpers need to impart the working knowledge and skills clients need to move forward. As Nelson-Jones (2005) puts it, "In the final analysis the purpose of using counseling skills is to enable clients to become more skilled in their own right. . . . Counselors are only skilled to the extent that they can be successful in skilling clients" (p. 14).

Just as doctors want their patients to learn how to prevent illness through good nutrition and healthy activities, just as dentists want their patients to engage in effective prevention activities, so skilled helpers want to see their clients not just managing this particular problem situation more effectively but also becoming more capable of managing subsequent problems in living more effectively. That is, helping at its best provides clients with tools to become more effective self-helpers. Therefore, although this book is about a process helpers can use to help clients, more fundamentally, it is about a problem-management and opportunity-development process that clients can use to help themselves. This process can help clients become more effective learns about problem management and opportunity development, better decision makers, and more responsible "agents of change" in their own lives. Chapter 2 provides an overview of this process.

Developing a working model of maturity or full human functioning. Pinsof (1995) says that the "alternative" to problem-centered therapy is "value-centered therapy organized not around presenting problems but around a definition of health, normality, or ideal functioning" (p. 3). Because I see no compelling reason to see value-centered helping as an "alternative," the approach taken here is an integration of both. A client's problem situations and unused opportunities are often related, directly or indirectly, to his or her degree or level of maturity. In order to help clients live more fully, helpers can benefit by having, at least in the back of their mind, a picture or conceptual model of what mature or optimal human functioning looks like (Sheldon, 2004).

One possible model is rooted in what I have called "social intelligence" (Egan, 1975) and what others (R. Bar-On, 1997; L. Bar-On & Parker, 2000; Goleman, 1995, 1998) have called "emotional intelligence" (EI). Perhaps *social-emotional intelligence* is a better term. Some authors understandably restrict EI to "the processes involved in the recognition, use, understanding, and management of one's own and others' *emotional states* [emphasis added] to solve emotion-laden problems and to regulate behavior" (Salovey, Brackett, & Mayer, 2004, p. I; see also Mayer & Salovey, 1997; Salovey & Mayer, 1990). Others (Bar-On, 1997; Bar-On & Parker, 2000; Goleman, 1995, 1998) see it as a broader set of cognitive and behavioral skills individuals need to live life fully: "For Bar-On, a clinical psychologist, EI becomes highly relevant since it answers the question 'Why are some individuals more able to succeed in life than others?'" (Neubauer & Freudenthaler, 2005, pp. 40–41). Nobel Laureate James Heckman (2006), discussing the individual, social, and economic necessity of investing in disadvantaged children as early as possible, talks about the importance of both cognitive and "noncognitive" skills such as motivation, perseverance, and self-restraint. EI focuses on such noncognitive skills.

The history of EI as a construct, including myriad disagreements over its nature (Salovey, Brackett, & Mayer, 2004; Schulze & Roberts, 2005), makes interesting reading, but does not serve our purpose here. Goleman's formulation of EI, despite its limitations, has face validity and common sense going for it. It stimulates our thinking about maturity. Perhaps we should call it the "maturity package." At this point we are interested in a thought-provoking paradigm rather than a finished scientific construct. The approach to EI taken here (see Box 1-1) draws on Egan's (1975) view of social intelligence and both Goleman's (1998) and Bar-On's (1997; Bar-On & Parker, 2000) formulation of EI. This is not the last word on social-emotional intelligence or maturity. You need to develop a schema that makes sense to you and your clients. Such a schema provides you with an optimal-human-functioning context that helps you understand why clients got into trouble in the first place and what they need to do to move on.

CHALLENGES FOR THE HELPING PROFESSIONS

The helping professions currently face many challenges that, directly or indirectly, affect helpers and their clients. Some of these challenges are named and briefly discussed here because they affect the context in which helping takes place. I find it helpful to understand the political, economic, and social "soup" in which the helping professions float. Every once in a while in restaurants people find strange things in their soup. Recently someone found a piece of a finger in her food. As the drama unfolded, some unsavory bits of human behavior were discovered. Inevitably helpers can expect to find both the savory and the unsavory in the political, economic, and social gumbo in which the helping professions simmer. There is not enough space in the introduction to this book to deal extensively with any of the issues outlined

Box 1-1 Maturity: Social-Emotional Intelligence

I. Mature people are self-managers. They know themselves, are in control of themselves, and get things done.

Self-Awareness. Self-managers know themselves without becoming preoccupied with themselves.

- They know their strengths and their limitations.
- They know how they experience emotions, how they tend to express them, and what impact this expression has on themselves and others.
- They understand and accept themselves and have a realistic sense of self-worth.

Self-Control. Positive self-control (not negative self-restriction) characterizes self-managers.

- They can be trusted because it is clear to others that they have standards of honest, integrity, and decency.
- They keep disruptive emotions and impulses in check.
- They find constructive way of coping with stress. They take responsibility for their actions and the consequences of their actions.
- They are flexible because they are open to new information, ideas, and ways of doing things.

Action. Self-managers have a bias toward action. They are doers rather than bystanders.

- They have life goals and pursue them.
- They take responsibility for their actions.
- They are assertive without being aggressive in expressing their ideas and needs.

(continued)

below. Rather, they are meant to pique your interest. There is a growing—and increasingly noisy—literature about all the issues touched on in this section.

Philosophical challenges. Epistemology is the part of philosophy that deals with human knowing. How do we know that we know? The average person might ask, "What kind of question is that? It's evident that we know. Life as we experience it would be impossible without knowing." This stance has not kept philosophers from musing on the mysteries of human knowing throughout the centuries. Helpers might well say, "So what? Let's move on," to all of this except for the fact that philosophy spills over into psychology. Epistemology makes a difference in our approach to clients.

To oversimplify complex issues for the sake of illustration, one philosophical assumption is that what we and our clients know is an objective reality separate from ourselves. This is the positivist assumption. Another assumption is that we and our clients in some way "create" our realities through our social interactions. This is the

(continued)

- They are interested in excellence rather than mediocrity.
- They manage problems and identify and develop opportunities.
- They persist in the face of obstacles.

II. Mature people handle relationships well. They know how to move creatively beyond themselves.

Empathy. They are aware of other people's feelings, needs, and concerns.
- They listen actively and without bias.
- They seek to understand and to communicate their understanding.
- They often anticipate the needs and concerns of others.

Communication. They strive to communicate well with others.
- They get their points across as clearly as possible.
- They are willing to challenge and be challenged constructively.
- They are willing to negotiate and resolve disagreements.

Interpersonal Relationships. Mature people prize solid relationships.
- They are intimate in responsible ways.
- Though capable of independent thought and action, they prize mutuality.
- They are open to influencing and being influenced, but without exercising or submitting to coercion.

The Wider World. They work at understanding the world in which they live.
- They are not parochial but have a sense of what the world is really like.
- Endowed with a sense of social responsibility, they seek to be a constructive member of their social groups.

constructivist assumption. As Ponterotto (2005, p. 29) puts it: "In marked contrast to positivism's naive realism (a single objective external reality), constructivism adheres to a relativist position that assumes multiple, apprehendable, and equally valid realities." Of course, it is not clear why constructivism is less "naive" than positivism. As we shall see later, there is perhaps something useful for helpers in both of these positions. Hansen (2004) suggests that helpers don't have to choose between supposedly irreconcilable systems. He proposes a "meta-epistemic system" that integrates the elements of the competing systems and illustrates it through a counseling case.

The temptation is to dismiss discussions like this as irrelevant to psychology and especially irrelevant to counseling. However, philosophical differences ultimately lead to different conceptions of and approaches to helping. Following a debate on philosophical assumptions in psychology in the *American Psychologist* (Cacioppo, Semin, & Berntson, 2004; Haig, 2005; Lau, 2005; Ramey & Chrysikou, 2005), Cacioppo, Berntson, and Semin (2005) cap a discussion by stating how important it is, "especially for students who hold the future of psychology in their hands, to take time to consider the effects of the philosophical perspective they inevitably bring to their scientific inquiries and perhaps even to revisit this choice periodically to ensure that the theoretical return on their scientific investment is optimized" (p. 348). This brings us to science.

The science challenge. Although the theory of relativity and quantum mechanics have revealed physical laws and phenomena that are both counterintuitive and difficult to understand, we tend to live our day-to-day lives in a more orderly Newtonian world. We read about this macro and micro messiness—curved space-time and particles that "recognize" and influence one another at a distance—but we do not experience it in everyday life, certainly not in the way we experience, say, gravity. On the other hand, psychology and the other social sciences deal with the behavior of human beings, which is often unpredictable and contrarian. For helpers, human messiness is not out of sight but center stage. If the subject matter of the social sciences—human behavior in all its messiness and unpredictability—differs radically from the subject matter of the physical or hard sciences, does it make sense to use the same research methodology or tools? Is the term "probability" as used in the social sciences the same as in the hard sciences?

Breckler (2004) sees science as both the process for accumulating knowledge—the *scientific method*—and the system for organizing it. But he admits that the process is ambiguous because it is not a single, well-defined process. The scientific method includes systematic observation, usually in the context of a model or theory, and the presentation, verification, and evaluation of, ideally, falsifiable hypotheses. But there are many different ways of doing each of these things, including quasi-experiments, natural experiments, mere observations, and qualitative investigations. "Aye, there's the rub," as Shakespeare's Hamlet would say. Purists in psychology opt for the traditional positivist experimental method—random assignment, independent variables, control groups, outcome measures, and the use of inferential statistics to test hypotheses. Yet, as part of a growing interest in qualitative research methods (Padgett, 2004), an entire issue of the APA *Journal of Counseling Psychology* (Haverkamp, Morrow, & Ponterotto, 2005) has been dedicated to qualitative and mixed methods (Tashakkori & Teddlie, 2003) in counseling psychology research.

Is strict positivism the way to go in the helping professions, or have the social sciences in their quest for respectability aped the approaches of the hard sciences instead of creating a different kind of science with a different set of research tools? The qualitative research movement in psychology (Camic, Rhodes, & Yardley, 2003; Denzin & Lincoln, 2000) can be seen as an effort to do just that. But there are dangers. Richard Feynman (1974), a Nobel Laureate in physics, thought that many things that were being passed off at the time as "social science" were more akin to certain superstitious practices of South Sea islanders—"cargo cult science" he called it. In his Caltech commencement address he outlined what he meant by research integrity and goes on to give examples of a lack of this kind of integrity in both "hard" and "soft" science research.

The evidence-based practice challenge. Throughout human history charlatans have taken advantage of people by huckstering products such as cure-all elixirs that have no real value. They do not work. Because helpers are not charlatans, the tools they use in helping clients—the helping relationship itself together with helping theories, models, methods, and skills—must work. And there must be evidence that they do work. In that sense all helping should be evidence-based. We will take a look at various evidence-based practices or treatments in Chapter 13, which deals with strategies for helping clients achieve their goals such as reducing the severity of post-traumatic stress disorder (PTSD) symptoms, dying well, healing a marriage, coping with prejudice, or launching a new career. But here a few words about evidence are in order.

It is one thing to say that helping should be evidence-based (Gibbs, 2003). It is quite another to answer such questions as: Whose evidence? What kind of evidence? Derived in what way? Verified in what way? With what kind of certainty? In relationship to what kind of probability? With what kind of applicability? Wampold (2003, p. 540) adds his own questions: "What is the purpose of collecting evidence? Who evaluates the evidence? What are the theoretical, historical, political, and cultural contexts in which the evidence is embedded? What decisions are made based on the evidence?" It is immediately evident that the positivists and constructivists mentioned above would not answer these questions in the same way. Philosophical assumptions—assumptions about the philosophy of science and the ambiguous nature of both science itself and scientific research—come back to haunt us.

The APA president described the evidence-based practice movement in U.S. society as a "juggernaut" for promoting accountability in medicine, psychology, education, and other professions. However, he has also suggested that the list of "empirically supported" treatments developed by APA Division 12 (Society of Clinical Psychology) was based on criteria that were too narrow: "Not taken into account were some of the broader strands of psychological research evidence (such as effectiveness research) and the other two pillars of what the Institute of Medicine (IOM) has defined as the foundation of evidence-based practice, namely clinical expertise and patient values" (Levant, 2005, p. 5).

The point here is that there is a philosophical, scientific, and professional debate going on about the very soul of helping. If you intend to be a professional helper, it is in your interest and the interest of your clients to become familiar with the debate, learn from it, and, if so moved, even dive into it. The first step is to become acquainted with the issues (Glicken, 2004; Silverman, 2005; Stout & Hayes, 2004; Westen, Novotny, & Thompson-Brenner, 2004). The literature is very extensive. It is also contentious. Every major book or article evokes a number of responses, so be prepared to sort through a great deal of disagreement. Norcross, Beutler, and Levant (2005) have edited a book in which experts explore nine fundamental questions in the debate about evidence-based practices. Each chapter ends with a dialogue—aptly named "convergence and contention"—among the contributors.

We do not live—at least not yet, if ever—in an evidence-based practice world. Think of governments and politics. Think of families. Think of the ways in which individuals construct and carry on their lives. Although medicine is making great strides in evidence-based practice, recent studies suggest that relatively few doctors are completely aboard the evidence-based express (Pfeffer & Sutton, 2006). Pfeffer and Sutton make it clear that things are even worse in the business world and call for an evidence-based movement in the ranks of managers: "Indeed, we would argue, managers are much more ignorant than doctors about which prescriptions are reliable—and they are less eager to find out. If doctors practiced medicine like many companies practice management, there would be more unnecessarily sick or dead patients and many more doctors in jail or suffering other penalties for malpractice" (p. 64). Evidence-based practice, then, is an ongoing professional struggle and journey— an essential journey, but one with lots of bumps and potholes.

The economics and politics challenge. Politics involves vying for scarce resources and promoting the ideology and policies that govern the distribution of these resources (Egan, 1994). Evidence-based practice is a trend that was started in the medical community (American Medical Association, 1992; Guyatt & Rennie, 2002) and has served that community well. It has quickly become important in an economic sense to the mental health professions because insurance companies have begun to offer preferential pay to organizations using evidence-based treatments. If your treatment is not on the evidence-based practice list, you may not be reimbursed. But some ask, What evidence is there that, say, a messy marriage, a problem in living, is in any way like a gall bladder gone bad? Is the medical model right for the helping professions? Duncan, Miller, and Sparks (2004) say emphatically no.

> The medical model, emphasizing diagnostic classification and evidence-based practice, has been transplanted wholesale into the field of human problems. Psychotherapy is almost exclusively described, researched, taught, practiced, and regulated in terms of the medical model's assumptions and practices. . . . [T]he fixation with nomenclature, categories, and diagnostic groupings is largely a waste of time. . . . They choke and smother alternative, hopeful ways of encouraging change and are based more on political and economic factors than science. (pp. 21–22)

So the cascade goes on from unchallengeable philosophical assumptions to the ambiguous nature of science to the competing definitions of evidence to politically and economically nuanced policy governing medical versus psychological approaches to practice—all finding their way into the relationship between client and helper. This book does not pretend to address—much less resolve—these debates, but they lurk in the background. If you intend to become a professional, you can't avoid them. The hope is that the commonsense problem-management process presented here will transcend a great deal of this contentiousness at the service of the client. For a peek at one view of politics in psychology see Wright and Cummings's (2005) collection of essays, *Destructive Trends in Mental Health*.

The "Does helping help?" challenge. This question must sound strange in a book on helping. However, ever since Eysenck (1952) questioned the usefulness of psychotherapy, there has been an ongoing debate as to the efficacy of both helping in general and the many different approaches to helping. Over the years some critics have expressed grave doubts about the very legitimacy of the helping professions themselves, even claiming that helping is a fraudulent process, a manipulative and malicious enterprise (see, for example, Furedi, 2004; Masson, 1988; Milton, 2002; Sommers & Satel, 2005). Masson went so far as to claim that in the United States, helping is a multibillion-dollar business that does no more than profit from people's misery. He also maintained that devaluing people is part and parcel of all therapy and that the helper's values and needs are inevitably imposed on the client. Although such criticisms are extreme, they should not be dismissed out of hand. What they say may be true of some forms of helping and of some helpers.

What evidence is there for the effectiveness of helping in general? There is a long history of outcome research. Hill and Corbett (1993) and Whiston and Sexton (1993) provide reviews that cover 50-year periods. There is a great deal of evidence showing that different kinds of helping, including counseling (Lambert & Cattani-Thompson, 1996), does help many people in many different situations. Hundreds of meta-analytical studies have been done over the past 20 years, and although some are, admittedly, quite crude, they still add up, in the eyes of many, to convincing evidence of the overall effectiveness of helping. And all this research and the methods used to determine successful outcomes are regularly updated (Lambert, Jasper, & White, 2005; Lambert & Ogles, 2004; Roth & Fonagy, 2004).

One possible sign of the maturing of the helping professions is the fact that the Surgeon General of the United States issued the first-ever report on mental health (see Satcher, 2000, for an executive overview). Even though many would say that the findings stated in this report have been known for some time, four are relevant here. First, mental health is fundamental to physical or overall health. Second, mental disorders are real health conditions. Third, the efficacy of mental health treatments are well documented. And fourth, a range of treatments exists for most mental disorders. To come to these conclusions, contributors, under the guidance of the Office of the Surgeon General, reviewed thousands of research studies and first-person accounts from individuals who had experienced mental disorders. However, I used the word "possible" in the first sentence of this paragraph because some would see the Surgeon General's endorsement as a sign that the medical profession had won and that the medical model would prevail.

What do clients think? Some studies ask clients whether they have been helped by counseling and psychotherapy and to what degree (Pekarik & Guidry, 1999). Client satisfaction studies have been widely used over the years and are highly regarded in many different mental health treatment centers (Bilbrey & Bilbrey, 1995; Lambert, Salzman, & Bickman, 1998). *Consumer Reports* (1994, 1995) published the results of a sophisticated large-scale survey project on client satisfaction with helping. The findings indicated the following:

- Clients believed that they had benefitted very substantially from psychotherapy.
- Psychotherapy alone did not differ in effectiveness from psychotherapy plus medication.
- No specific form of helping did better than any other for any particular kind of problem.
- Psychiatrists, psychologists, and social workers did not differ in their effectiveness as helpers.
- Long-term treatment produced appreciably better results than did short-term treatment.
- Clients whose choice of helper or length of therapy was limited by insurance or managed-care systems did not benefit as much as clients without those restrictions.

This study deals with the responses of real clients to questions about themselves, their helpers, processes used, and benefit received (see Seligman, 1995, for a discussion and critique of this study). Case closed? Hardly.

The *Consumer Reports* study has received a great deal of criticism on both theoretical and methodological grounds (Brock, Green, & Reich, 1998; Jacobson & Christensen, 1996). It has also been demonstrated that client satisfaction does not always mean that problems are being managed and opportunities developed (Pekarik & Guidry; 1999; Pekarik & Wolff, 1996): "Although often considered a traditional outcome measure, there are only low-to-moderate correlations between satisfaction and other measure of outcome. When traditional adjustment measures are clearly distinguished from satisfaction items, these correlations are especially low" (Pekarik & Guidry, 1999, p. 474).

Clients might be satisfied for a variety of reasons: for instance, because they like their counselors or because they feel less stressed. This does not mean that, in terms used in this book, they are managing problem situations or developing opportunities any better. The ideal, of course, is that clients feel satisfied because they have changed their lives for the better (Ankuta & Abeles, 1993).

And so doubts still persist. For instance, many outcome studies on helping are done in the lab or under lablike conditions. But lab results cannot be automatically compared to clinical-setting results. Real helping does not take place in a lab (see Henggeler, Schoenwald, & Pickrel, 1995; Nielsen et al., 2004; Weisz, Donenberg, Han, & Weiss, 1995). Furthermore, meta-analysis, the major tool used in efficacy studies, has itself come under criticism (Matt & Navarro, 1997).

Are there good helpers and bad helpers? Yes. There is indeed some evidence that therapy sometimes not only does not help but also actually makes things worse. That is, some helping leads to negative outcomes (see Mohr, 1995, and Strupp, Hadley, & Gomes-Schwartz, 1977, for reviews of the negative-outcome literature). Research shows that some of the factors associated with negative outcomes in helping are associated with clients, others with helpers. Sometimes clients who have severe interpersonal problems and severe symptomatology, who are poorly motivated, or who expect helping to be painless become more dysfunctional through therapy. Helpers who underestimate the severity of clients' problems, experience interpersonal difficulties with clients, use poor techniques, overuse any given technique, or disagree with clients over helping methodology can make things worse rather than better.

Finally, some helpers are incompetent. And even the competent and committed have their lapses. As noted by Luborsky and his associates (1986):

- There are considerable differences between therapists in their average success rates.

- There is considerable variability in outcome within the caseload of individual therapists.

- Variations in success rates typically have more to do with the therapist than with the type of treatment.

Although helping can and often does work, there is plenty of evidence that ineffective helping also abounds. Helping is a potentially powerful process that is all too easy to mismanage. It is no secret that because of inept helpers some clients get worse from treatment. Helping is not neutral; it is "for better or for worse." Albert Ellis (1984) claimed that inept helpers are either ineffective or inefficient. Even though the inefficient may ultimately help their clients, they use "methods that are often distinctly inept and that consequently lead these clients to achieve weak and unlasting results, frequently at the expense of enormous amounts of wasted time and money" (p. 24). Because studies on the efficacy of counseling and psychotherapy do not usually make a distinction between high-level and low-level helpers, and because the research on deterioration effects in therapy suggests that there is a large number of low-level or inadequate helpers, the negative results found in many studies are predictable.

What are we to conclude? Common sense tells us that some forms of helping actually help in the hands of good helpers. Personally, I have no doubt that in the hands of skilled and socially intelligent helpers, helping can do a great deal of good. Boisvert and Faust (2003), pooling the findings of leading international researchers who studied the effectiveness of helping, said that they were in agreement on the following:

- Therapy is helpful to the majority of clients.
- Most people achieve some change relatively quickly in therapy.
- People change more due to "common factors" than to "specific factors" associated with therapies.
- In general, therapies achieve similar outcomes.
- The relationship between the therapist and the client is the best predictor of treatment outcome.
- Most therapists learn more about effective therapy techniques from their experience than from the research.
- Approximately 10% of clients get worse as a result of therapy. (p. 511)

Norman Kagan (1973) long ago suggested that the basic issue confronting the helping professions is not validity—that is, whether helping helps or not—but reliability: "Not, can counseling and psychotherapy work, but does it work consistently? Not, can we educate people who are able to help others, but can we develop methods which will increase the likelihood that most of our graduates will become effective mental health workers as only a rare few do?" (p. 44). The question, then, is not "Does helping work?" but rather "How and under what conditions does it work?" The answer to the first question is easier than the answer to the second (see Bergin & Garfield, 1994).

Of course the typical client is not aware of the issues that are being discussed here. Often clients' problems are so pressing that they just want help. Therefore, "Let the practitioner beware." Become competent. Don't overpromise. Remain professionally self-critical. Keep your eye on results—that is, problem-managing and opportunity-developing outcomes for clients. Strive continually to become a better helper. Because helping in some generic sense "works," do not assume that it will always work with everyone. Finally, do not confuse difficult cases with impossible cases.

The common factors challenge. If, as the research findings above indicate, all therapies work equally well, then, as Alice in Wonderland said, "All should get prizes." However, if it is not the *specific* kind of therapy that produces success, then what does? One answer is *nonspecifics* or "common factors"—that is, factors that are common to all successful treatments. These factors have been teased out by a great deal of research, but, unsurprisingly, not everyone comes up with the same list and those that agree on a list differ in terms of how much each factor contributes to successful outcomes. Because that debate will not be resolved here, let's take a look at one early set. Lambert (1992) on empirical grounds suggested four sets of factors and estimated their contributions to therapeutic change: 40% due to client and extratherapeutic variables, 30% due to relationship factors, 15% due to expectancy and hope factors, and 15% due to techniques and therapeutic models. Empirically well-grounded or not, these results proved unpopular. These four sets are explored more fully in Hubble, Duncan, and Miller's (1999) very popular book on what works in therapy.

The paraprofessional helper/graduate education challenge. The fact that parapro-fessional and informal helpers do not go through a rigorous academic curriculum does not mean that they cannot be effective helpers. Indeed, studies have demon-strated that paraprofessional helpers can be as effective as professional helpers, and in some cases even more helpful (Durlak, 1979; Hattie, Sharpley, & Rogers, 1984; Tan, 1997). Studies worldwide have shown that this is not an unusual phenomenon. For instance, "minimally trained counselors" in Pakistan have helped community members significantly reduce levels of anxiety and depression (Ali, Rahbar, Naeem, & Gul, 2003). At the other end of the formal education and training scale, there is evidence (Atkins & Christensen, 2001; Beutler, Machado, & Neufeldt, 1994; Christensen & Jacobson, 1994; O'Donovan, Bain, & Dyck, 2005; Stein & Lambert, 1995) that questions the need for graduate training in psychology for those wanting to pursue careers as helpers. For discussion and debate on what graduate education in clinical psychology should look like in the 21st century see the special issue of the *Journal of Clinical Psychology* (Snyder & Elliott, 2005) on "visions of clinical psy-chology education."

Despite all the noise about the kind of training needed to become an effective helper, don't expect graduate programs to shut down soon. But the kind of studies mentioned above should challenge the profession to rethink its education and train-ing programs. The American Psychological Association ideal is that the Ph.D. clin-ical psychologist become a "scientist-practitioner." But there is evidence that this happens as the exception rather than the rule. Most become practitioners, hopefully good ones. For many students, however, the way research is taught does not inspire them to become scientists. And, as we have seen above, the science and research picture is not always a pretty one. I think there is another possibility. We need many more of what I call "translator-practitioners." Professionals in this role would fully understand the philosophical, scientific, research, evidence, and practice issues out-lined above and they would also be excellent practitioners. Their job would be to comb through the research, critique it, choose the best, and then turn it into the kinds of integrated frameworks, models, methods, and skills helpers need to serve their clients well. They would present their findings in language that both profes-sionals and paraprofessionals would understand (Bangert & Baumberger, 2005). The starting point in their efforts should be the client, not the profession, for without clients there would be no profession. They would pursue innovations that help clients and avoid fads that turn the profession into a business, and a lousy business at that.

The positive psychology challenge. Helping clients identify and develop unused potential and opportunities can be called a "positive psychology" goal. As guest edi-tors of the special Millennium Issue of the *American Psychologist,* Seligman and Csikszentmihalyi (2000) called for a better balance of perspectives in the helping professions. In their minds, too much attention is focused on pathology and too little on what they call "positive psychology": "Our message is to remind our field that psychology is not just the study of pathology, weakness, and damage; it is also the study of strength and virtue. Treatment is not just fixing what is broken; it is nurturing what is best" (p. 7). They and their fellow authors discuss such upbeat topics as:

- subjective well-being, happiness, hope, optimism (see Chang, 2001)
- a sense of vocation, developing a work ethic
- interpersonal skills, the capacity for love, forgiveness, civility, nurturance, altruism
- an appreciation of beauty and art
- responsibility, self-determination, courage, perseverance, moderation
- future mindedness, originality, creativity, talent
- a civic sense
- spirituality, wisdom

Obviously such an agenda supports a value-focused approach to helping. Seeing problem management as life-enhancing learning and treating all encounters with clients as opportunity-development sessions are part of the positive psychology approach. Seligman and his colleagues did not invent the content of positive psychology. Researchers like Albert Bandura, in his self-efficacy studies, have been exploring positive psychology issues for years without using the name. Seligman (1998), in setting up the Positive Psychology Network, had this to say:

> The network would support meetings, infrastructure, and the training of post-doctoral fellows who would rotate among investigators. The network would also seed collaborative research on foundational, meritorious projects. Each member of the network would be expected to generate the following "product" by the end of four years: to find a collaborator from within the network and together generate a major article, book, or externally funded research program within the field of Positive Psychology or Positive Social Science. The success of the network can be evaluated by the quality and visibility of these articles, books, and grant requests as well as by the spread of the field in the education and research focus of other scientists. So we will evaluate the success of the network by explicitly quantifying increased conventional funding, major conspicuous publications, new and tenured faculty, citation rate, and graduate and undergraduate course offerings, and the like in the field of Positive Psychology over the course of the four years of the network.

The Seligman and Csikszentmihalyi challenge has stimulated a great deal of theory, research, and debate. In 2000 the Templeton Foundation instituted The Templeton Positive Psychology Prize for excellent research in various areas of positive psychology (Azar, 2000). The initial awards were conferred for research on the beneficial effects of positive emotions, optimism and forward-thinking behavior, the ways positive emotions help people form and maintain relationships, and intellectual precocity and ways to help people realize their exceptional potential. For the past few years there has been an International Positive Psychology Summit. The Positive Psychology Network sponsors a positive psychology summer institute at the University of Pennsylvania. A substantial number of books (Carr, 2004; Seligman,

2004; Snyder & Lopez, 2005, 2006) and articles, including an issue of the *Review of General Psychology* (Simonton & Baumeister, 2005), have begun to appear. A note of caution. Positive psychology is not an "everything's going to be all right" approach to life and helping. Richard Lazarus (2000) puts it well:

> However, it might be worthwhile to note that the danger posed by accentu-ating the positive is that if a conditional and properly nuanced position is not adopted, positive psychology could remain at a Pollyanna level. Positive psychology could come to be characterized by simplistic, inspirational, and quasi-religious thinking and the message reduced to "positive affect is good and negative affect is bad." I hope that this ambitious and tantalizing effort truly advances what is known about human adaptation, as it should, and that it will not be just another fad that quickly comes and goes. (p. 670)

That said, Seligman and his colleagues have no intention of letting this happen. They do, support, and encourage the same kind of evidence-based research men-tioned above (Seligman, Steen, Park, & Peterson, 2005).

The term "positive psychology" will be used any number of times in this book. You, the reader, should take a critical look at how it is being used. The term should not be trivialized. Perhaps one of the main messages from the consideration of matu-rity, social-emotional intelligence, and all the different manifestations of positive psychology is that sometimes it is better to help clients *transcend* problems than to work through them. Helpers should have the tools to do just that.

The overextension challenge: Is helping for everyone? Just because, in the main, helping works, that does not mean that it is for everyone. Helpers see only a fraction of people with problems. Some don't come because they don't know how to get access to helpers. Others feel that they are not ready for change. Some don't acknowledge that they have problems. Some have "aversive expectations" of mental health services (Kushner & Sher, 1989) and stay clear. Most people muddle through without professional help. They pick up help wherever they find it. I often show a popular series of videotaped counseling sessions in which the client receives help from a number of helpers, each with a different approach to helping. Once the course participants have seen the tapes, I ask, "Does this client need counseling?" They all practically yell, "Yes!" Then I ask, "Well, if this client needs counseling, how many people in the world need counseling?" The question proves to be very sobering because the client in question is struggling with issues we all struggle with. She was one of what some would call "the worried well." Just because a person might well benefit from counseling, that does not mean that he or she "needs" counseling. Furthermore, many clients can "get better" in a variety of ways without help.

I once saw Melina Mercouri—singer, actress, freedom fighter, and Greek Minister of Culture during the '80s—on a TV talk show. Somehow or other the dis-cussion got around to psychotherapy. The host asked her, "Is psychotherapy big in Greece?" She threw her head back and laughed her infectious laugh and said, "Oh no, we're much too poor for that. We have to make do with our friends." This senti-ment is echoed in the Gold Star Wives of America Internet chat room (M. M. Phillips, 2005), where Iraq war widows "chat about everything from the grisly details of their husbands' deaths to the fond memories of their lives. They discuss spats with the in-laws and tips for raising children alone" (p. A6).

It is a waste of time and money to work with clients who, for whatever reason, don't want to grapple with their problem situations and develop their unused resources. Of course, it doesn't hurt to try, but the helper should know when to quit. In addition, clients with certain kinds of problems seem to be beyond help, at least at this stage of the development of the helping professions. Knowing when to help is important. So is knowing how much to help. Helping is an expensive proposition, both monetarily and psychologically. Even when it is "free," someone is paying for it through tax dollars, insurance premiums, or free-will offerings.

THE PROFESSIONAL CURRICULUM

"Beware the person of one book," we are told. For some people the central message of a book becomes a cause. Although causes empower people, including helping professionals, they can also cut them off from the community of practice the helping professions should be. They become closed to new ideas. Certainly all truth about helping cannot be found in one book. *The Skilled Helper* is not and cannot be "all that you've ever wanted to know about helping." It is part of the helping curriculum. Beyond a helping model and the skills that make it work, a practical curriculum is one that enables helpers to understand and work with their clients as effectively as possible in the service of problem management and opportunity development. The curriculum includes both working knowledge and skills.

Working knowledge is the translation of theory and research into the kind of *applied* understandings that enable helpers to work with clients. Although theory is important, it would be unfortunate if a helper were to grasp a theory but lose the person. *Skill* refers to the actual ability to deliver services. A fuller curriculum for training professional helpers includes, besides a working model of helping, many of the areas of working knowledge listed below. However, we face clients in their bio-psychological-social completeness. Because they are not divided up into the segments outlined below, helpers need to maintain an *integrative* perspective. Here are some possibilities for the "translator-practitioner."

Applied development psychology. It is likely that you will work with clients spread across the life span, from teens to the elderly. A working knowledge of the critical tasks, challenges, and crises at each stage of development is an essential part of professional competence (Newman & Newman, 2006; Sigelman & Rider, 2006). This includes a basic understanding of the impact of environmental factors such as culture and socioeconomic status on development (Zastrow, 2004). Dana Comstock (2005) has edited a counseling-specific developmental text that highlights diversity in the "critical contexts" that shape clients' lives and relationships.

The principles of cognitive therapy. Central to helping is developing a working knowledge of the principles of *cognitive psychology* as applied to helping (Beck, 1979; Beck, Freeman, & Davis, 2003), because the way people think and construct their worlds has a great deal to do with both getting into and getting out of trouble (Alloy & Riskind, 2005). As Beck noted, "In the broadest sense, cognitive therapy consists of all the approaches that alleviate psychological distress through the medium of correcting faulty conceptions and self-signals. The emphasis on thinking, however, should not obscure the importance of the emotional reactions which are generally the immediate source of distress" (1979, p. 214).

The framework of helping outlined in this book takes a cognitive-emotive-behavioral approach to counseling.

The principles of human behavior. If behavioral change is central to helping, then the ability to apply the principles of human behavior—what we know about such things as incentives, rewards, and punishment—to the helping process is a key competence. Wrestling with problem situations and undeveloped opportunities always involves incentives and rewards. Some writers look at behavior modification from the viewpoint of the professional helper (Miltenberger, 2001). Others focus on the principles of behavior from the viewpoint of the person who is trying to change his or her behavior, whether a client or someone interested in personal growth (Lee & Axelrod, 2005; L. K. Miller, 2006; Watson & Tharp, 2007).

Applied personality psychology. Applied personality psychology helps us understand in very practical ways what makes people "tick" and many of the ways in which individuals differ from one another. Brian Little (2001), noting the inordinate extensiveness of personality psychology, puts it this way: "Personality psychology seeks to integrate diverse influences on human conduct ranging from the genetic and neurophysiological underpinnings of traits to the historical contexts within which individual life stories can be rendered coherent. . . . The study of personality seeks to understand how individuals are like all other people, some other people, and no other person (to revise slightly the classic phrase of Kluckhohn and Murray, 1953, p. 53). It formulates theories about the nature of human nature, the role of individual differences, and the study of single cases" (p. 2). So, far from being a tool for stereotyping people, applied personality psychology helps counselors understand *this* person in *this* environmental context.

Abnormal psychology. Abnormal psychology can be described as a systematic understanding of the ways in which individuals get into cognitive, emotional, behavioral, and social trouble. Lately in the literature there has been less emphasis on different schools of thought on psychological disorders and more on the biological, psychological, cultural, social, familial, and even political factors that contribute to disordered human behavior (Barlow & Durant, 2005; Durand & Barlow, 2006). Integration is the order of the day. Psychological problems are not like a broken arm. The context is critical. Many object to traditional classifications of emotional disorders and their "diagnosis." The medical model has not proved useful for them or their clients.

That said, every problem in living has its own rich literature. So if you are counseling, say, substance-abuse clients, the literature has much to offer. Take smoking. It is not an innocent pastime. Many people who smoke don't like to think that it's a problem. "You have to die sometime anyway," a friend of mine would say. The research shows that nicotine dependence has terrible, even deadly, consequences. Dodgen (2005) starts by marshaling the evidence showing that nicotine is one of the more addictive, as well as destructive, substances of abuse. He then goes on to review and rate the effectiveness of available treatments and provides a step-by-step, 10-session assessment and treatment model that helpers can adapt to the need of their clients.

Applied social psychology. It is important for helpers to understand the ways in which clients act when they are in social settings. Some clients are "out of community" to their own detriment, whereas others are in community in dysfunctional ways. Social psychology deals with the behavior of groups and the influence of social factors on the individual and how individual behavior affects the group. Social psychology focuses on how individuals think about, influence, and relate to one another. Such topics as group pressure, conformity, compliance to authority, friendship, love, liking, stereotyping, prejudice, and aggression are the focus of research. These are issues that affect all of us and, of course, our clients. Breckler and Olson (2006) bring social psychology topics alive by involving the personal lives of their readers.

Diversity in all its forms. For some, diversity has come to mean multicultural diversity. I believe that this approach is both narrow and distorting. Some even see it as a new "force" in psychology. There has been such an explosion of writings on multicultural diversity in the helping professions in the last 10 years that I hesitate to list examples. Lena Hall (2005) has written a *dictionary* of multicultural psychology that is almost 200 pages long. Each entry has its own references for further study. Understanding diversity, including multicultural diversity, is extremely important for helpers. Because helpers deal with individual clients and not groups, getting a feeling for *this particular client* is essential. Stereotyping of all kinds is to be avoided, including the kind of stereotyping that comes from membership in a particular cultural or ethnic group.

Understanding specific populations. It is important for helpers to have an understanding of the needs and problems of *special populations*, such as the physically challenged, substance abusers, the homeless, the elderly, with which they work. For instance, for helpers who intend to work with the elderly, Aldwin and Gilmer (2004) provide comprehensive, multidisciplinary coverage of the physical aspects of aging, including age-related changes and disease-related processes, the demography of the aging population in the United States, theories of aging, and the promotion of optimal aging. Molinari and his associates (2003) and Knight (2004) discuss the kind of working knowledge and skills helpers need to work with older adults. Given the fact that more and more people are living longer and longer, it is anomalous that the elderly are so underrepresented in both the theory and research literature in counseling (Werth, Kopera-Frye, Blevins, & Bossick, 2003). There are increasing career opportunities for helpers drawn to working extensively with this population.

The helping professions. As you may have gathered by now, the helping professions are going through a period of turbulence. Because they focus on human behavior, perhaps turbulence is their normal state. So if you intend to enter one of the helping professions—for instance, clinical or counseling psychology, psychiatry, social work—you should know what you're getting into (Alle-Corliss & Alle-Corliss, 2006; Srebalus & Brown, 2001; Woodside & McClam, 2006). Dipping into the literature about the helping professions themselves can be quite eye-opening. It is also useful to know what job opportunities and career paths are opening up as the profession develops (Kuther, 2006). For instance, in the last 10 years there has been a great deal of growth in health care for helpers (R. T. Brown et al., 2002; Dobmeyer et al., 2003; Gray, Brody, & Johnson, 2005).

Personal and professional development. There is an excellent body of literature on the issues you will face personally in your journey to become a professional helper (Conyne & Bernak, 2005; Engels, 2004; Gladding, 2002; Hazler & Kottler, 2005;

Kaslow, 2004; Kottler, 2002; Orlinsky & Ronnestad, 2005; D. R. Peterson, 2003; M. Pope, 2006; Pope & Vasquez, 2005). There is wisdom in the subtitle of Hazler and Kottler's book, *The Emerging Professional Counselor: Student Dreams to Professional Realities*. The special issue of the *Journal of Clinical Psychology* (September, 2005) mentioned earlier begins with the personal accounts of the diverse roads a handful of notable therapists followed in becoming helpers (Farber, Manevich, Metzger, & Saypol, 2005; Norcross & Farber, 2005).

Along the way, helpers run into problems and often seek help themselves (Baker, 2003; Norcross, 2005b). Norcross notes that studies show that "personal therapy is an emotionally vital, interpersonally dense, and professionally formative experience that should be central to psychologists' development" (p. 840). You are headed into a dynamic profession that is constantly changing and you need to grow, develop, and change with it. Goldfried (2001) has edited a book dealing with how helpers evolve both professionally and personally. Helping is a tough, demanding profession. Burnout, a "syndrome of emotional exhaustion and cynicism" (Maslach & Jackson, 1981, p. 99), is not a rare occurrence, especially in these days of managed care (Rupert & Morgan, 2005). You have to learn how to take care of yourself (Skovholt, 2001). The American Counseling Association has even mounted a "campaign for counselor wellness" (Rollins, 2005).

Religion and spirituality. There is a growing body of literature in psychology dealing with religion and spirituality (Cashwell & Young, 2005; W. R. Miller, 1999; Richards & Bergin, 2005; see also the January 2003 *American Psychologist* for a series of articles on spirituality, religion, and health). Some researchers claim that religion and spirituality can be studied scientifically (W. R. Miller & Thoresen, 2003), whereas others contend that belief is belief and beyond the pale of science. The fact is, however, that many people in the world are religiously oriented. Often the problem situations they discuss have religious, spiritual, and moral implications. Therefore, books and articles like those cited above provide counselors with useful working knowledge.

Any given helper may have little interest in religion or spirituality, but if these are important for the client, then they are important (Hall, Dixon, & Mauzey, 2004). Sperry and Shafranske (2005) show how spirituality can be incorporated into many different approaches to psychotherapy. If you are not a Muslim and are soon to see a Muslim client for the first time, what do you do? In an engaging article entitled "Islam 101," Ali, Liu, and Humedian (2004) sketch out the religion and therapy implications and use a case to illustrate their points.

Helping and the law. When I was a student there were no courses and very few studies on helping and the law. We live in different times. There are at least two contact points between helping and the law. The first deals with the fact that many laws, at least theoretically, are in place to protect people from harm by others. So there are laws that protect clients and their rights in their interactions with their helpers (Sales, Miller, & Hall, 2005). The APA has been publishing a state-by-state series on the law and mental health professionals (Charlton, Fowler, & Ivandick, 2006). Ideally, counselors go beyond legal requirements in their efforts to be of service to their clients. However, because we live in such a litigious society, being ignorant of laws that protects clients can hurt you. Part VIII of *Psychologists' Desk Reference*

(Koocher, Norcross, & Hill, 2005) deals with the typical ethical and legal issues helpers can face. These articles constitute a good starting point. A second touching point refers to the contributions that the helping professions can make to legal decision making (English & Sales, 2005). Unfortunately, TV is filled with forensic psychology nonsense—you know, laughable "profilers" and "expert witnesses" that give you the creeps. But forensic psychology is a growing field (O'Donohue & Levensky, 2004; Walker & Shapiro, 2004; Weiner & Hess, 2005), and there are excellent forensic psychologists out there. A career choice for you?

In the end there is no such thing as the perfect professional curriculum to which all helpers would subscribe. Even if there were, it would be impossible for you to master all of it. Although much of this curriculum is touched on indirectly in the pages that follow, especially through the many examples offered, this book is not a substitute for such a curriculum.

MOVING FROM SMART TO WISE: MANAGING THE SHADOW SIDE OF HELPING

This book outlines a model of helping that is rational, linear, and systematic. What good is that, you well might ask, in a world that is often irrational, nonlinear, and chaotic? One answer is that rational models help clients bring much-needed discipline and order into their chaotic lives. Effective helpers do not apologize for using such models. But they also make sure that their humanity permeates their models.

More than intelligence is needed to apply the model well—smart is not smart enough. The helper who understands and uses the model together with the skills and techniques that make it work might well be smart, but he or she must also be wise. Effective helpers understand the limitations not only of helping models but also of helpers, the helping profession, clients, and the environments that affect the helping process. One dimension of wisdom is the ability to understand and manage these limitations, which in sum constitute what I call the *arational* dimensions, or the "shadow side," of life. The shadow side of helping can be defined as:

> *All those things that adversely affect the helping relationship, process, outcomes, and impact in substantive ways but that are not identified and explored by helper or client or even the profession itself.*

Helping models are flawed; helpers are sometimes selfish, lazy, and even predatory, and they are prone to burnout. Clients are sometimes selfish, lazy, and predatory, even in the helping relationship.

Indeed, if the world were completely rational, we would run out of clients. Not to worry, however, because many clients cause their own problems. People knowingly head down paths that lead to trouble. Life is not a straight road; often it is more like a maze. It often seems to be a contradictory process in which good and evil, the comic and the tragic, cowardice and heroism are inextricably intermingled. In the pages of this book the helping relationship and process are described and illustrated in very positive terms. They are described as they might be or even should be, not as they always are.

The Downside: The Messiness of Helping

We have already seen a cardinal example of the messiness of helping—the ongoing debate as to its efficacy and what makes for both effectiveness and efficiency. These are shadow-side issues in that they are hidden from clients and ignored by many practitioners. But there are plenty of others. Nevertheless, the ability to understand how helper, client, the relationship, and the helping process itself can go wrong is the first step toward managing the shadow side.

For instance, helpers' motives are not always as pure as they are portrayed in this book. Some helpers are not very committed even though they are in a profession in which success demands a high degree of commitment. Some helpers are competent; others are not. Incompetent helpers pass themselves off as professionals. Finally, although, like other people, helpers get into trouble, often they don't use the tools of their trade to get themselves out of trouble—"Don't do as I do; do as I say."

Clients often play games with themselves, their helpers, and the helping process. Helpers sometimes seduce their clients, and clients seduce their helpers, although not necessarily sexually. Hidden agendas are pursued by both helper and client. The helping relationship itself can end up as a conspiracy to do nothing. Clients, aided and abetted by their helpers, work on the wrong issues. Helpers fail to keep up with developments in the profession. They end up using helping methods that are not likely to benefit their clients. Helping is continued even though it is going nowhere. The list goes on.

Managing the shadow side is an exercise in integrity, social intelligence, and competence, not in cynicism. Clearly, not all these shadow-side things happen all the time. To assume that they do would be cynical. The shadow side is not usually an exercise in ill will. Few helpers set out to seduce their clients. Few clients set out to seduce their helpers. Helpers don't realize that they are incompetent. Clients don't realize that they are playing games. Most clients most of the time are well-intentioned, and most helpers do their best to put their own concerns aside to help their clients as best they can. But this does not mean that any given helper is always giving his or her best. Both clients and helpers have their temptations. Clients, as we shall see, have their blind spots that keep them mired in their problems, but helpers also have theirs.

All human endeavors have their shadow side. Companies and institutions are plagued with internal politics and are often guided by covert or vaguely understood beliefs, values, and norms that do not serve the best interests of the business, its employees, or its shareholders. If helpers don't know what's in the shadows, they are naive. If they believe that shadow-side realities win out more often than not, they are cynical. Helpers should be neither naive nor cynical. Rather, they should pursue a course of upbeat and compassionate realism.

Even the downside of the shadow side of helping can provide benefits. Consider the following analogy. The shadow side of helping is a kind of "noise" in the system. But scientists have discovered that sometimes a small amount of noise in a system, called "stochastic resonance," makes the system more sensitive and efficient (*Economist*, 1995). In the helping professions, noise in the guise of the debate around what makes helping both effective and efficient can ultimately benefit clients.

Despite the downside of the shadow side of helping, the tone of this book is unabashedly upbeat. However, helpers-to-be must not ignore the less palatable dimensions of the helping professions, including the less palatable dimensions of themselves. Therefore, throughout this book some of the common shadow-side realities that plague client, helper, and the profession itself are noted in the service of managing them. This book is by no means a treatise on the shadow side of helping. Rather its intent is to get helpers to begin to think about the shadow side of the profession. Wise helpers are idealistic without being naive. They also know the difference between realism and cynicism and opt for the former. They see the journey "from smart to wise" as a never-ending one. And they do not neglect the "smart" part of helping or of everyday living. They continually get better at, as John Ruscio subtitled his book *Critical Thinking in Psychology*, "separating sense from nonsense." This excellent book on critical thinking is a good starting point. There are signs that the helping professions are beginning to explore their shadow side. An example of that is the book *What Therapists Don't Talk About and Why* (Pope, Sonne, & Greene, 2006). The authors explore myths and taboos that stand in the way of effective helping.

The Upside: Common Sense and Wisdom in the Helping Professions

To my way of thinking, such things as wisdom and common sense are part of the "upside" shadows because they have not received a great deal of attention in the helping literature and they do not form part of the curriculum in helper-training programs. However, given what is happening in the positive psychology movement, perhaps this is beginning to change. Although helping is sometimes referred to as an art, the emphasis in the journals is on theory and research. In a way, the emphasis is on "smart" rather than "wise." Some writers and researchers, however, have focused on such things as wisdom, sagacity, street smarts, practical intelligence, and common sense in helping (for instance, Baltes & Staudinger, 2000; Hanna, 1994; Hanna & Ottens, 1995; Schmidt & Hunter, 1993; Sternberg, 1990; Sternberg, Wagner, Williams, & Horvath, 1995). This is part of the positive psychology approach to helping.

Take self-control, so essential for the kind of self-responsibility clients need to manage problems and develop opportunities. Although latter-day research has helped us understand more fully the dynamics of self-control, in truth, most of the techniques used to facilitate self-control have not been derived from formal theory and research. When it comes to self-control, the Bible, Koran, Talmud, and other books of wisdom contain much of what has been discovered through research studies (see Karoly, 1995, p. 273). Whereas some "smart" helpers might see the client as a "clinical entity subject to the templates and tools of the psychotherapeutic trade," wise helpers would see each client as a "vital, dynamic personage" (Hanna, 1994, p. 132) who needs to be helped to take advantage of the wisdom and common sense that is already within themselves, their societies, their cultures, their environments, and their helpers.

Helpers need to be wise, and part of their job is to impart wisdom, however indirectly, to their clients. Baltes and Staudinger (2000) define wisdom as "an expertise in the conduct and meaning of life" or "an expert knowledge system concerning the fundamental pragmatics of life" (pp. 124, 122). What is it that characterizes wisdom? Here are some possibilities (see Sternberg, 1990, 1998):

- self-knowledge; maturity
- knowledge of life's obligations and goals
- an understanding of cultural conditioning

- the guts to admit mistakes and the sense to learn from them
- a psychological and a human understanding of others; insight into human interactions
- the ability to "see through" situations; the ability to understand the meaning of events
- tolerance for ambiguity and the ability to work with it
- being comfortable with messy and ill-structured cases
- an understanding of the messiness of human beings
- openness to events that don't fit comfortably into logical or traditional categories
- the ability to frame a problem so that it is workable; the ability to reframe information
- avoidance of stereotypes
- holistic thinking; open-mindedness; open-endedness; contextual thinking
- "meta-thinking," or the ability to think about thinking and become aware about being aware
- the ability to see relationships among diverse factors; the ability to spot flaws in reasoning; intuition; the ability to synthesize
- the refusal to let experience become a liability through the creation of blind spots
- the ability to take the long view of problems
- the ability to blend seemingly antithetical helping roles—being one who cares and understands together with being one who challenges and "frustrates" (see Levin & Shepherd, 1974)
- an understanding of the spiritual dimensions of life

If social-emotional intelligence is about leading a good life, then wisdom is about excellence in living. As such, it focuses on knowing "how" (the procedural dimension) rather than merely knowing "what" (the factual dimension).

Overview of the Helping Model

RATIONAL PROBLEM SOLVING AND ITS LIMITATIONS

The problem-solving process is often described as a more or less straightforward natural and rational process of decision making. For instance, Yankelovich (1992) offered a seven-step process. Applied to helping, it looks something like this:

1. **Initial awareness.** Clients become aware of an issue or a set of issues. For instance, after a number of disputes over household finances, a couple develop a vague awareness dissatisfaction with the relationship itself.

2. **Urgency.** A sense of urgency develops, especially as the underlying problem situation—the dissatisfaction with the relationship itself—becomes more distressing. Even small annoyances are now seen in the light of overall dissatisfaction.

3. **Initial search for remedies.** Clients begin to look for remedies. However implicitly or perfunctorily, they explore different strategies for managing the problem situation. For instance, clients in difficult marriages begin thinking about complaining openly to their partners or friends, separating, getting a divorce, instituting subtle acts of revenge, having an affair, going to a marriage counselor, seeing a minister, unilaterally withdrawing from the relationship in one way or another, and so forth. The parties may try out one more of these remedies without evaluating their cost or consequences.

4. **Estimation of costs.** The costs of pursuing different remedies begin to become apparent. Someone in a troubled relationship might say to herself: "Being open and honest hasn't really worked. If I continue to put my cards on the table, I'll have to go through the agony of confrontation, denial, argument, counter accusations, and who knows what else." Or he might say, "Simply withdrawing from the relationship in small ways has been painful. What would I do if I were to go out on my own?" Or, "What would happen to the kids?" At this point clients often back away from dealing with the problem situation directly because there is no cost-free or painless way of dealing with it.

5. **Deliberation.** Because the problem situation does not go away, it is impossible to retreat completely. And so a more serious weighing of choices takes place. For instance, the costs of confronting the situation are weighed against the costs of merely withdrawing. Often, a kind of dialogue goes on in the client's mind

between steps 4 and 5. "I might have to go through the agony of a separation for the kids' sake. Maybe time apart is what we need."

6. **Rational decision.** An intellectual decision is made to accept some choice and pursue a certain course of action. "I'm going to bring all of this up to my spouse and suggest we see a marriage counselor." Or, "I'm going to get on with my life, find other things to do, and let the marriage go where it will."

7. **Rational-emotional decision.** However, a merely intellectual decision is often not enough to drive action. So the heart joins the head, as it were, in the decision. One spouse might finally say, "I've had enough of this! I'm leaving. It won't be comfortable, but it's better than living like this." The other might say, "It is unfair to both of us to go on like this; and it's certainly not good for the kids," and this drives the decision to seek help, even if it means going alone. Decisions driven by emotion and convictions are more likely to be translated into action.

Prochaska and Norcross (2002; Prochaska, Norcross, & DiClemente, 2005), seeing helping as facilitating client change, have developed a five-stage process of both client-mediated and helper-assisted change that many helpers find useful. The five stages are *precontemplation, contemplation, preparation, action,* and *maintenance.*

1. **Precontemplation.** In this stage the person is either unaware that he or she has a problem, or is only vaguely aware, or if aware, has no intention of doing anything about it ("it's your problem, not mine") even though others are aware of it and tell him about it ("with your heart the way it is, smoking's going to kill you, not later but sooner"). Resistance to recognizing or dealing with a problem is the hallmark of precontemplation.

2. **Contemplation.** In this stage the person knows that he or she has a problem, thinks seriously about doing something about it, but has yet to make a commitment to take action ("I know that smoking is doing me in and I've got to find a way, sooner or later, of giving it up"). The person is still ambivalent about change. The price of a desirable change still seems too high. Serious consideration of change is the central element of contemplation, but mere knowledge is usually not sufficient to motivate change.

3. **Preparation.** In this stage the person is on the verge of doing something about his or her problem situation or has already tried, unsuccessfully, to do something about it. However, failure is not a deterrent and the person still wants to move forward. Change is still possible and desirable enough to elicit some kind of action, however feeble or symbolic ("you know, I'm down to eight cigarettes a day"). In this stage the person may be doing such things as trying to avoid temptation, looking for understanding and support from family and friends, and planning or designing the change.

4. **Action.** In this stage individuals are involved in life-enhancing change and actually put time and effort into modifying their dysfunctional behavior. Change at this stage tends to be visible and is recognized by others. Modification of the undesirable behavior to an acceptable standard through systematic effort is the hallmark of action.

5. **Maintenance.** In this stage the person consolidates his or her gains and works to avoid relapse. This stage can go on for a long time, even a lifetime, for instance,

for the person who fights the urge to smoke every day. Of course, people are prone to relapses. There is a difference between a lapse and a relapse or a total collapse into the old way. The person can learn from lapses. A relapse or total collapse can means moving back to a previous stage or even starting the whole process over.

There are obvious similarities between these two processes. But there is nothing automatic about either of them. Four things should be noted. First, these stages and steps, however logically sequenced on paper, are often jumbled and intermingled in real-life problem-management situations. Second, these natural processes can be derailed at almost any point along the way. For instance, uncontrolled emotions spill out and destroy whatever progress has been made in healing a dysfunctional relationship, making a bad situation worse. Or the costs of managing the problem seem too high and so the process itself is put on the back burner. Third, decision making in difficult situations is seldom as rational as these processes might imply. As we shall see later, decision making, a process that has been called a journey as complex as life itself (B. Scott, 2000), has a deep shadow side. Fourth, the stages, steps, and processes described by both Yankelovich and Prochaska and his colleagues say little about the skills and methods that help people turn decisions into solution-focused action that produces problem-managing results.

The problem-management and opportunity-development process outlined in this chapter and developed in the rest of the book borrows from the natural processes outlined above, complements them with needed skills and techniques, suggests ways of helping clients turn decisions into action, focuses on solutions, that is, life-enhancing outcomes, provides ways of challenging backsliding, and, at its best, speeds up the entire process.

THE SKILLED HELPER MODEL: A PROBLEM MANAGEMENT/OPPORTUNITY DEVELOPMENT APPROACH TO HELPING

Because all approaches to helping must eventually help clients manage problem situations and develop unused opportunities, it seems logical to start with a flexible, humanistic, broadly based problem-management and opportunity-development model. Of those who write about problem-solving approaches, some use the term explicitly (Bedell & Lennox, 1997; Chang, D'Zurilla, & Sanna, 2004; D'Zurilla & Nezu, 1999, 2001; Elias & Tobias, 2002; Nezu, Nezu, Friedman, Faddis, & Houts, 1998), whereas others use some form of the process but not the name (Bertolino & O'Hanlon, 2002; Cormier & Nurius, 2003; Hill, 2004). General problem solving has a rich research history spanning some 100 years, but the research is scattered. But this does not belie its importance. Nobel Laureate Herbert Simon and his associates put it nicely:

> The work of managers, of scientists, of engineers, of lawyers . . . is largely work of making decisions and solving problems. It is work of choosing issues that require attention, setting goals, finding or designing suitable course of action, and evaluating and choosing among alternative actions. . . . Nothing is more important for the well being of society than that this work be performed effectively, that we address the many problems requiring attention at the national level . . . at the level of business organizations . . . and at the level of our individual lives. (1986, p. 1)

Problem solving applied "at the level of our individual lives," called, variously, *social problem solving* or *applied problem solving* or *personal problem solving* or *human problem solving,* has an evolving research base (see the research sections in Chang, D'Zurilla, & Sanna, 2004; D'Zurilla & Nezu, 1999; Heppner, Witty, & Dixon, 2004). Common sense suggests that problem-solving models, techniques, and skills are important for all of us, because all of us must grapple daily with problems of greater or lesser severity. Heppner's work over the past 20 years (for a review and summary, see Heppner, Witty, & Dixon, 2004) has focused on people's *appraisal* of their ability to cope with personal problems. Confidence in one's ability to solve problems, the guts to face up to them, and a sense of self-control (especially emotional self-control) and control of the problem situation lead in many ways to a richer life. O'Neil (2004), commenting on Heppner and his colleagues' review, says, "The core of their review is the authors' passion for helping people learn problem solving to improve their lives. Helping others solve problems is a passionate part of our mission as counseling psychologists" (p. 439).

Ask parents whether problem-management skills are important for their children, and they say "certainly." But ask where and how their children pick up these skills, they hem and haw. "Sometimes at home, but not always." Parents don't always see themselves as paragons of effective problem solving. "Maybe at school?" Yet review the curricula of our primary, secondary, and tertiary schools, and you will find little about problem solving that focuses on problems in living. Some say that formal courses in problem-solving skills are not found in our schools because such skills are picked up through experience. To a certain extent, that's true. However, if problem-management skills are so important, one may well wonder why society leaves the acquisition of these skills to chance. A problem-solving or problem-management mentality should be second nature to us. The world may be the laboratory for problem solving, but the skills needed to optimize learning in this lab should be taught. They are too important to be left to chance.

Institute a search and you will soon discover that there are dozens, if not hundreds, of models or approaches to helping, all of them claiming a high degree of success. Library shelves are filled with books on counseling and psychotherapy—books dealing with theory, research, and practice. Back in the 1980s it was estimated that there were between 250 and 400 different approaches to helping (see Herink, 1980; Karasu, 1986). Some of the approaches discussed then have, of course, fallen by the wayside, but many more have been added since. All are proposed with equal seriousness and all of which claim to lead to success. In the face of all this diversity, helpers, especially beginning helpers, need a basic, practical, working model of or approach to helping.

Because all approaches must eventually help clients manage problems and develop unused resources, the model or approach of choice in these pages is a flexible, humanistic, broadly based problem-management and opportunity-development—a model that is straightforward without ignoring the complexities of clients' lives or of the helping process itself. Indeed, because the problem-management and opportunity-development process outlined in this book is embedded in almost all approaches to helping, this model provides an excellent foundation for any "brand" of helping you eventually choose. This book provides the basics.

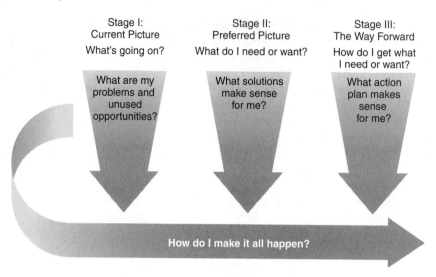

FIGURE 2-1
The Skilled Helper Model

The Stages and Tasks of the Helping Model

All worthwhile helping frameworks, models, or processes ultimately help clients ask and answer for themselves four fundamental questions:

- **What's going on?** What are the problems, issues, concerns, or undeveloped opportunities I should be working on? This is the client's *current picture*.

- **What do I need or want?** What do I want my life to look like? What changes would make me happier? This is the client's *preferred picture*.

- **How do I get there?** What do I need to do to make the preferred picture a reality? What *plan* will get me where I want to go? The plan outlines the *actions* I need to take to get what I need or want.

- **How do I make it all happen?** How do I turn planning and goal setting into the kind of action that leads to the *solutions, results, outcomes,* or *accomplishments* that constitute the preferred picture? How do I get going and stay going?

These four questions, turned into three logical "stages" and an implementation phase in Figure 2-1, provide the basic framework for the helping process. The term "stage" is placed in quotation marks because it has sequential overtones that are somewhat misleading. In practice the stages overlap and interact with one another as clients struggle to manage problems and develop opportunities. And, as we shall see, helping, like life itself, is not as logical as the models used to describe it.

An extended example is used to bring this process to life. The case, though real, has been disguised and simplified. It is not a session-by-session presentation. Rather, it illustrates ways in which one client was helped to ask and answer for himself the four fundamental questions outlined above. The client, Carlos, is voluntary, verbal, and, for the most part, cooperative. In actual practice, cases do not always flow as easily as this one. The simplification of the case, however, will help you see the main features of the helping process in action.

STAGE I: THE CURRENT PICTURE—"WHAT'S GOING ON?"—HELP CLIENTS CLARIFY THE KEY ISSUES CALLING FOR CHANGE

The *current picture* spells out the range of difficulties the client is facing. What are the problems, issues, concerns, and undeveloped opportunities with which Carlos needs to grapple? Here is a thumbnail sketch.

> Carlos, a man in his mid-twenties, has been working for a consulting firm for about a year. He is well-educated, though unlike some of his fellow workers, his undergraduate and MBA degrees are not from the "best" schools. He is bright, though practical rather than academic intelligence is his strong suit. He can be quite personable when he wants to. But he doesn't always want to. He comes from a family with traditional Hispanic cultural values. He can speak Spanish fairly well, but prefers not to, even at home.
>
> Trouble started when one of his colleagues, a consultant some twenty years older than Carlos, buttonholed him after a meeting one day and said: "I've seen you in action a few times. You know, you're your own worst enemy. You're headed for a fall and don't even know it. If I were you, I'd get some help." With that he walked away and Carlos had no further interaction with him.
>
> Though Carlos was first bothered by the incident, he sloughed it off. However, a few weeks later, while talking with one of his colleagues who had been at the firm for about three years, Carlos recounted the incident in a joking way. But his colleague didn't laugh. He merely said in a light-hearted manner, "Well, you never know, Carlos, there just might be something there."
>
> Through the company he has access to a couple of "developmental counselors." He is somewhat put off when he finds out that both of them are women, but he makes an appointment with one.

So here is a young man who has received a couple of shots across his bow. Although he certainly doesn't think that he needs help, he is unsettled enough to be willing to talk with someone.

The Three Tasks of Stage I

As we shall see in the chapters that follow, each stage is divided into three *tasks*. Like the stages themselves, the tasks are not steps in a mechanistic, "now do this" sense. And the tasks, like the stages, are interactive. In Stage I they are activities that help clients develop answers to three questions: What's going on in my life? What's really going on in my life? What should I be working on?

Task 1: Help clients tell their *stories*. Though helping is ultimately about solutions, some review of the problem or unused opportunity is called for. So Elena, Carlos's counselor, helps him tell his story. Through their dialogue, she helps him review what is happening in the workplace. Elena knows that if she can help Carlos get an undistorted picture of himself, his problems, and his unused opportunities, he will have a better chance of doing something about them. Her overall goal is to help Carlos manage the interpersonal dimensions of his work life better. His interpersonal style is both problem and opportunity.

Carlos gets over his initial reluctance and brings up a number of things that bother him. He feels that he is being discriminated against at work. He doesn't feel that he's on the fast track and he thinks that he should be. People at work don't understand and appreciate him. Though he lives with his parents because of financial reasons, he feels distant from his family. "We don't have that much in common any more. They don't want to be mainstream." He and his girlfriend are currently at odds. At one point he says, "I think I've moved beyond her anyway."

Task 2: Help clients develop *new perspectives* that help them reframe their stories. Counselors add great value when they can help their clients identify significant blind spots related to their problems and unused opportunities. Effectively challenged, blind spots yield to new perspectives that help clients think more realistically about problems, opportunities, and solutions.

It does not take Elena long to realize that Carlos has some significant blind spots. For instance, he tends to blame others for his problems. "They are keeping me back." Nor does he realize how self-centered he is. He gets angry when others stand in his way or don't cater to his needs and wants. Yet he is quite insensitive to anybody else's needs. His arrogant style rubs both colleagues and customers the wrong way. Without being brutal, Elena helps him see himself as others see him. She points out that being Hispanic and coming from the "wrong" school have little to do with his colleagues' hostility toward him. Elena, Hispanic herself, understands both Carlos's struggles and his excuses.

Task 3: Help clients achieve *leverage* by working on issues that make a difference. Many clients have a range of issues. In that case help them choose issues that will make a significant difference in their lives. If a client wants to work only on trivial things or does not want to work at all, then it might be better to defer counseling.

Carlos becomes more cooperative as it becomes clear to him that Elena has both his interests and those of the firm in mind. He sees her as "solid"—decent, businesslike, and not overly "psychological." He comes to realize that, although he has a number of concerns, he had better work on his interpersonal communication style and his relationships with both colleagues and clients. He also needs to do something about the "victim" mentality he has developed. With Elena's help he realizes that the flip side of his communication style problem is an enormous opportunity. He quickly sees that becoming a better communicator and relationship builder will help him in every social setting in life. Because this is a work setting and time is limited, Elena does not push him on the problems he's having at home or in his social life outside work. However, she suspects that the changes he makes at work might also apply outside.

STAGE II: THE PREFERRED PICTURE—"WHAT DO I WANT?"—HELP CLIENTS IDENTIFY, CHOOSE, AND SHAPE PROBLEM-MANAGING GOALS

In Stage II the counselors help clients explore and choose possibilities for a better future—a future in which key problem situations are managed and key opportunities developed. "What do you want this future to look like?" asks the helper. The client's answer constitute his or her "change agenda." Stage II focuses on outcomes.

Elena helps Carlos ask himself such questions as: What do I want? What do I need, whether I currently want it or not? What would my business life look like if it were more tolerable or—even better—more engaging and fulfilling?

Unfortunately, some approaches to problem solving or management skip Stage II. They move from the "What's wrong?" stage (Stage I) to a "What do I do about it?" stage (Stage III). As we shall see, however, helping clients discover what they want has a profound impact on the entire helping process. Stage II also has three tasks—that is, three ways of helping clients answer as creatively as possible the question "What do I need or want?"

The Three Tasks of Stage II

Task 1: Help clients use their imaginations to spell out *possibilities for a better future*. This often helps clients move beyond the problem-and-misery mind-set they bring with them and develop a sense of hope. Brainstorming possibilities for a better future can also help clients understand their problem situations better—"Now that I am beginning to know what I want, I can see my problems and unused opportunities more clearly."

Elena helps Carlos brainstorm goals that would help him repair some damaged relationships with both colleagues and clients and help him do something about his victim mentality. Carlos declares that he needs to become a "better communicator." Elena, pointing out that "becoming a better communicator" is a rather vague aspiration, asks him, "What do some of the good communicators you know look like?" Carlos comes up with a range of possibilities. "I'd give great presentations like Jeff." "I'd be a good problem solver like Sharon." "Tony listens a lot better than I do." "I'm not patient at all, but I like it when Abigail is patient with me." "Roger seems to have good relationships with everyone." And so forth. All of these become possibilities for a better communication style. She also gets him to explore further possibilities by asking him what a "repaired relationship" would look like both from his perspective and from the perspective of customers and colleagues he might have alienated.

Task 2: Help clients choose *realistic and challenging goals* that are real solutions to the key problems and unused opportunities identified in Stage I. Possibilities need to be turned into goals because helping is about solutions and outcomes. A client's goals, then, constitute his or her *agenda for change*. If goals are to be pursued and accomplished, they need to be, ideally, clear, related to the problems and unused opportunities the client has chosen to work on, substantive, realistic, prudent, sustainable, flexible, consistent with the client's values, and set in a reasonable time fame. Effective counselors help clients "shape" their agendas to meet these requirements.

Elena helps Carlos sort through some of the possibilities he has come up with. It becomes clear that changes in his interpersonal communication style would help him manage some problems and develop some opportunities at the same time. If he were to communicate well—in terms of both communication skills and the values that support relationship-building communication—he would come across more effectively and repair damaged relationships. But he needs more than skills. He needs to change his self-centered and poor-me attitude. Some of the possibilities Carlos comes up with in his brainstorming session with Elena can be put aside for the present. For instance, things like "becoming a terrific presenter" can wait.

Becoming a better communicator with upbeat interpersonal relationship values and attitudes is a substantial package. It includes Carlos's becoming good at the give-and-take of dialogue and the skills that make it work. These skills include visibly tuning in to others, active listening, thoughtfully processing what he hears, demonstrating understanding of the key points others are making, getting his own points across clearly, drawing others into the conversation, and the like. It also includes embracing the values that make conversations serve relationships—mutual respect, social sensitivity, emotional control, and collaboration.

Task 3: Help clients find the *incentives* that will help them *commit* themselves to their change agendas. The question clients must ask themselves is: "What am I willing to pay for what I need and want?" Without strong commitment, change agendas end up as no more than some "nice ideas." This does not mean that clients are not sincere in setting goals. Rather, once they leave the counseling session, they run into the demands of everyday life. The goals they set for themselves, however useful, face a great deal of competition. Counselors provide an important service when they help clients test their commitment to the better future embedded in the goals they choose.

Becoming a better communicator at the service of repairing and building relationships is hard work. Elena helps Carlos review the incentives he has for engaging in such work. One very strong one is this: He has to. His current interpersonal communication style will probably get him fired and prevent him from being successful in the future. Developing the values that should permeate dialogue is even harder work. It means undoing bad habits developed over years, which is difficult even when a client is committed to doing so. Elena does not dwell on Carlos's bad habits. Rather, she believes that embracing good habits, like showing interest in others and checking his understanding of what they have to say, will drive out his bad habits. If Carlos does all this work, the upside is enormous. Because communication is at the heart of everything he does, better communication skills and values will serve him well in every dimension of life. Elena does not find it difficult to help Carlos appreciate this attractive package of incentives. Positive psychology wins. In Carlos's case, developing opportunities is the main way of managing problems.

STAGE III: THE WAY FORWARD—"HOW DO I GET WHAT I NEED OR WANT?"—HELP CLIENTS DEVELOP STRATEGIES AND PLANS FOR ACCOMPLISHING GOALS

Stage III defines the actions that clients need to take in order to translate goals into problem-managing accomplishments. Stage III answers the question: "How do I get there?" It is about identifying and choosing action strategies and plans. Stage III, too, has three tasks that, in practice, intermingle with one another and with the tasks of the other stages.

The Three Tasks of Stage III
The three tasks of Stage III center around forging a practical plan for accomplishing problem-managing goals.

Task 1: Help clients review *possible strategies* to achieve goals. Stimulating clients to think of different ways of achieving their goals is usually an excellent investment of time. That said, clients should not leap into action. Hasty and disorganized action is often self-defeating. Complaints such as "I tried this and it didn't work. Then I tried that and it didn't work either!" are often signs of poor planning rather than of the impossibility of the task.

Elena helps Carlos explore different ways of becoming the "competent communicator" he wants to be in order to develop and foster satisfying work relationships and establish a better self-image with both colleagues and clients. To develop the skills he needs, he can read books, take courses at local colleges, attend courses for professionals, get a tutor or coach, or come up with his own approach. Elena helps Carlos brainstorm the possibilities. She also points out where he can get more information, but then let's him do his homework.

Task 2: Help clients choose the strategies that *best fit* their resources. Whereas Task 1 provides clients with a pool of possible strategies, Task 2 helps clients choose the action strategies that best fit their talents, resources, style, temperament, environment, and timetable.

With Elena's help, Carlos makes some choices. He chooses to attend an interpersonal-communication program for working professionals. Although more expensive than university-based programs, there is greater flexibility. This fits better with Carlos's rather hectic travel schedule. Second, Elena helps Carlos see that life is his lab. That is, every conversation is part of the program because every conversation is an opportunity to practice the skills he will be learning and demonstrate the attitudes that foster relationship building. To Carlos's credit, he adds that he needs to find a way of monitoring the degree that the values of effective dialogue are permeating his conversations. He decides to get a peer coach—a colleague who has both an excellent communication style and excellent relationships with both colleagues and clients. He finds someone who fits the bill and who is willing to help. He says to his colleague: "Be honest with me. Tell me the way it is."

Task 3: Help clients pull chosen strategies together into a *viable plan*. Help clients organize the actions they need to take to accomplish their goals. Plans are simply maps clients use to get where they want to go. A plan can be quite simple. Indeed, overly sophisticated plans are often self-defeating.

Carlos's plan is straightforward. He will begin the interpersonal-communication course within 2 weeks. He will use every conversation with colleagues and clients as his lab. At the end of each working day he will review the conversations he has had in terms of both skillfulness and values. He creates a checklist for himself that includes such questions as "How effectively did I listen? How clearly did I get my points across? How respectful and collaborative was I? How did I handle sensitive conversations such as those with colleagues and clients who had been turned off by my interpersonal style?" Time and schedules willing, he will meet with his peer coach once a week and with Elena once a month.

THE ACTION ARROW—"HOW DO I MAKE IT ALL HAPPEN?"—HELP CLIENTS IMPLEMENT THEIR PLANS

All three stages of the helping model sit on the "action arrow," indicating that clients need to act in their own behalf right from the beginning of the helping process. Stages I, II, and III are about planning for change, not constructive change itself. Planning is not action. Talking about problems and opportunities, discussing goals, and figuring out strategies for accomplishing goals is just so much blah, blah, blah without goal-accomplishing action. There is nothing magic about change; it is hard work. But, as we shall see in subsequent chapters, each stage and task of the process can promote problem-managing and opportunity-developing action right from the beginning.

Carlos, like most clients, runs into a number of obstacles as he tries to implement his plan. First, his travel schedule keeps conflicting with even the flexible communication-skills program in which he has enrolled. Because he finds the program very useful, he decides to use a tutor from the program to bring him up to speed whenever he cannot fit a session into his travel schedule. Although the company is paying for the program, he has to pay the tutor out of his own pocket. But it's worth it. He also finds that he is not very consistent in using the skills he is learning in the "lab of life." His progress is much slower than he thought it would be. Sometimes he loses heart. He often fails to do the evening self-evaluation sessions. Getting together with his peer coach proves to be almost impossible. And, although he is doing better with clients, he doesn't see much "repair" going on in his relationships with his colleagues. So in one of his discussions with Elena, he reviews the ups and downs of his program, and discusses his discouragement with how slowly some of his disenchanted colleagues are warming up to him.

He uses the discussion to reset the program. First, he agrees to see her for a half hour every other week. This provides him with an incentive to keep to the program. He wants to give her a good report. Second, he discusses his "bad attitude." Even though she said that the whole program would be a lot of work, he didn't realize just how much work. He finds that he can't do the program well without changing some basic attitudes and habits. It's not just a skills program. It cuts much deeper. He has to recommit himself to a deeper kind of attitudinal change. New attitudes look fine on paper. Developing them is another story. And so Carlos moves on—two small steps forward and a half step backward. He resets his schedule so he can get together with his peer coach. His sessions with both Elena and his peer coach help and he does make progress.

Figure 2-2 presents the full model in all its stages and tasks and their relationship to the action arrow. It includes two-way arrows between both stages and tasks to suggest the kind of flexibility needed to make the process work. As we shall see, every stage and task of the helping process has the ability to drive problem-managing action on the part of the client.

"HOW ARE WE DOING?"—ONGOING EVALUATION OF THE HELPING PROCESS

In the light of the "Does helping help?" discussion in Chapter 1, how do helpers using the problem-management and opportunity-development framework evaluate what is happening with each client? By making each case a "mini-experiment" in itself. In psychological research there has been a long history of what are called $N = 1$ or single-case research designs both to evaluate practice and conduct research (Blampied, 2000; Hilliard, 1993; Lundervold & Belwood, 2000; Persons, 1991; Valsiner, 1986). It's not enough to know that helping in general works. We have to know how well it is working in each case. Jay Lebow (2002) puts it this way: "A clinician can carry out with any individual client a method researchers call the 'single-case design'—which is simply a more formal and systematic way of documenting

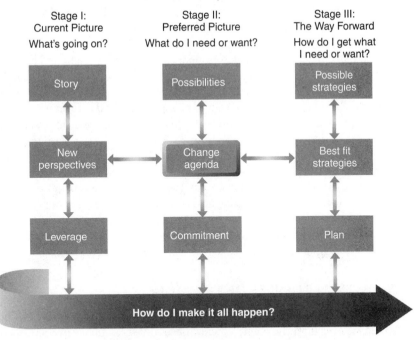

The Skilled-Helper Model

Stage I: Current Picture	Stage II: Preferred Picture	Stage III: The Way Forward
What's going on?	What do I need or want?	How do I get what I need or want?

FIGURE 2-2
The Helping Model Showing Interactive Stages and Steps

what he or she does anyway. . . . [A] therapist uses several questionnaires over the course of therapy to document the changes that occur, both in therapy and in the client's life" (p. 63). Bangert and Baumberger (2005) argue that although $N = 1$ is the most relevant design for practicing counselors, few are adept in using it. In one study (Hatfield & Ogles, 2004), only 37% of helpers indicated that they use some form of outcome assessment in practice.

In many helping models, evaluation is presented as the last step in the model. However, if evaluation occurs only at the end, it is too late. As Mash and Hunsley (1993) noted, early detection of what is going wrong in the helping process is needed to prevent failure. They claimed that an early-detection framework should be theory-based, ongoing, practical, and sensitive to whatever new perspectives might emerge from the helping process. The problem-management and opportunity-development model outlined in this chapter fills the bill. It is a tool to check progress throughout the helping process. As we shall see, it provides criteria for helper effectiveness, for client participation, and for assessing outcomes.

Because helpers and clients need to collaborate in this ongoing evaluation process, Elena works with Carlos in using the helping model as the evaluation framework. Carlos comes to appreciate Elena's skill in using the model. Once Carlos takes ongoing evaluation seriously, he begins to make progress. He gets feedback from the self-evaluation process that he begins to use more frequently, from the observations of his peer coach, and from his sessions with Elena. The ultimate feedback comes from goal accomplishment. In what ways and to what degree is he becoming a more competent communicator? How effectively are relationships with clients and colleagues being repaired? To what degree is he shedding his self-centered approach to relationships that contribute to his "others are out to get me" attitude?

There is evidence that soliciting ongoing feedback with respect to both the process and outcome of therapy makes it more effective and efficient and helps prevent therapist-client misunderstandings (Duncan, Miller, & Sparks, 2004; Lambert & Hawkins, 2004; Lambert et al., 2001). In Chapter 4 of their book, which emphasizes "practice-based evidence," Duncan, Miller, and Sparks demonstrate the value of two very short feedback forms—one on outcome, filled in and discussed at the beginning of a session (the Outcome Rating Scale focuses on what has happened in the client's life since the last session), and one on how the session itself has gone, administered at the end of the session (Session Rating Scale). These simple forms, which can be downloaded from their website (www.talkingcure.com), not only provide feedback for the helper but they also stimulate the discussion of important issues that might otherwise have been overlooked or silently set aside. Because their goal is to make therapy client-directed and outcome-informed, this feedback system is one way of encouraging clients to take the lead in the helping process.

FLEXIBILITY IN THE USE OF THE MODEL

There are many reasons why you need to use the helping model flexibly. The main one is this: Helping is for the client. Clients' needs take precedence over any model. That said, a number of points about flexibility need to be made.

First, clients start and proceed differently. Any stage or task of the helping process can be the entry point. For instance, Client A might start with something that he tried to do to solve a problem but that did not work—"I threatened to quit if they didn't give me a leave of absence, but it backfired. They told me to leave." The starting point is a failed strategy. Client B might start with what she believes she wants but does not have—"I need a boyfriend who will take me as I am. Joe keeps trying to redo me." Stage II is her entry point. Client C might start with the roots of his problem situation—"I don't think I've ever gotten over being abused by my uncle." Stage I is the entry point. Client D might announce that she really has no problems but is still vaguely dissatisfied with her life—"I don't know. Everyone tells me I've got a great life, but something's missing." The implication here is that she has not been seizing the kind of opportunities that could make her happy. Opportunity rather than problem is the starting point.

Second, clients engage in each stage and task of the model differently. Take clients' stories. Some clients spill out their stories all at once. Others "leak" bits and pieces of their story throughout the helping process. Still others tell only those parts that put them in a good light. Most clients talk about problems rather than opportunities. Because clients do not always present all their problems at once in neat packages, it is impossible to work through Stage I completely before moving on to Stages I, II, and III. It is not even advisable to do so. Some clients don't even understand their problems until they begin talking about what they want but don't have. Others need to engage in some kind of remedial action before they can adequately define the problem situation. That is, action sometimes precedes understanding. If some supposedly problem-solving action is not successful, then the counselor helps the client learn from it and return to the tasks of clarifying the problem or opportunity and then setting some realistic goals. Take the case of Woody.

> Woody, a sophomore in college, came to the student counseling services with a variety of interpersonal and somatic complaints. He felt attracted to a number of women on campus but did very little to become involved with them. After exploring this issue briefly, he said to the counselor, "Well, I just have to go out and do it." Two months later he returned and said that his experiment had been a disaster. He had gone out with a few women, but the chemistry never seemed right. Then he did meet someone he liked quite a bit. They went out a couple of times, but the third time he called, she said that she didn't want to see him any more. When he asked why, she muttered vaguely about his being too preoccupied with himself and ended the conversation. He felt so miserable he returned to the counseling center. He and the counselor took another look at his social life. This time, however, he had some experiences to probe. He wanted to explore this "chemistry thing" and his reaction to being described as "too preoccupied with himself."

Woody put into practice Weick's (1979) dictum that chaotic action is sometimes preferable to orderly inactivity. Once he acted, he learned a few things about himself. Some of these lessons proved to be painful, but he now had a better chance of examining his interpersonal style much more concretely.

Third, because the stages and tasks of the model intermingle, helpers will often find themselves moving back and forth in the model. Often two or more tasks or even two stages of the process merge into one another. For instance, clients can name parts of a problem situation, set goals, and develop strategies to achieve them in the same session. New and more substantial concerns arise while goals are being set, and the process moves back to an earlier, exploratory stage. Helping is seldom a linear event. In discussing a troubled relationship with a friend, one client said something like this:

> Every time I try to be nice to her, she throws it back in my face. So who says being more considerate is the answer? Maybe my problem is that I'm a wimp, not the self-centered jerk she makes me out to be. Maybe I'm being a wimp with you and you're letting me do it. Maybe it's time for me to start looking out for my own interests—you know, my own agenda rather than trying to make myself fit into everyone else's plans. I need to take a closer look at the person I want to be in my relationships with others.

In these few sentences the client mentions a failed action strategy, questions a previously set goal, hints at a new problem, suggests a difficulty with the helping relationship itself, offers, at least generically, a different approach to managing his problem, and recasts the problem as an opportunity to develop a more solid interpersonal style. Your challenge is to make sense of clients' entry points and guide them through whatever stage or task that will help move toward problem-managing and opportunity-developing action.

Of course, flexibility is not mere randomness or chaos. Focus and direction in helping are also essential. Letting clients wander around in the morass of problem situations under the guise of flexibility leads nowhere. The structure of the helping model is the very foundation for flexibility; it is the underlying "system" that keeps helping from being a set of random events. A helping model is like a map that helps you know, at any given moment, "where you are" with clients and what kinds of interventions would be most useful. Using the map metaphor, the stages and tasks of the model are orientation devices. At its best, it is a *shared* map that helps clients participate more fully in the helping process. They, too, need to know where they are going.

PROBLEM MANAGEMENT: A HUMAN UNIVERSAL

Given the diversity of clients, helping models should be vehicles of personal growth, not cultural domination (see Chapter 3 for more on cultural diversity). The advantage of problem-management and opportunity-development models of helping is that they are easily recognized across the world. At least that is my experience. Problem solving seems to be what McCrae and Costa (1997) call a "human universal" or what Norenzayan and Heine (2005), in a stimulating article, call a "psychological universal."

Many years ago, before presenting an earlier version of the helping process outlined in this book to some 300 college students and faculty members in Tanzania, I said, "All I can do is present to you the helping process I teach and use. You have to decide whether it makes sense in your own culture." At the end they said two things. First, the communication skills used in the helping process would have to be modified somewhat to fit their culture. Second, the problem-management helping process itself was very useful. This kind of cross-cultural validation is, as Norenzayan and Heine note, at the heart of universality: "A compelling case for universality can be made when a phenomenon is clearly identifiable in a large and diverse array of cultures" (2005, p. 769).

Since then, this scene has been repeated—in conferences and training events I and others have presented—over and over again on every continent. The model presented here spells out, in a flexible, step-by-step fashion, the way human beings tend to think about constructive change. The reason this process crosses cultures so easily is that its logic seems to be embedded in human beings. People don't so much learn the framework of the model. In essence it's already there. It is, to use Orlinsky and Howard's (1987) term, a "generic" model of helping. Of course, the process as outlined in these pages together with the skills and techniques that make it work still has to be adapted both to different cultural settings and to different individuals within those settings. This demands cultural sensitivity on the part of helpers. Universal psychological processes together with their cultural variations help make counseling both efficient and effective.

USING THE MODEL AS A "BROWSER": THE SEARCH FOR BEST PRACTICE

The claim in this book is this: Problem management and opportunity development is one of the principal processes—perhaps *the* principal process—underlying all successful counseling and psychotherapy. What is the novice to do in the face of the bewildering array of models and methods available? Even though there are only a handful of "major brands" of psychotherapy (see Capuzzi & Gross, 1999; Corey, 1996; Gilliland & James, 1997; Prochaska & Norcross, 2002; Sharf, 2003; Wachtel & Messer, 1997), choices need to be made.

Eclecticism. Many experienced helpers, even when they choose one specific school or approach to helping, often borrow methods and techniques from other approaches. Other helpers, without declaring allegiance to any particular school, stitch together their own approaches to helping. This borrowing and stitching is called "eclecticism" (Jensen, Bergin, & Greaves, 1990; Lazarus, Beutler, & Norcross, 1992; Prochaska & Norcross, 2000). In one study, some 40% of helpers said that eclecticism was their primary approach to helping (Milan, Montgomery, & Rogers, 1994). Effective eclecticism, however, must be more than a random borrowing of ideas and techniques from here and there. There must be some integrating framework to give coherence to the entire process; that is, to be effective, eclecticism must be systematic.

Problem management as underlying process. When any school, model, or eclectic mixture is successful, it is so precisely because it helps clients (1) identify and explore problem situations and unused opportunities, (2) determine what they need and want, (3) discover ways of getting what they need and want, and (4) translate what they learn into problem-managing action. That is, the problem-management and opportunity-development process outlined here underlies or is embedded in all approaches to helping because all approaches deal with constructive client change. Therefore, the model outlined in these pages together with the basic skills and techniques that make it work is a very useful starting point for novices no matter what school or approach or eclectic system they may ultimately choose or devise.

Using the model as a "browser." The helping model in this book can also be used as a tool—a "browser," to use an Internet term—for mining, organizing, and evaluating concepts and techniques that work for clients, no matter what their origin.

- **Mining.** First, helpers can use the problem-management model to mine any given school or approach, "digging out" whatever is useful without having to accept everything that is offered. The philosophy, communication skills, stages, and tasks of the model serve as tools for identifying methods and techniques that will serve the needs of clients.

- **Organizing.** Second, because the problem-management model is organized by stages and tasks, it can be used to organize the methods and techniques that have been mined from the rich literature on helping. For instance, a number of contemporary therapies have elaborated excellent techniques for helping clients identify blind spots and develop new perspectives on the problem situations they face. As we shall see, these techniques can be organized in the communication skills section dealing with challenge (Chapters 7 and 8).

- **Evaluating.** Because the problem-management model is pragmatic and focuses on outcomes of helping, it can be used to evaluate the vast number of helping techniques that are constantly being devised. The model enables helpers to ask in what way a technique or method contributes to the "bottom line," that is, to outcomes that serve the needs of clients.

The problem-management and opportunity-development model can serve these functions because it is an open-systems model, not a closed school. Although it takes a stand on how counselors may help their clients, it is open to being corroborated, complemented, and challenged by any other approach, model, or school of helping. The needs of clients, not the egos of model builders, must remain central to the helping process. Our clients deserve "best practice," whatever its source.

UNDERSTANDING AND DEALING WITH THE SHADOW SIDE OF HELPING MODELS

Besides the broad shadow-side themes mentioned in Chapter 1, there are a number of shadow-side pitfalls in the use of any helping model.

No model. Some helpers "wing it." They have no consistent, integrated model that has a track record of benefitting clients. Professional training programs often offer a wide variety of approaches to helping drawn from the "major brands" on offer. If helpers-to-be leave such programs knowing a great deal about different approaches but lacking an integrated approach for themselves, then they need to develop one quickly. The problem-management framework is a good place to start.

Fads. The helping professions are not immune to fads. A fad is an insight or a technique that would have some merit were it to be integrated into some overriding model or framework of helping. Instead it is marketed on its own as the central, if not the only meaningful, intervention needed. A fad need not be something new; it can be the "rediscovery" of a truth or a technique that has not found its proper place in the helping tool kit. Helpers become enamored of these ideas and techniques for a while and then abandon them. There will always be "hot topics" in helping. Note them and integrate whatever you find useful into a comprehensive approach to your clients. Many new approaches to helping make outrageous claims. Don't ignore them, but take the claims with a grain of salt and test the approach.

Rigid applications of helping models. Some helpers buy into a model early on and then ignore subsequent challenges or alterations to the model. They stop being learners. The "purity" of the model becomes more important than the needs of clients. Other helpers, especially beginners, apply a useful helping model too rigidly. They drag clients in a linear way through the model even though that is not what clients need. All of this adds up to excessive control. Effective models effectively used are liberating rather than controlling.

Virtuosity. A third form of ineptness is virtuosity. Some helpers tend to specialize in certain techniques and skills—exploring the past, assessment, goal setting, probing, challenging, and the like. Helpers who specialize not only run the risk of ignoring client needs but also often are not very effective even in their chosen specialties. For example, the counselor whose specialty is challenging clients is often an ineffective challenger. The reason is obvious: Challenge must be based on understanding. Challenge is just part of the picture, part of the hologram.

The antidote to all this is simple. Helpers need to become radically client-centered. Client-centered helping means that the needs of the client, not the models and methods of the helper, constitute the starting point and guide for helping. Therefore, flexibility is essential. In the end, helping is about solutions, results, outcomes, and impact rather than process. The values that drive client-centered helping are reviewed in Chapter 3.

THE HELPING RELATIONSHIP: VALUES IN ACTION

A Bias for Action as an Outcome-Focused Value

Values, Diversity, and Multiculturalism

 Understand and appreciate diversity

 Challenge whatever blind spots you may have

 Tailor your interventions in a diversity-sensitive way

 Work with individuals

A Working Charter: The Client-Helper Contract

Shadow-Side Realities in the Helping Relationship

 Ethical flaws

 Human tendencies in both helpers and clients

 Trouble in the relationship itself

 Vague and violated values

 Failure to share the helping process

 Flawed contracts

 Warring professionals

The Helping Relationship

Although theoreticians, researchers, and practitioners alike, not to mention clients, agree that the relationship between client and helper is important, there are significant differences as to how this relationship is to be characterized and played out in the helping process (Gaston, Goldfried, Greenberg, Horvath, Raue, & Watson, 1995; Hill, 1994; T. L. Sexton & Whiston, 1994; Weinberger, 1995). Some stress the *relationship* itself (see Bailey, Wood, & Nava, 1992; Cochran & Cochran, 2006; M. Kahn, 1990; E. W. Kelly, 1994; Patterson, 1985), whereas others highlight the *work* that is done through the relationship (Reandeau & Wampold, 1991) together with the *outcomes* to be achieved through the relationship (Horvath & Symonds, 1991).

The relationship itself. Patterson (1985) made the relationship itself central to helping. At that time he claimed that counseling or psychotherapy does not merely involve an interpersonal relationship; rather, it *is* an interpersonal relationship. E. W. Kelly (1994, 1997), in offering a humanistic model of counseling integration, argued that all counseling is distinctively human and fundamentally relational. Some traditional schools of psychotherapy indirectly emphasize the centrality of the helping relationship. For instance, in psychoanalytic or psychodynamic approaches, "transference"—the complex and often unconscious interpersonal dynamics between helper and client that are rooted in the client's and even the helper's past— is central (Gelso, Hill, Mohr, Rochlen, & Zack, 1999; Gelso, Kivlighan, Wise, Jones, & Friedman, 1997; Hill & Williams, 2000). Resolving these often murky dynamics is seen as intrinsic to successful therapeutic outcomes.

 In a different mode, Carl Rogers (1951, 1957), one of the great pioneers in the field of counseling, emphasized the quality of the relationship in representing the humanistic-experiential approach to helping (see Kelly, 1994, 1997). Rogers claimed that the unconditional positive regard, accurate empathy, and genuineness offered by the helper and perceived by the client were both necessary and often sufficient for therapeutic progress. Through this highly empathic relationship counselors, in his eyes, helped clients understand themselves, liberate their resources, and manage their lives more effectively. Rogers's work spawned the widely discussed client-centered approach to helping (Rogers, 1965). Unlike psychodynamic approaches, however, the empathic helping relationship was considered a facilitative condition, not a "problem" in itself to be explored and resolved.

The relationship as a means to an end. Others see the helping relationship as very important but still as a means to an end. In this view, a good relationship is practical because it enables client and counselor to do the work called for by whatever helping process is being used. The relationship is instrumental in achieving the goals of the helping process outlined in Chapter 1. Practitioners using cognitive and behavioral approaches to helping such as the manualized treatments discussed in Chapter 13, although sensitive to relationship issues (Arnkoff, 1995), tend toward the means-to-end view. Overstressing the relationship is a mistake because it obscures the ultimate goal of helping: clients better managing problem situations and developing life-enhancing opportunities. This goal won't be achieved if the relationship is poor, but if too much focus is placed on the relationship itself, both client and helper can be distracted from the real work to be done.

THE RELATIONSHIP AS A WORKING ALLIANCE

The term *working alliance*, first coined by Greenson (1967) and now used by advocates of different schools of helping, can be used to bring together the best of both the relationship-in-itself and relationship-as-means-to-achieving-desired-outcomes. Bordin (1979) defined the working alliance as the collaboration between the client and the helper based on their agreement on the goals and tasks of counseling. Bedi (2006) has suggested that the research community has given too much attention to therapists' views of the helping relationship to the neglect of clients' views. For instance, Ackerman and Hilsenroth (2003) presented a comprehensive examination of therapists' personal attributes and in-session activities that positively influence the therapeutic alliance. These attributes include flexibility, honesty, respect, trustworthiness, confidence, warmth, interest, and openness. Techniques that helped the alliance include exploration, reflection, noting past therapy success, accurate interpretation, facilitating the expression of affect, and paying attention to the client's experience. Bedi's study showed that clients appreciated the nature of the helping setting, helper's self-presentation and body language, nonverbal gestures, emotional support and care, honesty, validation, guidance, challenging, helper's education, helper's appreciation of client self-responsibility, and session administration. Although there is, predictably, considerable disagreement among practitioners as to what the critical dimensions of the working alliance are, how it operates, and what results it is to produce (see Hill & Nutt-Williams, 2000; Horvath, 2000; Weinberger, 1995), it is relatively simple to outline what it means in the context of the problem-management and opportunity-development process. Also, common sense suggests that helpers not get lost in the kind of detail outlined above.

The collaborative nature of helping. In the working alliance, helpers and clients are collaborators. Helping is not something that helpers do to clients; rather, it is a process that helpers and clients work through together. Helpers do not "cure" their patients.

Both have work to do in the problem-management and opportunity-development stages and tasks, and both have responsibilities related to outcomes. Outcomes depend on the competence and motivation of the helper and the client, and on the quality of their interactions. Helping is a two-person team effort in which helpers need to do their part and clients theirs. If either party refuses to play or plays incompetently, then the entire enterprise can fail.

The relationship as a forum for relearning. Even though helpers don't cure their clients, the relationship itself can be therapeutic. In the working alliance, the relationship itself is often a forum or a vehicle for social-emotional relearning (Mallinckrodt, 1996). Effective helpers model attitudes and behavior that help clients challenge and change their own attitudes and behavior. It is as if a client were to say to himself (though not in so many words), "She [the helper] obviously cares for and trusts me, so perhaps it is all right for me to care for and trust myself." Or, "He takes the risk of challenging me, so what's so bad about challenge when it's done well?" Or, "I came here frightened to death by relationships and now I'm experiencing a nonexploitative relationship that I cherish." Furthermore, protected by the safety of the helping relationship, clients can experiment with different behaviors during the sessions themselves. The shy person can speak up, the reclusive person can open up, the aggressive person can back off, the overly sensitive person can ask to be challenged, and so forth.

Clients can then transfer what they are learning to other social settings. It is as if a client might say to himself, "He [the helper] listens to me so carefully and makes sure that he understands my point of view even when he thinks I should reconsider it. My relationships outside would be a lot different if I were to do the same." Or, "I do a lot of stuff in the sessions that would make anyone angry. But she doesn't let herself become a victim of emotions, either her own or mine. And her self-control doesn't diminish her humanity at all. That would make a big difference in my life." The relearning dynamic, however subtle or covert, is often powerful. In sum, needed changes in both attitudes, emotional expression, and behavior often take place within the sessions themselves through the relationship.

Relationship flexibility. The idea that one kind of perfect relationship or alliance fits all clients is a myth. Different clients have different needs, and those needs are best met through different kinds of relationships and different modulations within the same relationship. One client may work best with a helper who expresses a great deal of warmth, whereas another may work best with a helper who is more objective and businesslike. Some clients come to counseling with a fear of intimacy. If helpers communicate a great deal of empathy and warmth right from the beginning, these clients might be put off. Once the client learns to trust the helper, stronger interventions can be used. Effective helpers use a mix of styles, skills, and techniques tailored to the kind of relationship that is right for each client (A. A. Lazarus, 1993; Mahrer, 1993). And they remain themselves while they do so.

If clients and helpers are collaborators, can we talk about who should take the lead? Does the one "who has something to say at any given moment" take the lead? Duncan, Miller, and Sparks (2004) answer that question in a way that many would consider radical:

> Data from forty years of outcome research provide strong empirical support for privileging the client's role in the change process (Hubble, Duncan, & Miller, 1999). In short, clients, not therapists, make therapy work. As a

result, therapy should be organized around their resources, perceptions, experiences, and ideas. . . . Therapists need only take direction from clients: following their lead; adopting their language, world view, goals, and ideas about the problem; and acknowledging their experiences with, and inclinations about, the change process. (pp. 11–12)

This view is radical because we have moved at least one step beyond Carl Rogers's client-centered approach. And, if up to now we have said that there is "no one right way" to relate to a client, we would have to change it to "there is no one right way to put the client in the driver's seat." Obviously thinking about the helping relationship is evolving. See Norcross (2002) and Norcross and Hill (2005) for a summary of research evidence related to helping relationships. See McWilliams (2005a, 2005b) and Norcross (2005) for an impassioned plea for preserving the humanity of therapists and therapeutic relationships in the face of our "culture's materialism, consumerism, appeals to vanity and greed, disdain of dependency and vulnerability, and abetment of narcissistic entitlement" (McWilliams, 2005a, p. 139).

THE VALUES THAT DRIVE THE HELPING RELATIONSHIP

One of the best ways to characterize a helping relationship is through the values that permeate and drive it. The relationship is the vehicle through which values come alive. Expressed concretely through working-alliance behaviors, values play a critical role in the helping process (Bergin, 1991; Beutler & Bergan, 1991; Kerr & Erb, 1991; Norcross & Wogan, 1987; Vachon & Agresti, 1992). Because it has become increasingly clear that helpers' values and clients' values interact over the course of the helping process, it is essential to consider the role of values in helping.

Culture, Personal Culture, and Values

Values are central to culture, but culture is more than values. The fuller notion of culture is, briefly, this: *shared beliefs and assumptions interact with shared values and produce shared norms that drive shared patterns of behavior.* Of course, culture is usually not applied directly to individuals but rather to societies, institutions, companies, professions, groups, families, and the like. However, counselors don't deal with societies but individuals and small groups of individuals such as families. So if we apply this basic culture framework to an individual, it goes something like this:

- Over the course of life individuals develop *assumptions and beliefs* about themselves, other people, and the world around them. For instance, Isaiah, a client suffering from posttraumatic stress disorder stemming from gang activity in his neighborhood and a brutal attack he suffered, has come to believe that the world is a heartless place.

- *Values*, what people prize, are picked up or inculcated along the path of life. Isaiah, because of dangers he encounters in his community, has come to value or prize personal security.

- Assumptions and beliefs, interacting with values, generate *norms*, the "dos and don'ts" we carry around inside ourselves. For Isaiah one of these is, "Don't trust people. You'll get hurt."

- These norms drive *patterns of behavior* and these patterns of behavior constitute, as it were, the *bottom line* of personal or individual culture—"the way I live my life." For Isaiah this means not taking chances with people. He's a loner.

Because no individual is an island, personal cultures do not develop in a vacuum. The beliefs, values, and norms people develop are greatly influenced by the groups to which they belong. That said, individuals within any given culture can and often do personalize the beliefs, values, and norms of the cultures in which they live. People within the same culture tailor these beliefs, values, and norms in different ways (Massimini & Delle Fave, 2000). Individuals are not cultural carbon copies. Individuals from the same culture often differ widely in their personal cultures. Effective helpers come to understand both the cultural background of their clients and the personal culture of each individual client. For instance, Isaiah has many of the cultural characteristics of his family, his ethnic group, his neighborhood, his school, and his socioeconomic class, but he is not a carbon copy of any of these cultures. His mix is unique.

Because patterns of behavior constitute the "bottom line" of culture, a popular definition of societal, institutional, and familial culture is "the way we do things here." This definition applied to the individual client is "the way I choose to live my life." Helpers, too, although influenced by the cultures of the various helping professions, have their personal cultures as helpers, that is, "the way I do helping." Inevitably, the helper's personal-professional culture interacts with the client's for better or for worse. The focus in this chapter is on some of the values traditionally found in the helping professions together with the norms these values generate. Indirectly, this entire book focuses on the culture of helping—that is, the beliefs, values, and norms that can and should drive the helping process.

The Pragmatics of Values

Values are not just ideals. They are also a set of practical criteria for making decisions. As such, they are drivers of behavior. For instance, a helper might say to himself or herself during a session with a difficult client something like this:

> The arrogant, I'm-always-right attitude of this client needs to be challenged. How I challenge her is important, because I don't want to damage our relationship. I value genuineness and openness. Therefore, I can challenge her by describing her behavior and the impact it has on me and might have on others and I can do so respectfully, that is, without belittling her.

Working values help counselors make decisions on how to proceed. Helpers without a set of working values are adrift. Those who don't have an explicit set of values have an implicit or "default" set that may or may not serve the helping process. Therefore, reviewing the values that drive your behaviors as a helper is not optional.

Helping-related values, like your other values, cannot be handed to you on a platter. Much less can they be shoved down your throat. Therefore, this chapter is meant to stimulate your thinking about the values that should drive helping. In the final analysis, as you sit with your clients, only those beliefs, values, and norms that you have made your own will make a difference in your helping behavior.

This does not mean that you will invent a set of values different from everyone else's. Tradition is an important part of value formation, and we all learn from the rich tradition of the helping professions. And so, in the following pages, five major values from the tradition of the helping professions—respect, empathy, genuineness, client empowerment, and a bias toward action—are translated into a set of norms. Respect is the *foundation* value; empathy is the value that *orients* helpers in every interaction with their clients; genuineness is the what-you-see-is-what-you-get *professional* value; client empowerment is the value that highlights *self-responsibility*; a bias toward action is an *outcome-focused* value. These values serve as a starting point for your reflection on the values that should drive the helping process. Don't just swallow them: Analyze, reflect on, and debate them. Come up with your own values package.

Respect as the Foundation Value

Respect for clients is the foundation on which all helping interventions are built. Respect is such a fundamental concept that, like most such concepts, it eludes definition. The word comes from a Latin root that includes the idea of "seeing" or "viewing." Indeed, respect is a particular way of viewing oneself and others. If it is to make a difference, respect cannot remain just an attitude or a way of viewing others. Here are some norms that flow from the interaction between a belief in the dignity of the person and the value of respect.

Do no harm. This is the first rule of the physician and the first rule of the helper. Yet some helpers do harm either because they are unprincipled or because they are incompetent. Helping is not a neutral process—it is for better or for worse. In a world in which such things as child abuse, wife battering, and exploitation of workers are much more common than we care to think, it is important to emphasize a nonmanipulative and nonexploitative approach to clients.

Become competent and committed. Master whatever model of helping you use. Get good at the basic problem-management and opportunity-development framework outlined in this book and the skills that make it work. There is no place for the "caring incompetent" in the helping professions. It would be great to say that everyone who graduates from some kind of helping training program is not only competent but also increases his or her competence over his or her career. Unfortunately, this is not the case.

Make it clear that you are "for" the client. The way you act with clients will tell them a great deal about your attitude toward them. Your manner should indicate that you are "for" the client, that you care for him or her in a down-to-earth, nonsentimental way. It is as if you are saying to the client, "Working with you is worth my time and energy." Respect is both gracious and tough-minded. Being for the client is not the same as taking the client's side or acting as the client's advocate. "Being for" means taking clients' points of view seriously even when they need to be challenged. Respect often involves helping clients place demands on themselves. Of course, this kind of "tough love" in no way excludes appropriate warmth toward clients.

Assume the client's goodwill. Work on the assumption that clients want to work at living more effectively, at least until that assumption is proved false. As we shall see in Chapter 9, the reluctance and resistance of some clients, particularly involuntary clients, is not necessarily evidence of ill will. Respect involves entering clients' world to understand their reluctance and a willingness to help clients work through it.

Do not rush to judgment. You are not there to judge clients or to shove your values down their throats. You are there to help them identify, explore, and review and challenge the consequences of the values they have adopted. Let's say that a client during the first session says somewhat arrogantly, "When I'm dealing with other people, I say whatever I want when I want. If others don't like it, well, that's their problem. My first obligation is to myself, being the person I am." Irked by the client's attitude, a helper might respond judgmentally by saying, "You've just put your finger on the core of your problem! How can you expect to get along with people with this kind of self-centered philosophy?" However, another counselor, taking a different approach, might respond, "So being yourself is one of your top priorities and being totally frank is, for you, part of that picture." The first counselor rushes to judgment; the second neither judges nor condones. At this point she merely tries to understand the client's point of view and let him know that she understands—even if she realizes that this point of view needs to be reviewed and challenged later.

Keep the client's agenda in focus. Helpers should pursue their clients' agendas, not their own. Here are three examples of helpers who lost clients because of lack of appreciation of their agendas. One helper recalled, painfully, that he lost a client because he had become too preoccupied with his theories of depression rather than the client's painful depressive episodes. Another helper who dismissed as either trivial or irrelevant a client's bereavement over a pet that had died was dumbfounded and crushed when the client made an attempt on her own life. The loss of the pet was the last straw in a life that was spiraling downward. A third helper, a white male who prided himself on his multicultural focus in counseling, went for counseling himself when a Hispanic client quit therapy, saying, perhaps somewhat unfairly, as he was leaving, "I don't think you're interested in me. You're more interested in Anglo-Hispanic politics."

Cantor (2005) has written a brief overview of clients' rights in psychotherapy in which she broadens helpers' obligations: "It is the obligation of treating psychologists to respect the rights of patients *and*, to the limits of their ability, to assist patients to press third-party payers and other entities to do likewise" (p. 181, emphasis added). This hints at the kind of social justice agenda that some espouse in the helping professions (C. C. Lee, 2006; the November 2004 issue of *The Counseling Psychologist* is devoted to training counseling psychologists as "social justice agents"). There is disagreement in the community of helpers, not about social justice, but about institutional versus individual obligations in pursuing a social-justice agenda and what forms such an agenda should take.

Empathy as a Primary Orientation Value
Although empathy is a rich concept in the helping professions, it has been a confusing one (see Bohart & Greenberg, 1997, and Duan & Hill, 1996, for overviews). Different theoreticians and researchers have defined it in different ways. Some see it as a *personality trait*, a disposition to feel what other people feel or to understand others "from the inside," as it were. In this view some people are by nature more empathic than others. Others see empathy, not as a personality trait, but as a situation-specific *state* of feeling for and understanding of another person's experiences. The implication is that helpers can learn how to bring about this state in themselves because it is so useful in the counseling process. Still others, building on the state

approach, have focused on empathy as a *process* with stages. For instance, Barrett-Lennard (1981) identified three phases—empathic resonance, expressed empathy, and received empathy. Carl Rogers (1975) talked about sensing a client's inner world and communicating that sensing.

The Nature of Empathy

In this chapter empathy is seen as a basic *value* that informs and drives *all* helping behavior. Empathy as a communication skill is discussed and illustrated in Chapter 5.

A rich concept. A number of authors look at empathy from a value point of view and talk about the behaviors that flow from it. Sometimes their language is almost lyrical. For instance, Kohut (1978) said, "Empathy, the accepting, confirming, and understanding human echo evoked by the self, is a psychological nutrient without which human life, as we know and cherish it, could not be sustained" (p. 705). In this view, empathy is a value, a philosophy, or a cause with almost religious overtones. Because empathy seems to be in rather short supply, it might be safer to say that life is fuller because of mutual empathy. Covey (1989), naming empathic communication one of the "seven habits of highly effective people," said that empathy provides those with whom we are interacting with "psychological air" that helps them breathe more freely in their relationships. Goleman (1995, 1998) puts empathy at the heart of emotional intelligence. It is the individual's "social radar" through which he or she senses others' feelings and perspectives and takes an active interest in their concerns.

Can a helper understand a client who is very different from him or her? Can a normal person understand a person with bipolar disorder? The practical answer is yes (Hatcher et al., 2005). That is, clients working with helpers who espouse the value of empathy as delineated here and who have the communication competence described in Chapter 5 to express their understanding feel understood. Answering the philosophical question behind the question is another matter.

The importance of empathy begins early in life. The WAVE Trust, an international charity dedicated to advancing public awareness of the root causes of violence and the means to prevent and reduce it in society, commissioned research that came up with an extraordinary finding: "Empathy is the single greatest inhibitor of the development of propensity to violence. Empathy fails to develop when parents or prime carers fail to attune with their infants" (Hosking & Walsh, 2005, p. 20). In the report, empathy is defined:

> where the observed experiences of others come to affect our own thoughts and feelings in a caring fashion. Empathy entails the ability to step outside oneself emotionally and to be able to suppress temporarily one's own perspective on events to take another's. (p. 20)

To "attune" to a child means "attempting to respond to his or her needs, particularly emotionally, resulting, resulting in the child's sense of being understood, cared for, and valued" (p. 20).

A key helping value. Empathy as a value is a radical commitment on the part of helpers to understand clients as fully as possible in three different ways. First, empathy is a commitment to work at understanding each client from *his or her point of view* together with the feelings surrounding this point of view and to communicate this understanding whenever it is deemed helpful. Second, it is a commitment to understand individuals in and through the *context* of their lives. The social settings, both large and small, in which they have developed and currently "live and move and have their being" provide routes to understanding. Third, empathy is also a commitment to understand the *dissonance* between the client's point of view and reality. But, as Goleman (1995, 1998) notes, there is nothing passive about empathy. Empathic helpers respectfully communicate the kinds of understanding outlined in this paragraph to their clients and generally take an active interest in their concerns. Of course, empathy is a critical value when interacting with clients from different cultures (Wang et al., 2003).

Genuineness as a Professional Value

Like respect, helper genuineness refers to both a set of attitudes and a set of counselor behaviors. Some writers call genuineness "congruence." Genuine people are at home with themselves and therefore can comfortably be themselves in all their interactions. Being genuine has both positive and negative implications; it means doing some things and not doing others.

Do not overemphasize the helping role. Genuine helpers do not take refuge in the role of counselor. Ideally, relating at deeper levels to others and helping are part of their lifestyle, not roles they put on or take off at will. This keeps them far away from being patronizing and condescending. Years ago Gibb (1968, 1978) suggested ways of being "role-free." He said that helpers should learn how to do the following:

- Express directly to another whatever they are presently experiencing.
- Communicate without distorting their own messages.
- Listen to others without distorting the messages they hear.
- Reveal their true motivation in the process of communicating their messages.
- Be spontaneous and free in their communications with others rather than use habitual and planned strategies.
- Respond immediately to another's need or state instead of waiting for the "right" time or giving themselves enough time to come up with the "right" response.
- Manifest their vulnerabilities and, in general, the "stuff" of their inner lives.
- Live in and communicate about the here and now.
- Strive for interdependence rather than dependence or counter dependence in their relationships with their clients.
- Learn how to enjoy psychological closeness.
- Be concrete in their communications.
- Be willing to commit themselves to others.

By this Gibb did not mean that helpers should be "free spirits," inflicting themselves on others. Indeed, "free spirit" helpers can even be dangerous. Freedom from role means that counselors should not use the role or façade of counselor to protect themselves, to substitute for competence, or to fool the client in other ways.

Be spontaneous. Many of the behaviors suggested by Gibb are ways of being spontaneous. Effective helpers, although being tactful as part of their respect for others, do not constantly weigh what they say to clients. They do not put a number of filters between their inner lives and what they express to others. On the other hand, being genuine does not mean verbalizing every thought to the client.

Avoid defensiveness. Genuine helpers are nondefensive. They know their own strengths and deficits and are presumably trying to live mature, meaningful lives. When clients express negative attitudes toward them, they examine the behavior that might cause the client to think negatively, try to understand the clients' points of view, and continue to work with them. Consider the following example:

CLIENT: I don't think I'm really getting anything out of these sessions at all. I still feel drained all the time. Why should I waste my time coming here?

HELPER A: If you were honest with yourself, you'd see that you are the one wasting time. Change is hard and you keep putting it off.

* * * *

HELPER B: Well, that's your decision.

Helpers A and B are both defensive, though in different ways. It is more likely that the client will react to their defensiveness than that she will move forward.

HELPER C: So from where you're sitting, there's no payoff for being here. Just a lot of dreary work and nothing to show for it.

Helper C centers on the experience of the client, with a view to "resetting the system" and helping her explore her responsibility for making the helping process work. Because genuine helpers are at home with themselves, they can allow themselves to examine negative criticism honestly. Counselor C, for instance, would be the most likely of the three to ask himself or herself whether he or she is contributing to the apparent stalemate.

Be open. Genuine helpers are capable of deeper levels of self-disclosure even within the helping relationship. They do not see self-disclosure as an end in itself, but they feel free to reveal themselves, even in deeper ways, when and if it is appropriate. Being open also means that the helper has no hidden agendas: "What you see is what you get." Helper self-disclosure is discussed in Chapter 8.

Client Empowerment as a Responsibility-Focused Value

The second goal of helping, outlined in Chapter 1, deals with empowerment—that is, helping clients identify, develop, and use resources that will make them more effective agents of change both within the helping sessions themselves and in their everyday lives (Strong, Yoder, & Corcoran, 1995). The opposite of empowerment is dependency (Abramson, Cloud, Kesse, & Keese, 1994; Bornstein & Bowen, 1995), deference (Rennie, 1994), and oppression (McWhirter, 1996). Because helpers are often experienced by clients as relatively powerful people and because even the most egalitarian and client-centered of helpers do influence clients, it is necessary to come to terms with social influence in the helping process.

Helping as a social-influence process. People influence one another every day in every social setting of life. E. R. Smith and Mackie (2000) consider it one of eight basic principles needed to understand human behavior. William Crano (2000) suggests that "social influence research has been, and remains, the defining hallmark of social psychology" (p. 68). Parents influence each other and their kids. In turn they are influenced by their kids. Teachers influence students and students influence teachers. Bosses influence subordinates and vice versa. Team leaders influence team members, and members influence both one another and the leader. The world is abuzz with social influence. It could not be otherwise. However, social influence is a form of power and power too often leads to manipulation and oppression.

It is not surprising, then, that helping as a social-influence process has received a fair amount of attention in the helping literature (Dorn, 1986; Heppner & Claiborn, 1989; Heppner & Frazier, 1992; Houser, Feldman, Williams, & Fierstien, 1998; W. T. Hoyt, 1996; McCarthy & Frieze, 1999; McNeill & Stolenberg, 1989; Strong, 1968, 1991; Tracey, 1991). Helpers can influence clients without robbing them of self-responsibility. Even better, they can exercise their trade in such a way that clients are, to use a bit of current business jargon, "empowered" rather than oppressed both in the helping sessions themselves and in the social settings of everyday life. With empowerment, of course, comes increased self-responsibility.

Imagine a continuum. At one end lies "directing clients' lives" and at the other "leaving clients completely to their own devices." Somewhere along that continuum is "helping clients make their own decisions and act on them." Most forms of helper influence will fall somewhere in between the extremes. Preventing a client from jumping off a bridge moves, understandably, to the controlling end of the continuum. On the other hand, simply accepting and in no way challenging a client's decision to put off dealing with a troubled relationship because he or she is "not ready" moves toward the other end. As Hare-Mustin and Marecek (1986) noted, there is a tension between the right of clients to determine their own way of managing their lives and the therapist's obligation to help them live more effectively.

Norms for empowerment and self-responsibility. Helpers don't self-righteously "empower" clients. That would be patronizing and condescending. In a classic work, Freire (1970) warned helpers against making helping itself just one more form of oppression for those who are already oppressed. Effective counselors help clients discover, develop, and use the untapped power within themselves. Here, then, is a range of empowerment-based norms, some adapted from the work of Farrelly and Brandsma (1974).

Start with the premise that clients can change if they choose. Clients have more resources for managing problems in living and developing opportunities than they—or sometimes their helpers—assume. The helper's basic attitude should be that clients have the resources both to participate collaboratively in the helping process and to manage their lives more effectively. These resources may be blocked in a variety of ways or simply unused. The counselor's job is to help clients identify, free, and cultivate these resources. The counselor also helps clients assess their resources realistically so that their aspirations do not outstrip their resources.

Do not see clients as victims. Even when clients have been victimized by institutions or individuals, don't see them as helpless victims. The cult of victimhood is already growing too fast in society. Even if victimizing circumstances have diminished a client's degree of freedom—the abused spouse's inability to leave a deadly relationship, for example—work with the freedom that is left.

Don't be fooled by appearances. One counselor trainer in a meeting with his colleagues dismissed a reserved, self-deprecating trainee with the words, "She'll never make it. She's more like a client than a trainee." Fortunately, his colleagues did not work from the same assumption. The woman went on to become one of the program's best students. She was accepted as an intern at a prestigious mental-health center and was hired by the center after graduation.

Share the helping process with clients. Clients, like helpers, can benefit from maps of the helping process. Helping should not be a "black box" for them. Clients have a right to know what they are getting into (Heinssen, 1994; Heinssen, Levendusky, & Hunter, 1995; Hunter, 1995; Manthei & Miller, 2000). How to clue clients into the helping process is another matter. Helpers can simply explain what helping is all about. A simple pamphlet outlining the stages and steps of the helping process can be of great help, provided that it is in language that clients can readily understand. Just what kind of detail will help will differ from client to client. Obviously, clients should not be overwhelmed by distracting detail from the beginning. Nor should highly distressed clients be told to contain their anxiety until helpers teach them the helping model. Rather, the details of the model can be shared over a number of sessions. There is no one right way. In my opinion, however, clients should be told as much about the model as they can assimilate.

Help clients see counseling sessions as work sessions. Helping is about client-enhancing change. Therefore, counseling sessions deal with exploring the need for change, determining the kind of change needed, creating programs of constructive change, engaging in change "pilot projects," and finding ways of dealing with obstacles to change. This is work, pure and simple. This search for and implementation of solutions can be arduous, even agonizing, but it can also be deeply satisfying, even exhilarating. Helping clients develop the "work ethic" that makes them partners in the helping process can be one of the helper's most formidable challenges. Some helpers go so far as to cancel counseling sessions until the client is "ready to work." Helping clients discover incentives to work is, of course, less dramatic and hard work in itself.

Become a consultant to clients. Helpers can see themselves as consultants hired by clients to help them face problems in living more effectively. Consultants in the business world adopt a variety of roles. They listen, observe, collect data, report observations, teach, train, coach, provide support, challenge, advise, offer suggestions, and even become advocates for certain positions. But the responsibility for running the business remains with those who hire the consultant. Therefore, even though some of the activities of the consultant can be seen as quite challenging, the decisions are still made by managers. Consulting, then, is a social-influence process, but it is a collaborative one that does not rob managers of the responsibilities that belong to them. In this respect, it is a useful analogy to helping. The best clients, like the best managers, learn how to use their consultants to add value in managing problems and developing opportunities.

Accept helping as a natural, two-way influence process. Tyler, Pargament, and Gatz (1983) moved a step beyond the consultant role in what they called the "resource collaborator role." Seeing both helper and client as people with defects, they focused on the give-and-take that should characterize the helping process. In their view, either client or helper can approach the other to originate the helping process. The two have equal status in defining the terms of the relationship, in originating actions within it, and in evaluating both outcomes and the relationship itself. In the best case, positive change occurs in both parties.

Helping is a two-way street. Clients and therapists change one another in the helping process. Even a cursory glance at helping reveals that clients can affect helpers in many ways. For instance, Wei-Lian has to correct Timothy, his counselor, a number of times when Timothy tries to share his understanding of what Wei-Lian has said. For instance, at one point, when Timothy says, "So you don't like the way your father forces his opinions on you," Wei-Lian replies, "No, my father is my father and I must always respect him. I need to listen to his wisdom." The problem is that Timothy has been inadvertently basing some of his responses on his own cultural assumptions rather than on Wei-Lian's. When Timothy finally realizes what he is doing, he says to Wei-Lian, "When I talk with you, I need to be more of a learner. I'm coming to realize that Chinese culture is quite different from mine. I need your help."

Focus on learning instead of helping. Although many see helping as an education process, it is probably better characterized as a learning process. Effective counseling helps clients get on a learning track. Both the helping sessions themselves and the time between sessions involve learning, unlearning, and relearning. Howell (1982) gave us a good description of learning when he said that "learning is incorporated into living to the extent that viable options are increased" (p. 14). In the helping process, learning takes place when options that add value to life are opened up, seized, and acted on. If the collaboration between helpers and clients is successful, clients learn in very practical ways. They have more "degrees of freedom" in their lives as they open up options and take advantage of them. This is precisely what counseling helped Carlos (see Chapter 2) to do. He unlearned, learned, relearned, and acted on his learnings.

Do not see clients as overly fragile. Neither pampering nor brutalizing clients serves their best interests. However, many clients are less fragile than helpers make them out to be. Helpers who constantly see clients as fragile may well be acting in a self-protective way. Driscoll (1984) noted that early in the helping process, too many helpers shy away from doing much more than listening. The natural deference many clients display early in the helping process (Rennie, 1994)—including their fear of criticizing the therapist, understanding the therapist's frame of reference, meeting the perceived expectations of the therapist, and showing indebtedness to the therapist—can send the wrong message to helpers. Clients early on may be fearful of making some kind of irretrievable error. This does not mean that they are fragile. Reasonable caution on your part is appropriate, but you can easily become overly cautious. Driscoll suggested that helpers intervene more right from the beginning—for instance, by reasonably challenging the way clients think and act and by getting them to begin to outline what they want and are willing to work for.

There are cultural differences in the ways people approach self-responsibility. Consider Taiwanese university students. They tend to see help seeking as a sign of weakness, even as a cause for shame; informal is more acceptable than formal help; the situation has to be quite serious before they think of seeking help; and they are reluctant to seek help from strangers (Lin, 2002). Typical convictions about dealing with problem situations are found in such sayings as "God helps people who help themselves" and "Solve problems on my own" (p. 51). It is also important to note that this kind of self-reliance differs greatly from the kind of individualism found in Western societies such as Australia and the United States. You will find different convictions about self-responsibility in other cultures. In the end, of course, what matters is *this* client's approach to self-responsibility. Another rich source for understanding and developing self-responsibility is the growing literature on *conscientiousness* (Bogg & Roberts, 2004; B. W. Roberts, Walton, & Bogg, 2005).

A Bias for Action as an Outcome-Focused Value

The overall goal of helping clients become more effective in problem management and opportunity development was mentioned in Chapter 1. Another goal is to help clients become more effective "agents" in the helping process and in their daily lives—*doers* rather than mere reactors, *preventers* rather than fixers, *initiators* rather than followers.

> Lawrence was liked by his superiors for two reasons. First, he was competent—he got things done. Second, he did whatever they wanted him to do. They moved him from job to job when it suited them. He never complained. However, as he matured and began to think more of his future, he realized that there was a great deal of truth in the adage, "If you're not in charge of your own career, no one is." After a session with a career counselor, he outlined the kind of career he wanted and presented it to his superiors. He pointed out to them how this would serve both the company's interests and his own. At first they were taken aback by Lawrence's assertiveness, but then they agreed. Later, when they seemed to be sidetracking him, he stood up for his rights. Assertiveness was his bias for action.

The doer is more likely to pursue "stretch" rather than merely adaptive goals in managing problems. The doer is also more likely to move beyond problem management to opportunity development.

Action and discretionary change. If clients are to become more effective agents in their lives, they need to understand the difference between discretionary and nondiscretionary change. Nondiscretionary change is mandated change. If the courts say to a divorced man negotiating visiting rights with his children, "You can't have visiting rights unless you stop drinking," then the change is nondiscretionary. There will be no visiting rights without the change. In contrast, a man and wife having difficulties with their marriage are not under the gun to change the current pattern. Change here is discretionary. "IF you want a more productive relationship, THEN you must change in the following ways."

The fact of discretionary change is central to mediocrity. If we don't *have* to change, very often we don't. We need merely review the track record of our New Year's resolutions. Unfortunately, in helping situations, clients often see change as discretionary. They may talk about it as if it were nondiscretionary, but deep down a great deal of "I don't really have to change" pervades the helping process. "Other people should change; the world should change. But I don't have to." This is not cynical. It's the way things are. The sad track record of discretionary change is not meant to discourage you but to make you more realistic about the challenges you face as a helper and about the challenges you help your clients face.

A pragmatic bias toward client action on your part—rather than merely talking about action—is a cardinal value. Effective helpers tend to be active with clients and see no particular value in mere listening and nodding. They engage clients in a dialogue. During that dialogue, they constantly ask themselves, "What can I do to raise the probability that this client will act on her own behalf intelligently and prudently?" I know a man who years ago went "into therapy" (as "into another world") because, among other things, he was indecisive. Over the years he became engaged several times to different women and each time broke it off. So much for decisiveness.

Real-life focus. If clients are to make progress, they must "do better" in their day-to-day lives. The focus of helping, then, is not narrowly on the helping sessions and client-helper interactions themselves, but on clients' managing their day-to-day lives more effectively. A friend of mine in his early days as a helper exulted in the "solid relationship" he was building with a client until in the third interview she stopped, stared at him, and said, "You're really filled with yourself, aren't you? But, you know, we're not getting anywhere." He had become so lost in relationship building that he forgot about the client's pressing everyday concerns. In working with difficult clients, Hanna (2003) has developed a model of change that highlights the importance of clients as "active agents" (p. 11). The behaviorist view of human beings, he suggests, has overemphasized mechanistic and deterministic factors in human behavior. This view shortchanges clients. As Bohart and Tallman (1999) have put it, "Clients in varying degrees can solve their own problems, come up with their own ideas, and actively contribute to the therapy process" (p. xiv). The view taken here is that they not only can, but they must.

Values, Diversity, and Multiculturalism

Although dealing knowledgeably and sensitively with diversity and that particular form of diversity called multiculturalism is part of both respect and empathy and is related to genuineness and client empowerment, it is given special attention here because of the emphasis currently being placed on diversity both in the workplace and in the helping professions. There has been an explosion of literature on diversity and multiculturalism over the past few years. A simple search on the website of an Internet bookseller yielded over 16,000 entries. A dictionary of multicultural psychology (L. E. Hall, 2005) runs more than 170 pages. So I hesitate to cite references to books and articles that should be read. There is both an upside and a downside to this avalanche. One plus is that helpers are forced to take a look at the blind spots they may have about diversity and culture and to take a deeper look at the world around us. One minus is that multiculturalism runs the danger of becoming a fad or even an industry. Let's start with a simple example.

Sue Smith, a Midwestern American, is married to Patrick Lee, an immigrant from Singapore. They are having problems. Many clients come to helpers because they are having difficulties in their relationships with others or because relationship difficulties are part of a larger problem situation. Therefore, understanding clients' different approaches to developing and sustaining relationships is important. Guisinger and Blatt (1994) put this in a broader multicultural perspective: "Western psychologies have traditionally given greater importance to self-development than to interpersonal relatedness, stressing the development of autonomy, independence, and identity as central factors in the mature personality. In contrast, women, many minority groups, and non-Western societies have generally placed greater emphasis on issues of relatedness" (p. 104). In this case Sue is deep into the development of autonomy, independence, and her identity as a successful working woman. Patrick runs a successful small website development business. Guisinger and Blatt go on to point out that both interpersonal relatedness and self-definition are essential for maturity. Helping Sue and Patrick, individuals from different cultures, achieve the "right balance" depends on understanding what the "right balance" means in any given culture.

Because multiculturalism and other forms of diversity are politically and professionally charged, we can expect controversy and debate. Although most would agree that diversity and multiculturalism should take their rightful place in psychology and helping, there is a great deal of disagreement about what "rightful place" means and how this task is to be accomplished. Add this to the challenges outlined in Chapter 1. Because psychology is an inexact science and helping is a profession filled with ambiguities, we need to learn how to live with differences and remain civil in doing so. There is no on right way. In the meantime, here are a few guidelines for developing a style in counseling that honors the best in the diversity and multicultural traditions.

Understand and appreciate diversity. Although clients have in common their humanity, they differ from one another in a whole host of ways—abilities, accent, age, attractiveness, color, developmental stage, disabilities, economic status, education, ethnicity, fitness, gender, group culture, health, national origin, occupation, personal culture, personality variables, politics, problem type, religion, sexual orientation, social status—to name some of the major categories. We differ from one another in hundreds of ways. And who is to say which differences are key? This presents several challenges for helpers. For one, it is essential that helpers understand clients and their problem situations contextually. For instance, a life-threatening illness might be one kind of reality for a 20-year-old and quite a different reality for an 80-year-old. Homelessness is a complex phenomenon. A homeless client with a history of drug abuse who has dropped out of graduate school is far different from a drifter who hates homeless shelters and resists every effort to get him to go to one.

Although it is true that over time helpers can come to understand a great deal about the characteristics of the populations with whom they work—for instance, they can and should understand the different development tasks and challenges that take place over the life span, and if they work with the elderly, they can and should grow in their understanding of the challenges, needs, problems, and opportunities of the aged—still, it is impossible to know everything about every population. Read a hefty abnormal psychology text. It soon becomes clear that it is impossible to become an expert on every syndrome discussed there. This impossibility becomes even more dramatic when the combinations and permutations of characteristics are taken into consideration. How could an African American, middle-class, highly educated, younger, urban, Episcopalian, female psychologist possibly understand a poor, unemployed, homeless, middle-aged, uneducated, lapsed-Catholic male, born of migrant workers, the father a Mexican, the mother a Polish immigrant? Indeed, how can anybody fully understand anybody else? If the legitimate principles relating to diversity were to be pushed too far, no one would be able to understand and help anybody else.

Because appreciating diversity is so important, there might be a tendency to assume it or find it where it does not exist. Consider the psychological differences between women and men. Janet Hyde (2005) challenges the "vast psychological differences model" found in the popular media. She posits a "gender similarities hypothesis," based on a review of 46 meta-analytic studies, which contends that females and males "are similar on most, but not all, psychological variables. That is, men and women, as well as boys and girls, are more alike than they are different" (p. 581). She goes on to claim that distortions about gender differences comes at a cost: "It is time to consider the costs of overinflated claims of gender differences. Arguably, they cause harm in numerous realms, including women's opportunities in the workplace, couple conflict and communication, and analyses of self-esteem problems among adolescents" (p. 590). Her study may ignite a brush fire, if not a bonfire, of both discussion and controversy. Overemphasizing diversity can lead to needless conflict. The underlying lesson is clear. Appreciate diversity but be careful how you assess and handle it.

Challenge whatever blind spots you may have. Because helpers often differ from their clients in many ways, there is often the challenge to avoid diversity-related blind spots that can lead to inept interactions and interventions during the helping process. For instance, a physically attractive and extroverted helper might have blind spots with regard to the social flexibility and self-esteem of a physically unattractive and introverted client. Much of the literature on diversity and multiculturalism targets such blind spots. Counselors would do well to become aware of their own cultural values and biases. They should also make every effort to understand the worldviews of their clients. Helpers with diversity blind spots are handicapped. Helpers should, as a matter of course, become aware of the key ways in which they differ from their clients and take special care to be sensitive to those differences.

Tailor your interventions in a diversity-sensitive way. Both a practical understanding of diversity and self-knowledge need to be translated into appropriate interventions. The way a Hispanic helper challenges a Hispanic client may be inappropriate for a white client and vice versa. The way a younger helper shares his own experience with a younger client might be inappropriate for an older client and vice versa. Client self-disclosure, especially more intimate disclosure, might be relatively easy for a person from one culture, let us say North American culture, but very difficult for a client from another, let us say Asian or British culture. In this case, interventions that call for intimate self-disclosure may be seen as inappropriate by such a client.

But personal cultures within the same social culture differ. Even though a client may be from a culture that is more open to self-disclosure, he or she may be frightened to death by it. Therefore, with clients who come from a culture that has a different perception of self-disclosure or with any client who finds it difficult, it might make more sense, after an initial discussion of the problem situation in broad terms, to move to what the client wants instead of what he or she currently has (that is, Stage II: helping clients determine outcomes) rather than to the more intimate details of the problem situation. Once the helping relationship is on firmer ground, the client can move to the work he or she sees as more intimate or demanding.

Work with individuals. The diversity principle is clear: The more helpers understand the broad characteristics, needs, and behaviors of the populations with whom they work—African Americans, Caucasian Americans, diabetics, the elderly, the drug addict, the homeless, you name it—the better positioned they are to adapt these broad parameters and the counseling process itself to the individuals with whom they work. But, whereas diversity focuses on differences both between and within groups—cultures and subcultures, if you will—helpers interact with clients as individuals. As Satel (1996) pointed out: "Psychotherapy can never be about celebrating racial diversity because it is not about groups; it is about individuals and their infinite complexity" (p. A14). Your clients are individuals, not cultures, subcultures, or groups. Remember that category traits can destroy understanding as well as facilitate it.

Of course, individuals often have group characteristics, but they do not come as members of a homogeneous group because there are no homogeneous groups. One of the principal learnings of social psychology is this: There are as many differences, and sometimes more, within groups as between groups (see Weinrach & Thomas, 1996, pp. 473–474). This middle-class black male is this individual. This poor Asian woman is this person. In a very real sense, a conversation between identical twins is a cross-cultural event because they are different individuals with differences in personal assumptions, beliefs, values, norms, and patterns of behavior. Genetics and group culture account for commonalities among individuals, but personhood and personal cultures emphasize each person's uniqueness. Finally, valuing diversity is not the same as espousing a splintered, antagonistic society in which one's group membership is more important than one's humanity. On the other hand, valuing individuality is not the same as espousing a "society of one." This would make counseling and other forms of human interaction impossible.

Take Sean, a client you are seeing for the second time. He is very bright, well spoken, gay, Hispanic, poorly educated, lower-middle-class, slight of build, indifferent to his Catholic heritage, a churchgoer from habit, underemployed, good-looking, honest, and at sea because he feels "defeated." At his age, 26, life should be opening up, but he feels that it is closing down. He feels trapped. Understand this individual in any way you can, but work with Sean.

An extended discussion of diversity and multiculturalism in helping can be found in my booklet *Skilled Helping Around the World* (2006)—but don't expect to find the final answer there either.

A WORKING CHARTER: THE CLIENT-HELPER CONTRACT

Both implicit and explicit contracts govern the transactions that take place between people in a wide variety of situations, including marriage (in which some but by no means all of the provisions of the contract are explicit) and friendship (in which the provisions are usually implicit). If helping is to be a collaborative venture, then both parties must understand what their responsibilities are. Perhaps the term *working charter* is better than *contract*. It avoids the legal implications of the latter term and connotes a cooperative venture.

To achieve these objectives, the working charter should include, generically, the issues that have been covered in Chapters 1 through 3—that is, (a) the nature and goals of the helping process, (b) an overview of the helping approach together with some idea of the techniques to be used, (c) a sense of the flexibility built into the process, (d) how this process will help clients achieve their goals, (e) relevant information about yourself and your background, (f) how the relationship is to be structured and the kinds of responsibilities both you and the client will have, (g) the values that will drive the helping process, and (h) procedural issues. "Procedural issues" refers to the nuts and bolts of the helping process, such things as where sessions will be held and how long they will last. Procedural limitations should also be discussed—for instance, ground rules about whether the client can contact the helper between sessions. "Ordinarily we won't contact each other between sessions, unless there is some kind of emergency or we prearrange it for a particular purpose." Manthei and Miller (2000) have written a practical book for clients on the elements of a working charter. There is evidence that charters also work with the seriously mentally ill (Heinssen, Levendusky, & Hunter, 1995).

SHADOW-SIDE REALITIES IN THE HELPING RELATIONSHIP

There are common flaws in the working alliance that remain in the shadows either because they are not dealt with effectively by the helping professions themselves or because individual helpers are inept at addressing them with clients. Here are some of them (and don't forget the book *What Therapists Don't Talk About and Why*, mentioned in Chapter 1).

Ethical flaws. Little has been said about ethics in the helping process so far, not because it is not important, but because it is so important. There is a vast literature on ethical responsibilities in the helping professions (see Bersoff, 1995; Canter, Bennett, Jones, & Nagy, 1994; Claiborn, Berberoglu, Nerison, & Somberg, 1994; Corey, Corey, & Callanan, 1997; Cottone & Claus, 2000; Fisher & Younggren, 1997; Keith-Spiegel, 1994; Lowman, 1998). There is also a growing literature on ways in which helpers violate their ethical responsibilities. Because this area is too vast and too important to be given summary treatment here, helpers-to-be are urged to make this part of their professional development program.

Human tendencies in both helpers and clients. Neither helpers nor clients are usually heroic figures. They are human beings with all-too-human tendencies. For instance, helpers find clients attractive or unattractive. There is nothing wrong with this. However, they must be able to manage closeness in therapy in a way that furthers the helping process (R. S. Schwartz, 1993). They must deal with both positive and negative feelings toward clients lest they end up doing silly things. They may have to fight the tendency to be less challenging with attractive clients or not to listen carefully to unattractive clients. Clients, too, have their tendencies. Some have unrealistic expectations of counseling (Tinsley, Bowman, & Barich, 1993), whereas others trip over their own distorted views of their helpers. In such cases helpers have to manage both expectations and the relationship.

Very often these human tendencies on the part of both client and helper are not center-stage in awareness. Rather they constitute a subtext within the relationship. Unskilled helpers can get caught up in both their own and their clients' games. The working alliance breaks down. Skilled helpers, on the other hand, understand the shadow side of both themselves and their clients and manage them. Tools that helpers need to challenge themselves and invite their clients to do the same are discussed in Chapters 7 and 8.

Trouble in the relationship itself. The helping relationship might be flawed from the beginning. That is, the fit or chemistry between helper and client might not be right. But, for a variety of reasons, it is not easy for a helper to say, "I don't think I'm the one for you." On the other hand, high-level helpers can work with a wide variety of clients. They create their own chemistry. They make the relationship work.

One coach/counselor in a work setting was asked to work with a very bright manager whose interpersonal style left much to be desired. But the relationship was troubled from the start. Early on it become clear to the coach that his client expected him to "say good things" about him to senior managers. The client also had a tendency to play "mind" games with the coach, saying things like, "I wonder what's going on in your mind right now. I bet you're thinking things about me that you're not telling me." Managing expectations and managing the relationship proved to be hard work. However, he knew enough about the company to realize that "style" was an issue for the senior team. Because the client was bright and innovative, promotion was a distinct possibility, but because of his style promotion was probably "his to lose." The coach remained respectful and empathetic, but challenged "the crap." This shocked the client because he had always been able to "win" in his encounters with subordinates and peers. He stopped playing games and eventually realized that becoming better at interpersonal relations had only an upside.

Even if the relationship starts off on the right foot, it can deteriorate (Arnkoff, 1995; Omer, 2000). In fact, some deterioration is normal. Kivlighan and Shaughnessy (2000; see also Ackerman & Hilsenroth, 2003, and Stiles et al., 2004) talk about the "tear-and-repair" phenomenon. Many therapeutic relationships start well, get into trouble, and then recover. Experienced helpers are not surprised by this. However, some helping relationships get caught up in what Binder and Strupp (1997) call "negative process." They suggest that the ability of therapists to establish and maintain a good alliance has been overestimated. Hostile interchanges between helpers and clients are common in all treatment models. When impasses and ruptures in the relationship take place, ineffective helpers get bogged down. One study showed that both helpers and clients too often lack both the skill and the will for repair (Watson & Greenberg, 2000).

Factors associated with relationship breakdowns include "a client history of interpersonal problems, a lack of agreement between therapists and clients about the tasks and goals of therapy, interference in the therapy by others, transference, possible therapist mistakes, and therapist personal issues" (Hill, Nutt-Williams, Heaton, Thompson, & Rhodes, 1996, p. 207). If impasses and ruptures are not addressed, premature termination often takes place. When this happens, helpers predictably blame clients: "She wasn't ready," "He didn't want to work," "She was impossible," and so forth. Often the real problem is that such helpers fail to create the right chemistry.

Vague and violated values. Helpers do not always have a clear idea of what their values are. Or the values they say they hold—that is, their espoused values—do not always coincide with their actions. Values too often remain "good ideas" and are not translated into specific norms that drive helping behavior. For instance, even though helpers value self-responsibility in their clients, they see clients as helpless, make decisions for them, and direct rather than guide. Often they do so out of frustration. Expediency leads them to compromise their values and then rationalize their compromises. "I blew up at a client today, but he really deserved it. It probably did more good than my unappreciated patience." I bet.

Failure to share the helping process. When it comes to sharing the helping process itself, some counselors are reluctant to let the client know what the process is all about. Of course, helpers who "fly by the seat of their pants" can't tell clients what it's all about because they don't know what it's all about themselves. Still others seem to think that knowledge of helping processes is secret or sacred or dangerous and should not be communicated to the client, even though there is no evidence to support such beliefs (Dauser, Hedstrom, & Croteau, 1995; Somberg, Stone, & Claiborn, 1993; Sullivan, Martin, & Handelsman, 1993; Winborn, 1977).

Flawed contracts. There is an extensive shadow side to both explicit and implicit contracts. Even when a contract is written, the contracting parties interpret some of its provisions differently. Over time they forget what they contracted to and differences become more pronounced. These differences are seldom discussed. In counseling, the helper-client contract has been, traditionally, implicit, even though the need for more explicit structure has been discussed for years (Proctor & Rosen, 1983). Because of this, clients' expectations may differ from helpers' expectations (Benbenishty & Schul, 1987). Implicit contracts are not enough, but they still abound (Handelsman & Galvin, 1988; Weinrach, 1989; Woody, 1991).

Warring professionals. There are not just debates but also conflicts close to internecine wars in the helping professions. For instance, the debate on the "correct" approach to diversity and multiculturalism brings out some of the best and some of the worst in the helping community. Accusations, however subtle or blatant, of cultural imperialism on the one side and "political correctness" on the other fly back and forth. As has been noted, the debate on whether or how the helping professions should take political stands or engage in social engineering generates more heat than light. No significant article is published about any significant dimension of counseling without a barrage of often testy replies.

What happened to learning from one another and integration? The search for the truth gives way at times to the need to be right. It is not always clear how all of this serves the needs of clients. Indeed, clients are often enough left out of the debate. Just as many businesses today are reinventing themselves by starting with their customers and markets, so the helping professions should continually reinvent themselves by looking at helping through the eyes of clients.

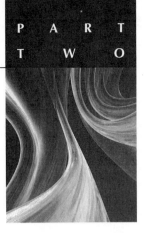

THE THERAPEUTIC DIALOGUE

In Part Two, the basic communication skills needed to be an effective helper are reviewed and illustrated. These skills are integrated under the title of "The Therapeutic Dialogue," meaning the helping dialogue, the problem-management dialogue, the opportunity-development dialogue. Dialogue is at the heart of the communication between helper and client. Chapter 4 includes an overview of both interpersonal communication and dialogue together with the first two basic communication skills. The first, often called "attending," is now called, more pragmatically, *visibly tuning in* to clients. This skill focuses on the helper's empathic presence to the client. The second skill is *active listening.* Helpers visibly tune in, not only to demonstrate their solidarity with their clients, but also to listen to and understand what their clients are saying both directly and indirectly. Chapter 5 deals with the skill of *empathy,* that is, the communication skill expression of the value of empathy discussed in Chapter 3, through which helpers both share and check out their understanding with clients. Chapter 6 explores the skills of *probing* and *summarizing* that counselors use to help clients explore issues more thoroughly. Chapters 7 and 8 deal with the skills and wisdom helpers need to invite clients to *challenge* dysfunctional forms of thinking, emotional expression, and action. Chapter 9, the last in Part Two, focuses on the kinds of *reluctance and resistance* found in many clients together with the strength and *resilience* they have, however deeply buried, to move beyond reluctance and resistance and manage problem situations and develop unused resources and opportunities more creatively.

COMMUNICATION: THE SKILLS OF TUNING IN AND ACTIVELY LISTENING TO CLIENTS

Hearing the slant or spin: Tough-minded listening and processing

Musing on what's missing

LISTENING TO ONESELF: THE HELPER'S INTERNAL CONVERSATION

THE SHADOW SIDE OF LISTENING TO CLIENTS

Forms of Distorted Listening

Filtered listening

Evaluative listening

Stereotype-based listening

Fact-centered rather than person-centered listening

Sympathetic listening

Interrupting

Myths About Nonverbal Behavior

THE SHADOW SIDE OF COMMUNICATION SKILLS—PART 1

THE IMPORTANCE OF DIALOGUE IN HELPING

Conversations between helpers and their clients should be a therapeutic or helping *dialogue* (Paré & Lysack, 2004; Seikkula & Trimble, 2005). Interpersonal communication competence means not only being good at the individual communication skills outlined in this and following chapters, but also marshaling them at the service of dialogue. There are four requirements for true dialogue (Egan, 2006):

- **Turn taking.** Dialogue is interactive. You talk, then I talk. In counseling this means that, generally speaking, monologues on the part of either client or helper don't add value. On the other hand, turn taking opens up the possibility for mutual learning. Helpers learn about their clients and base their interventions on what they come to understand through the give-and-take of the dialogue. Clients come to understand themselves and their concerns more fully and learn how to face up to the challenge their problems and opportunities present.

- **Connecting.** What each person says in the conversation should be connected in some way to what the other person has said. The helper's comments should be connected to the client's remarks and, ideally, vice versa. Helper and client need to engage each other if their working alliance is to be productive.

- **Mutual influencing.** In true dialogue the parties are open to being influenced by what the other person has to say. This echoes the social-influence dimension of counseling discussed in Chapter 3. Helpers certainly influence their clients, and the best helpers learn from and are influenced by the clients. Therefore, counselors need to be open-minded and help their clients be open to new learning.

- **Cocreating outcomes.** Good dialogue leads to outcomes that benefit both parties. As we have seen, counseling is about results, accomplishments, outcomes. The job of the counselor is neither to tell clients what to do nor merely to leave them to their own devices. The counselor's job to act as a catalyst for the kind of problem-managing dialogue that helps clients find their own answers. In true dialogue, neither party should know exactly what the outcome will be. If you know what you're going to tell a client or if the client has already made up his or her mind what he or she is going to say and do, the two of you may well have a conversation, but it is probably not a dialogue. Only clients can change themselves. Helpers facilitate change through effective dialogue.

There is another requirement for true dialogue in the sense that it is being used here. The values outlined in Chapter 3 must permeate and drive client-helper conversations. Exploitative dialogue is a contradiction in terms.

Although individual communication skills are a necessary part of communication competence, dialogue is the integrating mechanism. Individual skills are the building blocks for effective dialogue. The communication skills outlined in these chapters are not special skills peculiar to helping. Rather, they are extensions of the kinds of skills all of us need in our everyday interpersonal transactions (Adler, Proctor, & Towne, 2005; DeVito, 2004). Ideally, helpers-to-be would enter training programs with this basic set of interpersonal communication skills in place, and training would simply help them adapt the skills to the helping process.

Unfortunately, this is often not the case. Training or retraining in communication skills is the norm. These communication skills need to become "second nature" to helpers. People like Bob Carkhuff (1987), Allen Ivey (Ivey & Ivey, 2007), and Carl Rogers (1951, 1957, 1965) have been trailblazers in developing and humanizing communication skills and integrating them into the helping process. Their influence is seen throughout this book. The manual that accompanies this text, *Exercises in Helping Skills*, provides opportunities for extensive practice in all the communication skills discussed in these chapters.

It would also be helpful if clients had the communication skills outlined here and the ability to weave them into constructive dialogues with their helpers. Once more, this is not the case. In fact, many clients are in trouble precisely because they do not know how to establish and maintain healthy interpersonal relationships, which are nourished by effective communication. In the following sections, you will find suggestions for helping clients with poor communications engage in dialogue, starting with the first skill: visibly tuning in to clients.

I. VISIBLY TUNING IN: THE IMPORTANCE OF EMPATHIC PRESENCE

During some of the more dramatic moments of life, simply being with another person is extremely important. If a friend of yours is in the hospital, just your being there can make a difference, even if conversation is impossible. Similarly, being with a friend who has just lost his wife can be very comforting to him, even if little is said. Your empathic presence is comforting. Most people appreciate it when others pay attention to them. By the same token, being ignored is often painful: The averted face is too often a sign of the averted heart. Given how sensitive most of us are to others' attention or inattention, it is paradoxical how insensitive we can be at times about paying attention to others.

Helping and other deep interpersonal transactions demand a certain robustness or intensity of presence. Visibly tuning in to others contributes to this presence. It is an expression of empathy that tells clients that you are with them, and it puts you in a position to listen carefully to their concerns. Your attention can be manifested in both physical and psychological ways. Because nonverbal behavior can play an important part in empathic communication, let's start by briefly exploring nonverbal behavior as a channel of communication.

Nonverbal Behavior as a Channel of Communication

Over the years both researchers and practitioners have come to appreciate the importance of nonverbal behavior in counseling (Andersen, 1999; Ekman, 1992, 1993; Ekman & Friesen, 1975; Ekman & Rosenberg, 1998; Grace, Kivlighan, Jr., & Kunce, 1995; Hickson & Stacks, 1993; Highlen & Hill, 1984; Knapp & Hall, 1992; McCroskey, 1993; Mehrabian, 1972, 1981; Norton, 1983; Richmond & McCroskey, 2000; Riggio & Feldman, 2005; Russell, 1995; Russell, Fernandez-Dols, & Mandler, 1997; for a wealth of information about nonverbal behavior, see the following Internet site: http://www3.usal.es/~nonverbal/introduction.htm, last updated in 2003, but filled with useful references). Highlen and Hill suggested that nonverbal behaviors regulate conversations, communicate emotions, modify verbal messages, provide important messages about the helping relationship, give insights into self-perceptions, and provide clues that clients (or counselors) are not saying what they are thinking. This area has taken on even more importance because of the multicultural nature of helping.

The face and body are extremely communicative. We know from experience that even when people are together in silence, the atmosphere can be filled with messages. Sometimes clients' facial expressions, bodily motions, voice quality, and physiological responses communicate more than their words do.

Studies of nonverbal behavior should not be overinterpreted. Taken together, however, these studies do highlight pervasiveness and importance of nonverbal behavior in human communication. The following factors, on the part of both helpers and clients, play an important role in the therapeutic dialogue:

- **bodily behavior,** such as posture, body movements, and gestures
- **eye behavior,** such as eye contact, staring, eye movement
- **facial expressions,** such as smiles, frowns, raised eyebrows, and twisted lips
- **voice-related behavior,** such as tone of voice, pitch, volume, intensity, inflection, spacing of words, emphases, pauses, silences, and fluency
- **observable autonomic physiological responses,** such as quickened breathing, blushing, paleness, and pupil dilation
- **physical characteristics,** such as fitness, height, weight, and complexion
- **space,** that is, how close or far a person chooses to be during a conversation
- **general appearance,** such as grooming and dress

People constantly "speak" to one another through their nonverbal behavior. Effective helpers learn this "language" and how to use it effectively in their interactions with their clients. They also learn how to "read" relevant messages embedded in the nonverbal behavior of their clients.

Helpers' Nonverbal Behavior

Before you begin interpreting the nonverbal behavior of your clients (discussed later in this chapter), take a look at yourself. You speak to your clients through all the nonverbal categories outlined above. At times your nonverbal behavior is as important as, or even more important than, your words. Your nonverbal behavior influences clients for better or for worse. Clients read in your nonverbal behavior cues that indicate the quality of your presence to them. Attentive presence can invite or encourage them to trust you, open up, and explore the significant dimensions of their problem situations.

Half-hearted presence can promote distrust and lead to clients' reluctance to reveal themselves to you. Clients may misinterpret your nonverbal behavior. For instance, you may be comfortable with the space between you and your client, but it is too close for the client. Or remaining silent might in your mind mean giving a client time to think, but the client might feel embarrassed. Part of listening, then, is being sensitive to clients' reactions to your nonverbal behavior.

Effective helpers are mindful of, but not preoccupied with, the stream of nonverbal messages they send to clients. Reading your own bodily reactions is an important first step. For instance, if you feel your muscles tensing as the client talks to you, you can say to yourself, "I'm getting anxious here. What's going on? And what nonverbal messages indicating my discomfort am I sending to the client?" Again, you probably would not use these words. Rather you read the signals your body is sending you without letting them distract you from your client.

You can also use your body to censor instinctive or impulsive messages that you feel are inappropriate. For instance, if the client says something that instinctively angers you, you can control the external expression of the anger (for instance, a sour look) in order to give yourself time to reflect. Such self-control is not phony because your respect for your client takes precedence over your instinctive reactions. Not dumping your annoyance or anger on your clients through nonverbal behavior is not the same as denying it. Becoming aware of it is the first step in dealing with it.

In a more positive vein, you can "punctuate" what you say with nonverbal messages. For instance, Denise is especially attentive when Jennie talks about actions she could take to do something about her problem situation. She leans forward, nods, and say "uh-huh." She uses nonverbal behavior to reinforce, let's say, Jennie's intention to act constructively in renewing contact with a couple of key friends.

On the other hand, don't become preoccupied with your body and the qualities of your voice as a source of communication. Rather, learn to use your body instinctively as a means of communication. Being aware of and at home with nonverbal communication can reflect an inner peace with yourself, the helping process, and your clients. Your nonverbal behavior should enhance rather than stand in the way of your working alliance with your clients.

Although the skills of being visibly tuned in can be learned, they will be phony if they are not driven by the attitudes and values such as respect and empathy discussed in Chapter 3. Your mind set—what's in your heart—is as important as your visible presence. If you are not actively interested in the welfare of your client or if you resent working with a client, subtle or not-so-subtle nonverbal clues will color your behavior. I once mentioned to a doctor my concerns about an invasive diagnostic procedure he intended to use. The doctor said the right words to reassure me, but his physical presence and the way he rushed his words said, "I've heard this dozens of times. I really don't have time for your concerns. Let's get on with this." His words were right but the real message was in the nonverbal messages that accompanied his words.

Guidelines for Visibly Tuning in to Clients

There are certain key nonverbal skills you can use to visibly tune in to clients. These skills can be summarized in the acronym SOLER. Because communication skills are particularly sensitive to cultural differences, care should be taken in adapting what follows to different cultures. What follows, however is only a framework.

S: Face the client *Squarely*. That is, adopt a posture that indicates involvement. In North American culture, facing another person squarely is often considered a basic posture of involvement. It usually says, "I'm here with you; I'm available to you." Turning your body away from another person while you talk to him or her can lessen your degree of contact with that person. Even when people are seated in a circle, they usually try in some way to turn toward the individuals to whom they are speaking. The word *squarely* here should not be taken too literally. "Squarely" is not a military term. The point is that your bodily orientation should convey the message that you are involved with the client. If, for any reason, facing the person squarely is too threatening, then an angled position may be more helpful. The point is not inches and angles but the quality of your presence. Your body sends out messages whether you like it or not. Make them congruent with what you are trying to do.

O: Adopt an *Open* posture. Crossed arms and crossed legs can be signs of lessened involvement with or availability to others. An open posture can be a sign that you're open to the client and to what he or she has to say. In North American culture, an open posture is generally seen as a nondefensive posture. Again, the word *open* can be taken literally or metaphorically. If your legs are crossed, this does not mean that you are not involved with the client. But it is important to ask yourself, "To what degree does my present posture communicate openness and availability to the client?" If you are empathic and open-minded, let your posture mirror what is in your heart.

L: Remember that it is possible at times to *Lean* toward the other. Watch two people in a restaurant who are intimately engaged in conversation. Very often they are both leaning forward over the table as a natural sign of their involvement. The main thing is to remember that the upper part of your body is on a hinge. It can move toward a person and back away. In North American culture, a slight inclination toward a person is often seen as saying, "I'm with you, I'm interested in you and in what you have to say." Leaning back (the severest form of which is a slouch) can be a way of saying, "I'm not entirely with you" or "I'm bored." Leaning too far forward, however, or doing so too soon, may frighten a client. It can be seen as a way of placing a demand on the other for some kind of closeness or intimacy. In a wider sense, the word *lean* can refer to a kind of bodily flexibility or responsiveness that enhances your communication with a client. And bodily flexibility can mirror mental flexibility.

E: Maintain good *Eye* contact. In North American culture, fairly steady eye contact is not unnatural for people deep in conversation. It is not the same as staring. Again, watch two people deep in conversation. You may be amazed at the amount of direct eye contact. Maintaining good eye contact with a client is another way of saying, "I'm with you; I'm interested; I want to hear what you have to say." Obviously, this principle is not violated if you occasionally look away. Indeed, you have to if you don't want to stare. But if you catch yourself looking away frequently, your behavior may give you a hint about some kind of reluctance to be with this person or to get involved with him or her. Or it may say something about your own discomfort. In other cultures, however, too much eye contact, especially with someone in a position of authority, is out of order. I have learned much about the cultural meaning of eye contact from my Asian students and clients.

R: Try to be relatively *Relaxed* or natural in these behaviors. Being relaxed means two things. First, it means not fidgeting nervously or engaging in distracting facial expressions. The client may wonder what's making you nervous. Second, it means becoming comfortable with using your body as a vehicle of personal contact and expression. Your being natural in the use of these skills helps put the client at ease.

A counselor trained in the *Skilled Helper* model was teaching counseling to visually impaired students in the Royal National College for the Blind. Most of her clients were visually impaired. However, she wrote this about SOLER:

> In counseling students who are blind or visually impaired, eye contact has little or no relevance. However, attention on voice direction is extremely important, and people with a visual impairment will tell you how insulted they feel when sighted people are talking to them while looking somewhere else.
>
> I teach SOLER as part of listening and attending skills and can adapt each letter of the acronym [to my visually impaired students] with the exception of the E. . . . After much thought, I would like to change your acronym to SOLAR, the A being for "Aim," that is, aim your head and body in the direction of your client so that when they hear your voice, be it linguistically or paralinguistically, they know that you are attending directly to what they are saying. (personal communication)

This underscores the fact that people are more sensitive to how you orient yourself to them nonverbally than you might imagine. Anything that distracts from your "being there" can harm the dialogue. The point to be stressed is that a respectful, empathic, genuine, and caring mind-set might well lose its impact if the client does not see these internal attitudes reflected in your external behaviors.

In the beginning you may become overly self-conscious about the way you visibly tune in, especially if you are not used to being attentive. Still, the guidelines just presented are just that—guidelines. They should not be taken as absolute rules to be applied rigidly in all cases. Box 4-1 summarizes, in question form, the main points

Box 4-1   Questions on Visibly Tuning In

- What are my attitudes toward this client?
- How would I rate the quality of my presence to this client?
- To what degree does my nonverbal behavior indicate a willingness to work with the client?
- What attitudes am I expressing in my nonverbal behavior?
- What attitudes am I expressing in my verbal behavior?
- To what degree does the client experience me as effectively present and working with him or her?
- To what degree does my nonverbal behavior reinforce my internal attitudes?
- In what ways am I distracted from giving my full attention to this client? What am I doing to handle these distractions? How might I be more effectively present to this person?

related to being visibly tuned in to clients. Turn to the *Exercises in Helping Skills* for opportunities to "practice" the skill of visibly tuning in. Every conversation you have is an opportunity to practice.

II. Active Listening: The Foundation of Understanding

Visibly tuning in to clients is not, of course, an end in itself. We tune in both mentally and visibly in order to listen to what clients have to say—their stories, complaints, points of view, intentions, proposals, decisions, and everything else. Listening carefully to a client's concerns seems to be a concept so simple to grasp and so easy to do that one may wonder why it is given such explicit treatment here. Nonetheless, it is amazing how often people fail to listen to one another. Full listening means listening actively, listening accurately, and listening for meaning. Listening is not merely a skill. It is a rich metaphor for the helping relationship itself—indeed, for all relationships. I will attempt to tap some of that richness here.

Active listening plays a key role in all human service endeavors. Take the doctor-patient relationship, for instance. Patients have two central concerns about their doctors—their medical competence and their ability to relate and communicate. Listening, then, is an important medical skill. Hippocrates told aspiring doctors to "listen to the patient, and the patient will tell you what is wrong." Today, even though doctors use sophisticated high-tech diagnostic methods together with a hands-on approach in making their diagnoses, listening to patients is still an extremely important part of health care. If Hippocrates were living, I'm sure that he would still be giving the same advice to doctors.

The Accreditation Council for Graduate Medical Education (ACGME) obviously agrees (see http://www.acgme.org, last retrieved here in February 2006). This body is responsible for the accreditation of post-M.D. medical training programs within the United States. Its goal is to use state-of-the-art concepts from education and health care in the accreditation of residency programs.

In an initiative called the "Outcome Project," ACGME has identified six general competencies important to the practice of medicine, one of which is "interpersonal and communication skills." The goal of the Outcome Project is to base programs' accreditation status on how well they educate residents and prepare them for the practice of medicine. Medical schools, including Harvard, are gearing up to offer communication skills programs based on ACGME's new competency standards (Landro, 2005). For some, this is long overdue. For instance, Graybar and Leonard (2005) decry what they see as a decline in the importance of listening to patients in mental health settings. They believe that this is due to "the abuses of managed care, the marketing and misuse of psychotropic medications, and the growth of brief, manualized, empirically supported treatments (ESTs)" (p. 1).

Inadequate Listening

Effective listening is not a state of mind, like being happy or relaxed. It's not something that "just happens." It's an activity. In other words, effective listening requires work. Let's first take a look at the opposite of active listening. All of us have been, at one time or another, both perpetrators and victims of the following forms of inactive or inadequate listening.

Nonlistening. Sometimes we go through the motions of listening but are not really engaged. At times we get away with it. Sometimes we are caught. "What would you do?" Jennifer asks her colleague, Kieran, after outlining a problem the university counseling center is having with the structure of the sales group. Embarrassed, Kieran replies, "I'm not sure." Staring him down, she says, "You haven't been listening to a word I've said." For whatever reason, he had tuned her out. Obviously no helper sets out not to listen, but even the best can let their mind wander as they listen to the same kind of stories over and over again, forgetting that the story is unique to *this* client.

Partial listening. This is listening that skims the surface. The helper picks up bits and pieces, but not necessarily the essential points the client is making. For instance, Janice's client, Dean, is talking to her about a date that went terribly wrong. Janice only half listens. It seems that Dean is not that interesting. Dean stops talking and looks rather dejected. Janice tries to pull together the pieces of the story she did listen to. Her attempt to express understanding has a hollow ring to it. Dean pauses and then switches to a different topic. Inadequate listening helps neither understanding nor relationships.

Tape-recorder listening. What clients look for from listening is not the helper's ability to repeat their words. A tape recorder could do that perfectly. People want more than physical presence in human communication; they want the other person to be present psychologically, socially, and emotionally. Sometimes helpers fail to visibly tune in and listen; they are totally present. Clients pick up on signs of nonlistening and lack of total presence. How many times have you heard someone exclaim, "You're not listening to what I'm saying!" When the person accused of not listening answers, almost predictably, "I *am* too listening; I can repeat everything you've said," the accuser is not comforted. Usually clients are too polite or cowed or preoccupied with their own concerns to say anything when they find themselves in that situation. But it is a shame if your auditory equipment is in order, but you are elsewhere. Your clients want *you*, a live counselor, not a tape recorder.

Rehearsing. Picture Sid, a novice counselor, sitting with Casey, a client. At one point in the conversation when Casey talks about some "wild dreams" he is having, Sid says to himself, "I know very little about dreams; I wonder what I'm going to say?" Sid stops listening and begins rehearsing his response. Even when experienced helpers begin to mull over how they will respond to the client, they stop listening. On the other hand, effective helpers listen intently to clients and to the themes and core messages embedded in what clients are saying. They are never at a loss in responding. They don't need to rehearse. And their responses are much more likely to help clients move forward in the problem-management process. When the client stops speaking, they often pause to reflect on what he or she just said, and then speak. Pausing says, "I'm still mulling over what you've just said. I want to respond thoughtfully." They pause because they have listened.

Empathic Listening: Listening to Clients' Stories and Their Search for Solutions

The opposite of inactive or inadequate listening is empathic listening, listening driven by the value of empathy. Empathic listening centers on the kind of attending, observing, and listening—the kind of "being with"—needed to develop an understanding of clients and their worlds. Although it might be metaphysically impossible to actually get "inside" the world of another person and experience the world as he or she does, it is possible to approximate this.

Carl Rogers (1980) talked passionately about basic empathic listening—being with and understanding the other—even calling it "an unappreciated way of being" (p. 137). He used the word *unappreciated* because in his view few people in the general population developed this "deep listening" ability and even so-called expert helpers did not give it the attention it deserved. Here is his description of empathic listening, or "being with":

> It means entering the private perceptual world of the other and becoming thoroughly at home in it. It involves being sensitive, moment by moment, to the changing felt meanings which flow in this other person, to the fear or rage or tenderness or confusion or whatever that he or she is experiencing. It means temporarily living in the other's life, moving about in it delicately without making judgments. (p. 142)

Such empathic listening is selfless because helpers must put aside their own concerns to be fully with their clients. Of course, Rogers pointed out that this deeper understanding of clients remains sterile unless it is somehow communicated to them. Although clients can appreciate how intensely they are attended and listened to, they and their concerns still need to be understood. Empathic listening leads to empathic understanding, which leads to empathic responding.

Empathic participation in the world of another person obviously admits of degrees. As a helper, you must be able to enter clients' worlds deeply enough to understand their struggles with problem situations or their search for opportunities with enough depth to make your participation in problem management and opportunity development valid and substantial. If your help is based on an incorrect or invalid understanding of the client, then your helping may lead him or her astray. If your understanding is valid but superficial, then you might miss the central issues of the client's life.

The following case will be used to help you develop a better behavioral feel for empathic listening.

Jennie, an African American college senior, was raped by a "friend" on a date. She received some immediate counseling from the university Student Development Center and some ongoing support during the subsequent investigation. But even though she was raped, it turned out that it was impossible for her to prove her case. The entire experience—both the rape and the investigation that followed—left her shaken, unsure of herself, angry, and mistrustful of institutions she had assumed would be on her side (especially the university and the legal system). When Denise, a middle-aged and middle-class African American social worker who was a counselor for a health maintenance organization (HMO), first saw her a couple of years after the incident, Jennie was plagued by a number of somatic complaints, including headaches and gastric problems. At work, she engaged in angry outbursts whenever she felt that someone was taking advantage of her. Otherwise she had become quite passive and chronically depressed. She saw herself as a woman victimized by society and was slowly giving up on herself.

Denise is a pro, so she doesn't have much of a problem with inadequate listening. She is an empathic listener par excellence. Also Jennie's story and what she's going through right now are engaging. Denise knows she has to be with Jennie every step of the way.

Focused listening. In many ways helping is a "talking game." Therefore, the kind and quality of talk are both crucial. Listening at its best is both focused *and* unbiased.

Two forms of focus are offered here. First, the problem-management helping model itself, because it is not theory- or school-focused, helps counselors organize what they are hearing without prejudice. Problem-management and opportunity-development dialogue is at the heart of helping. Helpers listen intently to clients' stories to help them search for solutions.

The second aid to focused listening is a particular view of *personality*. Pervin's (1996) definition of personality is "the complex organization of cognitions [thoughts], affects [emotion], and behaviors that gives direction and pattern (coherence) to the person's life" (p. 414). Rasmussen (2005), author of one of the books in a personality-guided therapy series (Everly & Lating, 2004; Farmer & Nelson-Gray, 2005; Harper, 2004; Magnativa, 2005), adds the notion of "activating event" (p. 4) to Pervin's triad. Because clients tell their stories in terms of their experiences (activating events), thoughts, behaviors, and emotions, we as helpers are listening to their personalities at work, as it were. More precisely, we are listening to both individual thoughts and *patterns* of thinking, to both individual behaviors and *patterns* of behaving, and to both individual emotions and *patterns* of emotionality—all stimulated by both internal and external experiences or events.

Clients' stories, then, tend to be mixtures of clients' experiences, thoughts, emotions, and behaviors.

- Clients talk about their *experiences*—that is, what happens to them. If a client tells you that she was fired from her job, she is talking about her problem situation as an experience. Jennie, of course, talked about being raped, belittled, and ignored.

- Clients talk about the way they *think*, what goes on in their head. Jennie shares her points of view about the rape and its aftermath. She thinks that the color of her skin worked against her.

- Clients talk about their *behavior*—that is, what they do or refrain from doing. If a client tells you that he smokes and drinks a lot, he is talking about his external behavior. If a different client says that she spends a great deal of time daydreaming, she is talking about her internal behavior. Jennie talked about pulling away from her family and friends after the rape investigation.

- Clients talk about their *affect*—that is, the feelings, emotions, and moods that arise from or are associated with their experiences and both internal and external behavior. If a client tells you how depressed she gets after fights with her fiancé, she is talking about the mood associated with her experiences and behavior. Jennie talked about her shame, her feelings of betrayal, and her anger.

Of course, thoughts, actions, and emotions are interrelated in the day-to-day lives of clients. And so they mix them together in telling their stories. Consider this example. A client is talking to a counselor in the personnel department of a large company. The client says in a very agitated way: "I've just had one of the lousiest days of my life." At this point the counselor knows that something went wrong and that the client feels bad about it, but she knows relatively little about the specific experiences, thoughts, and behaviors that has made the day such a horror for the client. However, the client continues.

Toward the end of the day my boss yelled at me in front of some of my colleagues for not landing an order from a new customer [an experience]. I lost my temper [emotion] and yelled right back at him [behavior]. He blew up and fired me on the spot [an experience for the client]. I really think that he's a jerk and the company should not tolerate people like him in supervisory positions [a thought, a point of view]. And now I feel awful [emotion] and am trying to find out if I really have been fired and, if so, if I can get my job back [behavior]. I have every intention to fight this; it's unjust [a thought, a resolve, a point of view].

Now the counselor knows a great deal more about the problem situation. Problem situations are much clearer when they are spelled out as *specific* experiences, thoughts, behaviors, and feelings related to specific situations. Because clients spend so much time telling their stories, a few words about each of these elements are in order.

Listening to clients' experiences.   Most clients spend a fair amount of time, sometimes too much time, talking about what happens *to* them.

- "My wife doesn't understand me."
- "My ulcer acts up when family members argue."
- "My boss doesn't give me feedback."
- "I get headaches a lot."

It is of paramount importance to listen to and understand clients' experiences. However, because experiences often dwell on what other people do or fail to do, experience-focused stories at times smack a bit of passivity. The implication is that others—or the world in general—are to blame for the client's problems.

- "She doesn't do anything all day. The house is always a mess when I come home. No wonder I can't concentrate at work."
- "He tells his little jokes, and I'm always the butt of them. He makes me feel bad about myself most of the time."

Some clients talk about experiences that are internal and out of their control.

- "These feelings of depression come from nowhere and seem to suffocate me."
- "I just can't stop thinking of him."

The last statement sounds like an action, but it is expressed as an experience. It is something happening to the client, at least to the client's way of thinking.

One reason that some clients fail to manage the problem situations of their lives is that they are too passive or see themselves as victims, adversely affected by other people, by the immediate social settings of life such as the family, by society in its larger organizations and institutions such as government or the workplace, by cultural prescriptions, or even by internal forces. They feel that they are no longer in control of their lives or some dimension of life. Therefore, they talk extensively about these experiences.

- "Company policy discriminates against women. It's that simple."
- "The economy is booming but the kind of jobs I want are already taken."
- "No innovative teacher gets very far around here."

Of course, some clients *are* treated unfairly; they are victimized by the behaviors of others in the social and institutional settings of their lives. Although they can be helped to cope with victimization, full management of their problem situations demands changes in the social settings themselves. A counselor helped one client to cope with a brutal husband, but ultimately the courts had to intervene to keep him at bay.

For other clients, talking constantly about experiences is a way of avoiding responsibility: "It's not my fault. After all, these things are happening to me." Sykes (1992), in his book *A Nation of Victims*, was troubled by the tendency of the United States to become a "nation of whiners unwilling to take responsibility for our actions." Whether his statement is true or not, counselors must be able to distinguish "whiners" from those who are truly being victimized. Helpers need to honor clients' experiences without getting lost in them.

Listening to clients' thoughts and patterns of thinking. A lot goes on in clients' heads. Common ways in which clients share their thinking include sharing *points of view*, stating *intentions*, declaring *decisions*, and offering *proposals* and *plans*.

Clients' points of view. As clients tell their stories, explore possibilities for a better future, set goals, make plans, and review obstacles to accomplishing these plans, they often share their points of view. A point of view is a client's personal estimation of something. A full point of view includes the point of view itself, the reasons for it, an illustration to bring it to life, and some indication of how open the client might be to modifying it. There is no expectation that anyone else need adopt the point of view. That would be a form of selling or persuasion. But, realistically, the implication often is: "I think this way. Why don't other people think this way?"

For instance, Aurora, an 80-year-old woman, is talking to a counselor about the various challenges of old age. At one point she says,

> My sister in Florida [85-years-old]—Sis, we call her—is sick. She's probably dying, but she wants to stay at home. She's asked me to come down there and take care of her. I think that's asking too much. I could use some help myself these days. But she keeps calling.

Aurora's point of view is that her sister's request is not realistic because she herself needs some help to get by. But her sister persists in trying to persuade her to change her mind.

Points of view reveal clients' beliefs, values, attitudes, and convictions. Clients may share their points of view about everything under the sun. You will need to listen to and understand the ones that are relevant to their problem situations or undeveloped opportunities. Let's return briefly to Jennie and Denise.

> JENNIE: You just can't trust the system. They're not going to help. They take the easy way out. I don't care which system it might be. Church, government, the community, sometimes even the family. They're not going to give you much help.

Denise listens carefully to Jennie's point of view and realizes how much it is influencing her behavior. In Jennie's case it's easy to see where the point of view comes from. But Denise also knows that at some point she needs to challenge Jennie's point of view because it may be one of the things that is keeping her locked in her misery. Points of view have power.

Clients' intentions, proposals, and plans. Clients provide a window into their thinking when they state intentions, offer proposals, or make a case for certain courses of action. Consider Lydia. She is a single parent of two young children and a member of the "working poor." Her wages don't cover her expenses. She has no insurance. The father of her children has long disappeared. She says to a social worker,

> I've been thinking of quitting my job. I'm making the minimum wage and with travel expenses and all I just can't make ends meet. I spend too much time traveling and don't see enough of my kids. Friends look after them when I'm gone, but that's hit and miss and puts a burden on them. You know if I go on welfare I could make almost as much. And then I could pick up jobs that would pay me cash. I've got friends who do this. I believe I could make ends meet. My kids and I would be better off. And I wouldn't be hassled as much.

Lydia is making a case, not announcing a decision. The case includes what she wants to do (quit her job and move into the "alternate" work economy), the reasons for doing it (the inadequacy of her current work situation, the need to make ends meet), and the implications for herself and her children (she'd be less hassled and her kids would see more of her and be better off).

Clients' decisions. Clients talk about decisions they are making or have already made. A client might say, "I've decided to stop drinking. Cold turkey." Or, "I'm tired of being alone. I'm going to join a dating service." Decisions usually have implications for the decision maker and for others. The client who has decided to quit drinking cold turkey has his work cut out for him, but there are implications for his spouse. For instance, she's used to coping with a drunk, but now she may have to learn how to cope with this "new person" in the house. Commands, instructions, and even hints are, in a way, decisions about other people's behavior. A client might say to her helper, "Don't bring up my ex-husband any more. I'm finished with him."

Sharing a decision fully means spelling out the decision itself, the reasons for the decision, the implications for self and others, and some indication as to whether the decision or any part of it is open to review. For instance, Jennie, in talking with Denise about future employment, says in a rather languid tone of voice, "I'm not going to get any kind of job where I have to fight the race thing. Or the woman thing. I'm tired of fighting. I only get hurt. I know that this limits my opportunities, but I can live with that." Note that this is more than a point of view. Jennie is more or less saying, "I've made up my mind." She notes the implication for herself—a limitation of job opportunities—and an implication for Denise might be, "So that's the end of it. Don't try to convince me otherwise." Denise hears the message and the implied command. However, she believes that some of Jennie's messages need challenging. Decisions can be tricky. Often enough, *how* they are delivered says a great deal about the decision itself. Given the rather languid way in which Jennie delivers her decision, Denise thinks that it might not be Jennie's final decision. This is something that has to be checked out. A dialogue with Jennie about the reasons for her decision and a review of its implications are possible routes for a challenge.

Listening to clients' behaviors and patterns of behavior. All of us do things that get us into trouble and fail to do things that will help us get out of trouble or develop opportunities. Clients are no different.

- "When he ignores me, I get to work thinking of ways of getting back at him."
- "Whenever anyone gets on my case for having a father in jail, I let him have it. I'm not taking that kind of crap from anyone."

- "Even though I feel the depression coming on, I don't take the pills the doctor gave me."
- "When I get bored, I find some friends and go get drunk."
- "I have a lot of sexual partners and have unprotected sex whenever my partner will let me."

Some clients talk freely about their experiences, what happens to them, but seem more reluctant to talk about their behaviors. One reason for this is that they can't talk about behaviors without bringing up issues of personal responsibility.

Listening to the client's feelings, emotions, and moods. Feelings, emotions, and moods constitute a river that continually runs through us—peaceful, meandering, turbulent, or raging—often beneficial, sometimes dangerous, seldom neutral. They are certainly an important part of clients' problem situations and undeveloped opportunities (Greenberg, 2002; Greenberg & Paivio, 1997; Plutchik, 2001, 2003). Some have complained that psychologists—both researchers and practitioners—don't take emotions seriously enough. For instance, anger is an ubiquitous and extremely important emotion, but "it has been oddly neglected by the clinical community" (Norcross & Kobayashi, 1999, p. 275—this is the opening article of a series of articles on anger in a special section of the *Journal of Clinical Psychology, 55*, March 1999).

But that was in 1999. Today there are plenty of signs that things have changed. In 2001, a new American Psychological Association journal, *Emotion*, entered the scene because of the recognition that emotion is fundamental to so much of human life. The journal includes articles ranging from the so-called "softer" side of psychology through hard-nosed molecular science. There is a rich history of scholarly books and articles written about emotion (for instance, see Griffiths, 1997, with its rich set of references). On the other hand, popular self-help books written by professionals have proved useful to both clients and practitioners for years (McKay & Dinkmeyer, 1994; McKay, Davis, & Fanning, 1997). These very practical books tend to take a positive psychology approach to the experience, regulation, and use of emotion in everyday life. Inevitably, there is the commercialization of emotion highlighted by a veritable industry focused on "anger management" (D. Seligman, 2003). Some of the programs are excellent; some are trash. Roffman (2004) rejects the notion of anger as a "thing-to-be-managed" (p. 161) and looks on anger as a resource. Anyway, there is more to emotional life than anger.

Recognizing key feelings, emotions, and moods (or the lack thereof) is very important for at least three reasons. First, they pervade our lives. There is an emotional tone to just about everything we do. Feelings, emotions, and moods pervade clients' stories, points of view, decisions, and intentions or proposals. Second, they greatly affect the quality of our lives. A bout of depression can stop us in our tracks. A client who gets out from under the burden of self-doubt breathes more freely. Third, feelings, emotions, and moods are drivers of our behavior. As Lang (1995) pointed out, they are "action dispositions" (p. 372). Clients driven by anger can do desperate things. On the other hand, enthusiastic clients can accomplish more than they ever thought they could. The good news is that we can learn how to tune ourselves in to our clients' and our own feelings, emotions, and moods (Machado, Beutler, & Greenberg, 1999) at the service of discovering how to regulate them.

Understanding the role of feelings, emotions, and moods in clients' problem situations and their desire to identify and develop opportunities is central to the helping process. Emotions highlight learning opportunities.

- "I've been feeling pretty sorry for myself ever since he left me." This client learns that self-pity constricts her world and limits problem-managing action.

- "I yelled at my mother last night and now I feel very ashamed of myself." Shame may well be a wake-up call in this client's relationship with his mother.

- "I've been anxious for the past few weeks, but I don't know why. I wake up feeling scared and then it goes away but comes back again several times during the day." Anxiety has become a bad habit for this client. It is self-perpetuating. What can the client do to break through the vicious circle?

- "I finally finished the term paper that I've been putting off for weeks and I feel great!" Here emotion becomes a tool in this client's struggle against procrastination.

The last item in this list brings up an important point. In the psychological literature, negative emotions tend to receive more attention than positive emotions. Now work is under way to study positive emotions and their beneficial effects. There are indications that we can use positive emotions to promote both physical and psychological well-being (Salovey, Rothman, Detweiler, & Steward, 2000). There are indications that emotions can free up psychological resources, act as opportunities for learning, and promote health-related behaviors. In managing problems and developing opportunities, social support plays a key role. As Salovey and his colleagues note, clients are more likely to elicit social support if they manifest a positive attitude toward life. Potential supporters tend to shun clients who let their negative emotions get the best of them.

Of course, clients often express feelings without talking about them. When a client says, "My boss gave me a raise and I didn't even ask for one!" you can feel the emotion in her voice. A client who is talking listlessly and staring down at the floor may not say, in so many words, "I feel depressed." A dying person may express feelings of anger and depression without talking about them. Other clients feel deeply about things but do their best to hold their feelings back. But effective helpers can usually pick up on clues or hints, whether verbal or nonverbal, that indicate the feelings and emotions rumbling inside.

Clients' stories, points of view, decisions, and expressed intentions or proposals for action are permeated by feelings, emotions, and moods. Your job is to listen carefully to the ways in which they affect, color, and give meaning to words they are using. The meaning is not just in the words. It's in the full package.

Listening for strengths, opportunities, and resources. If you listen only for problems, you will end up talking mainly about problems. And you will shortchange your clients. Every client has something going for him or her. Your job is to spot clients' resources and help them invest these resources in managing problem situations and opportunities. If it is true that people generally use only a fraction of their potential (Maslow, 1968), then there is much to be tapped. For instance, a counselor is working with a 65-year-old, successful businessman who, with his wife, has raised three children. The children are well-educated and successful in their own right. The man is having difficulty coping with some health problems. The counselor learns that the man was one of a group of poor inner-city boys in a longitudinal study. The boys had a mean IQ of 80 and a lot of social disadvantages (see Vaillant, 2000). As the counselor listens to the man's story, he hears a history of resilience. The counselor helps him review the strategies he used to cope as he was growing up. Energized by this, the man says, "I never gave up then. Why should I start giving up now?"

One section of the positive psychology movement focuses on strengths, especially strengths that clients have but fail to use as they struggle with problem situations (Aspinwall & Staudinger, 2003; Peterson & Seligman, 2004). Taken together, these writings constitute another approach to formulating the kind of model of human maturity mentioned in Chapter 1. Just the list of the strengths examined in a book on "positive psychological assessment," edited by Lopez and Snyder (2003), gives the reader a lift—hope, optimism, self-efficacy, problem-solving, internal locus of control, creativity, wisdom, courage, positive emotions, self-esteem, love, emotional intelligence, forgiveness, humor, gratitude, faith, morality, coping, and well-being. Though we might long for a world in which striving for these virtues was a priority, listening for hints of any or all of these capabilities in our clients is a first step. Finally, all of these writers are committed to developing a *science* of human strengths.

Listening to clients' nonverbal messages and modifiers. Carton, Kessler, and Pape (1999) showed that the ability of people to read nonverbal messages is one factor in establishing and maintaining relationships. So once we have an understanding of our own nonverbal "speech," we can turn to an exploration of clients' nonverbal behavior. Clients send a steady stream of clues and messages through their nonverbal behavior. Helpers need to learn how to read these messages without distorting or overinterpreting them. For instance, when Denise says to Jennie, "It seems that it's hard talking about yourself," Jennie says, "No, I don't mind at all." But the real answer is probably in her nonverbal behavior, for she speaks hesitatingly while looking away and frowning. Reading such cues helps Denise understand Jennie better. Our nonverbal behavior has a way of "leaking" messages about what we really mean. The very spontaneity of nonverbal behaviors contributes to this leakage even in the case of highly defensive clients. It is not easy for clients to fake nonverbal behavior (Wahlsten, 1991).

Besides being a channel of communication in itself, such nonverbal behavior as facial expressions, bodily motions, and voice quality often modify and punctuate verbal messages in much the same way that periods, question marks, exclamation points, and underlining punctuate written language. All the kinds of nonverbal behavior mentioned earlier in this chapter can punctuate or modify verbal communication in the following ways:

- **Confirming or repeating.** Nonverbal behavior can confirm or repeat what is being said verbally. For instance, once when Denise responds to Jennie with just the right degree of understanding—she hits the mark—not only does Jennie say, "That's right!" but also her eyes light up (facial expression), she leans forward a bit (bodily motion), and her voice is very animated (voice quality). Her nonverbal behavior confirms her verbal message.

- **Denying or confusing.** Nonverbal behavior can deny or confuse what is being said verbally. When challenged by Denise, Jennie denies that she is upset, but her voice falters a bit (voice quality) and her upper lip quivers (facial expression). Her nonverbal behavior carries the real message.

- **Strengthening or emphasizing.** Nonverbal behavior can strengthen or emphasize what is being said. When Denise suggests to Jennie that she ask her boss what he means by her "erratic behavior," Jennie says in a startled voice, "Oh, I don't think I could do that!" while slouching down and putting her face in her hands. Her nonverbal behavior underscores her verbal message.

- **Adding intensity.** Nonverbal behavior often adds emotional color or intensity to verbal messages. When Jennie tells Denise that she doesn't like to be confronted without first being understood and then stares at her fixedly and silently with a frown on her face, Jennie's nonverbal behavior tells Denise that her feelings are intense.

- **Controlling or regulating.** Nonverbal cues are often used in conversation to regulate or control what is happening. Let's say that in a group counseling session, Nina looks at Tom and gives every indication that she is going to speak to him. But he looks away. Nina hesitates and then decides not to say anything. Tom has used a nonverbal gesture to control her behavior.

In reading nonverbal behavior—"reading" is used here instead of "interpreting"—caution is a must. We listen to clients in order understand them, not to dissect them. But merely noticing nonverbal behavior is not enough. Trainees can learn how to identify useful nonverbal messages, clues, and modifiers by watching videotaped interactions, including their own interactions (Costanzo, 1992). Once you develop a working knowledge of nonverbal behavior and its possible meanings, the next step is practice.

Because nonverbal behaviors can often mean a number of things, how can you tell which meaning is the real one? The key is the context in which they take place. Effective helpers listen to the entire context of the helping interview and do not become overly fixated on details of behavior. They are aware of and use the nonverbal communication system, but they are not seduced or overwhelmed by it. Sometimes novice helpers will fasten selectively on this or that bit of nonverbal behavior. For example, they will make too much of a half-smile or a frown on the face of a client. They will seize upon the smile or the frown and, in overinterpreting it, lose the person.

Putting it all together. When clients talk about their concerns, they mix all forms of discourse—thoughts, stories, experiences, emotions, actions, decisions in progress, points of view, proposed actions—together. For example, what follows came out through dialogue in one of Jennie's sessions with Denise. For the sake of illustration, it is presented here in summary form in Jennie's words. She is talking much more animatedly and maintains much more eye contact with Denise than she usually does.

A couple of weeks ago I met a woman at work who has a story similar to mine. We talked for a while and got along so well that we decided to meet outside of work. I had dinner with her last night. She went into her story in more depth. I was amazed. At times I thought I was listening to myself! Because she had been hurt, she was narrowing her world down into a little patch so that she could control everything and not get hurt anymore. I saw right away that I'm trying to do my own version of the same thing. I know you've been telling me that, but I haven't been listening very well. Here's a woman with lots going for her and she's hiding out. As I came back from dinner I said to myself you've got to change.

So, Denise, I want to revisit two areas we've talked about—my work life and my social life. I don't want to live in the hole I've dug for myself. I could see clearly some of the things she should do. So here's what I want to do. I want to engage in some little experiments in broadening my social life. Starting with my family. And I want to discuss the kind of work I want without putting all the limitations on it. I want to start coming out of the hole I'm in. And I want to help my new friend do the same.

Everything is here. Or elements of everything—a story about her new friend, including experiences, actions, and feelings; points of view about her new friend; decisions about where she wants her life to go; proposals about experiments in her social life and her relationship with her friend. The point is this: Developing frameworks for listening can help you zero in on the key messages your clients are communicating and help you identify and understand the feelings, emotions, and moods that go with them. Of course, there is no need to go overboard on listening. Remember that you are a human being listening to a human being, not a vacuum cleaner indiscriminately sweeping up every scrap of information. Quality, not quantity, is important. This brings us to the next topic.

PROCESSING WHAT YOU HEAR: THE THOUGHTFUL SEARCH FOR MEANING

As we listen, we process what we hear. The trick is to become a thoughtful processor. As we shall see a bit further along, there are many less-than-thoughtful ways of processing clients' stories, points of view, and messages. But first, what does thoughtful processing look like?

Identifying key messages and feelings. Denise listens to what Jennie has to say early on about her past and present experiences, actions, and emotions. She listens to Jennie's points of view and the decisions Jennie has made or is in the process of making. She listens to Jennie's intentions and proposals. Jennie tells Denise about an intention gone awry and the emotions that went with it: "When the investigation began, I had every intention of pushing my case, because I knew that some of the men on campus were getting away with murder. But then it began to dawn on me that people were not taking me seriously because I was an African American woman. First I was angry, but then I just got numb. . . ."

Later, Jennie says, "I get headaches a lot now. I don't like taking pills, so I try to tough it out. I have also become very sensitive to any kind of injustice, even in movies or on television. But I've stopped being any kind of crusader. That got me nowhere." As Denise listens to Jennie speak, questions based on the listening frameworks outlined here arise in the back of her mind:

- "What are the main points here?"
- "What experiences and actions are most important?"
- "What themes are coming through?"
- "What is Jennie's point of view?"
- "What is most important to her?"
- "What does she want me to understand?"
- "What decisions are implied in what she's saying?"
- "What is she proposing to do?"

If helpers think that everything that their clients say is key, then nothing is key. In the end, helpers make a clinical judgment as to what is key. Of course, as we shall see later, helpers have ways of checking their understanding. Denise doesn't distract herself from Jennie by asking the above questions of herself directly, but they are part of her active listening and symbolize her interest in Jennie's world. What Denise has learned from both theory and experience over the years constitutes the basis for the clinical judgments she makes.

Understanding clients through context. People are more than the sum of their verbal and nonverbal messages. Listening, in its deepest sense, means listening to clients themselves as influenced by the contexts in which they "live, move, and have their being." As mentioned earlier, it is important to interpret a client's nonverbal behavior in the context of the entire helping session. It is also essential to understand clients' stories, points of view, and messages and the emotions that permeate them through the wider context of their lives. Tiedens and Leach (2004), in their edited book *The Social Life of Emotions,* develop the theme that emotions cannot be understood independently of the social relationships and groups in which they occur. All the things that make people different—culture, personality, personal style, ethnicity, key life experiences, education, travel, economic status, and the other forms of diversity discussed in Chapter 3—provide the context for the client's problems and unused opportunities. Key elements of this context become part of the client's story, whether they are mentioned directly or not. Effective helpers listen through this wider context without being overwhelmed by the details of it.

Here is my version of a simple context-stage-approach-style model (McAuliffe & Ericksen, 1999) for helping counselors think about their clients in context.

- **Context.** This deals with background, the circumstances of the client's life. What circumstances surround the client and how do these circumstances affect the way the client understands and deals with her problems and opportunities?
- **Stage.** This deals with developmental stage. What age-related psychosocial tasks and challenges is the client currently facing and how does the way he goes about these tasks affect the problem situation or opportunity?
- **Approach.** This deals with the client's approach to coming to know and make sense of the world about him or her. How does the client go about constructing meaning, including such things as determining what is important and what is right?
- **Style.** This deals with personality. How does the client's personality style and temperament affect his understanding of himself and his approach to the world?

This is, of course, but one framework. You need to discover for yourself the contextual frameworks that can help you understand your clients as "people-in-systems" (Egan & Cowan, 1979). And contextual frameworks need to be updated. For instance, Arnett (2000) provides a new emerging-adulthood framework for understanding the developmental tasks and challenges from the late teens through the 20s in societies that allow young people of independent role exploration during those years, whereas at the other end of the spectrum Qualls and Abeles (2000) reframe the challenges of growing older and debunk popular misconceptions about aging.

A new graduate program at the University of Michigan, called Personality and Social Contexts, explores the contexts in which people live out their lives and how these contexts influence their lives (DeAngelis, 2005).

To return to our previous example, Denise tries to understand Jennie's verbal and nonverbal messages, especially the core messages, in the context of Jennie's life. As she listens to Jennie's story, Denise says to herself right from the start something like this:

> Here is an intelligent African American woman from a conservative Catholic background. She has been very loyal to the church because it proved to be a refuge in the inner city. It was a gathering place for her family and friends. It provided her with a decent primary and secondary school education and a shot at college. She did very well in her studies. Initially college was a shock. It was her first venture into a predominantly white and secular culture. But she chose her friends carefully and carved out a niche for herself. Because studies were much more demanding, she had to come to grips with the fact that, in this larger environment, she was, academically, closer to average. The rape and investigation put a great deal of stress on what proved to be a rather fragile social network. Her life began to unravel. She pulled away from her family, her church, and the small circle of friends she had at college. At a time when she needed support the most, she cut it off. After graduation she continued to stay "out of community." Now she is underemployed as a secretary in a small company. This does little for her sense of personal worth.

Denise listens to Jennie *through* this context without assuming that it need define Jennie. The helping context is also important. Denise needs to be sensitive about how Jennie might feel about talking to a woman who is quite different from her and also needs to understand that Jennie might well have some misgivings about the helping professions.

In sum, Denise tries to pull together the themes she sees emerging in Jennie's story and tries to see these themes in context. She listens to Jennie's discussion of her headaches (experiences), her self-imposed social isolation (behaviors), and her chronic depression (feelings) against the background of her social history—the pressures of being religious in a secular society at school, the problems associated with being an upwardly mobile African American woman in a predominantly white male society. Denise sees the rape and investigation as social, not merely personal, events. She listens actively and carefully, because she knows that her ability to help depends, in part, on not distorting what she hears. She does not focus narrowly on Jennie's inner world, as if Jennie could be separated from the social context of her life. Finally, although Denise listens to Jennie through the context of Jennie's life, she does not get lost in it. She uses context both to understand Jennie and to help her manage her problems and develop her opportunities more fully.

Hearing the slant or spin: Tough-minded listening and processing. This is the kind of listening needed in order to help clients explore issues more deeply and to identify blind spots that need to be challenged. Skilled helpers not only listen to clients' stories, points of view, decisions, intentions, and proposals but also to any slant or spin that clients might give their stories. Although clients' visions of and feelings about themselves, others, and the world are real and need to be understood, their perceptions are sometimes distorted.

For instance, if a client sees herself as ugly, her experience of herself as ugly is real and needs to be listened to and understood. If her experience does not square with the facts—if she is, in fact, beautiful—then this, too, must be listened to and understood. If a client sees himself as above average in his ability to communicate with others when, in reality, he is below average, his experience of himself needs to be listened to and understood, but reality cannot be ignored. Tough-minded listening includes detecting the gaps, distortions, and dissonance that are part of the client's experienced reality.

> Denise realizes from the beginning that some of Jennie's understandings of herself and her world are not accurate. For instance, in reflecting on all that has happened, Jennie remarks that she probably got what she deserved. When Denise asks her what she means, she says, "My ambitions were too high. I was getting beyond my place in life." This is the slant or spin Jennie gives to her career aspirations. It is one thing to understand how Jennie might put this interpretation on what has happened; it is another to assume that such an interpretation reflects reality. To be client-centered, helpers must first be reality-centered.

Of course, helpers need not challenge clients as soon as they hear any kind of distortion. Rather, they note gaps and distortions, choose key ones, and challenge them when it is appropriate to do so (see Chapters 7–8).

Musing on what's missing. Clients often leave key elements out when talking about problems and opportunities. Having frameworks for listening can help you spot important things that are missing. For instance, they tell their stories but leave out key experiences, behaviors, or feelings. They offer points of view but say nothing about what's behind them or their implications. They deliver decisions but don't give the reasons for them or spell out the implications. They propose courses of action but don't say why they want to head in a particular direction, what the implications are for themselves or others, what resources they might need, or how flexible they are. As you listen, it's important to note what they put in and what they leave out.

For instance, when it comes to stories, clients often leave out their own behavior or their feelings. Jennie says, "I got a call from an old girlfriend last week. I'm not sure how she tracked me down. We must have chatted away for 20 minutes. You know, catching up." Because Jennie says this in a rather matter-of-fact way, it's not clear how she felt about it at the time or feels now. Nor is there any indication of what she might want to do about it—for instance, stay in touch.

In another session Jennie says, "I was talking with my brother the other day. He runs a small business. He asked me to come and work for him. I told him no. . . . By the way, I have to change the time of our next appointment. I forgot I've got a doctor's appointment." Denise notes the experience (being offered a job) and Jennie's behavior or reaction (a decision conveyed to her brother refusing the offer). But Jennie leaves out the reasons for her refusal or the implications for herself or her brother or their relationship and moves to another topic.

Note that this is not a search for the "hidden stuff" that clients are leaving unsaid. We all leave out key details from time to time. Rather, because Denise understands what full versions of stories, points of view, and messages look like, and she notes what parts are missing. She then uses her clinical judgment—a large part of which is common sense—to determine whether or not to ask about the missing parts. For instance, when she asks Jennie why she refused her brother point-blank, Jennie says, "Well, he's a good guy and I'd probably like the work, but this is no time to be getting mixed up with family." This is one more indication of how restricted Jennie has allowed her social life to become. It may be that she has determined that support from her family is out of bounds. In Chapter 6, you will find ways of helping clients fill out their stories with essential but missing details related to stories, points of view, and messages.

LISTENING TO ONESELF: THE HELPER'S INTERNAL CONVERSATION

The conversation helpers have with themselves during helping sessions is the "internal conversation." To be an effective helper, you need to listen not only to the client but also to yourself. Granted, you don't want to become self-preoccupied, but listening to yourself on a "second channel" can help you identify both what you might do to be of further help to the client and what might be standing in the way of your being with and listening to the client. It is a positive form of self-consciousness.

I remember when this second channel did not work very well for me. A friend of mine who had been in and out of mental hospitals for a few years and whom I had not seen for over 6 months showed up unannounced one evening at my apartment. He was in a highly excited state. A torrent of ideas, some outlandish, some brilliant, flowed nonstop from him. I sincerely wanted to be with him as best I could, but I was very uncomfortable. I started by more or less naturally following the guidelines of tuning in, but I kept catching myself at the other end of the couch on which we were both sitting with my arms and legs crossed. I think that I was defending myself from the torrent of ideas. When I discovered myself almost literally tied up in knots, I would untwist my arms and legs, only to find them crossed again a few minutes later. It was hard work being with him. In retrospect, I realize I was concerned for my own security. I have since learned to listen to myself on the second channel a little better. When I listen to my nonverbal behavior as well as my internal dialogue, I interact with clients better.

Helpers can use this second channel to listen to what they are "saying" to themselves, their nonverbal behavior, and their feelings and emotions. These messages can refer to the helper, the client, or the relationship.

- "I'm letting the client get under my skin. I had better do something to reset the dialogue."
- "My mind has been wandering. I'm preoccupied with what I have to do tomorrow. I had better put that out of my mind."
- "Here's a client who has had a tough time of it, but her self-pity is standing in the way of her doing anything about it. My instinct is to be sympathetic. I need to talk to her about her self-pity, but I had better go slow."
- "It's not clear that this client is interested in changing. It's time to test the waters."

The point is that this internal conversation goes on all the time. It can be a distraction or it can be another tool for helping. In one study, Fauth and Williams (2005) found that helpers' internal conversations were "generally helpful rather than hindering from both the trainee and student-client perspectives" (p. 443). The client, too, is having his or her internal conversation. One intriguing study (Hill, Thompson, Cogar, & Denman, 1993) suggested that both client and therapist are more or less aware of the other's "covert processes." This study showed that even though helpers knew that clients were having their own internal conversations and left things unsaid, they were not very good at determining what those things were. At times there are verbal or nonverbal hints as to what the client's internal dialogue might be. Helping clients move key points from their internal conversations into the helping dialogue is a key task and will be discussed in Chapter 8.

THE SHADOW SIDE OF LISTENING TO CLIENTS

Listening as described here is not as easy as it sounds. Obstacles and distractions abound. Some relate to listening generally. Others relate more specifically to listening to and interpreting clients' nonverbal behavior.

Forms of Distorted Listening

As you will see from your own experience, the following kinds of distorted listening permeate human communication. They also insinuate themselves at times into the helping dialogue. Sometimes more than one kind of distortion contaminates the helping dialogue. They are part of the shadow side because helpers never intend to engage in these kinds of listening. Rather, helpers fall into them at times without even realizing that they are doing so. But they stand in the way of the kind of openminded listening and processing needed for real dialogue.

Filtered listening. It is impossible to listen to other people in a completely unbiased way. Through socialization we develop a variety of filters through which we listen to ourselves, others, and the world around us. As Hall (1977) noted: "One of the functions of culture is to provide a highly selective screen between man and the outside world. In its many forms, culture therefore designates what we pay attention to and what we ignore. This screening provides structure for the world" (p. 85). We need filters to provide structure for ourselves as we interact with the world. But personal, familial, sociological, and cultural filters introduce various forms of bias into our listening and do so without our being aware of it.

The stronger the cultural filters, the greater the likelihood of bias. For instance, a white, middle-class helper probably tends to use white, middle-class filters in listening to others. Perhaps this makes little difference if the client is also white and middle class, but if the helper is listening to an Asian client who is well-to-do and has high social status in his community, to an African American mother from an urban ghetto, or to a poor white subsistence farmer, then the helper's cultural filters might introduce bias. Prejudices, whether conscious or not, distort understanding. Like everyone else, helpers are tempted to pigeonhole clients because of gender, race, sexual orientation, nationality, social status, religious persuasion, political preferences, lifestyle, and the like. Helpers' self-knowledge is essential. This includes ferreting out the biases and prejudices that distort listening.

Evaluative listening. Most people, even when they listen attentively, listen evaluatively. That is, as they listen, they are judging what the other person is saying as good/bad, right/wrong, acceptable/unacceptable, likable/unlikable, relevant/irrelevant, and so forth. Helpers are not exempt from this universal tendency. The following interchange takes place between Jennie and a friend of hers. Jennie recounts it to Denise as part of her story.

JENNIE: Well, the rape and the investigation are not dead, at least not in my mind. They are not as vivid as they used to be, but they are there.

FRIEND: That's the problem, isn't it? Why don't you do yourself a favor and forget about it? Get on with life, for God's sake!

Evaluative listening gives way to advice giving. It might well be sound advice, but the point here is that Jennie's friend listens and responds evaluatively. Clients should first be understood, then, if necessary, challenged or helped to challenge themselves. Evaluative listening, translated into advice giving, will just put clients off. Indeed, a judgment that a client's point of view, once understood, needs to be expanded or transcended or that a pattern of behavior, once listened to and understood, needs to be altered can be quite useful. That is, there are productive forms of evaluative listening. It is practically impossible to suspend judgment completely. Nevertheless, it is possible to set one's judgment aside for the time being in the interest of understanding clients, their worlds, their stories, their points of view, and their decisions "from the inside."

Stereotype-based listening. I remember my reaction to hearing a doctor refer to me as the "hernia in 304." We don't like to be stereotyped, even when the stereotype has some validity. The very labels we learn in our training—paranoid, neurotic, sexual disorder, borderline—can militate against empathic understanding. Books on personality theories provide us with stereotypes: "He's a perfectionist." We even pigeonhole ourselves: "I'm a Type A personality." Though in this case the stereotype is often used as an excuse.

In psychotherapy, diagnostic categories can take precedence over the clients being diagnosed. Helpers forget at times that their labels are interpretations rather than understandings of their clients. You can be "correct" in your diagnosis and still lose the person. In short, what you learn as you study psychology may help you to organize what you hear, but it may also distort your listening. To use terms borrowed from Gestalt psychology, make sure that your client remains "figure"—in the forefront of your attention—and that models and theories about clients remain "ground"—knowledge that remains in the background and is used only in the interest of understanding and helping this unique client.

Fact-centered rather than person-centered listening. Some helpers ask clients many informational questions, as if clients would be cured if enough facts about them were known. It's entirely possible to collect facts but miss the person. The antidote is to listen to clients contextually, trying to focus on themes and key messages. Denise, as she listens to Jennie, picks up what is called a "pessimistic explanatory style" theme (Peterson, Seligman, & Vaillant, 1988). Clients with this style tend to say, directly or indirectly, about unfortunate events such things as, "It will never go away," "It affects everything I do," and "It is my fault." Denise knows that the research indicates that people who fall victim to this style tend to end up with poorer health than those who do not. There may be a link, she hypothesizes, between Jennie's somatic complaints (headaches, gastric problems) and this explanatory style. This is a theme worth exploring.

Sympathetic listening. Because most clients are experiencing some kind of misery and because some have been victimized by others or by society itself, helpers tend to feel sympathy for them. Sometimes these feelings are strong enough to distort the stories that clients are telling. Consider this case.

Liz was counseling Ben, a man who had lost his wife and daughter to a tornado. Liz had recently lost her husband to cancer. As Ben talked about his own tragedy during their first meeting, she wanted to hold him. Later that day she took a long walk and realized how her sympathy for Ben had distorted what she heard. She heard the depth of his loss, but, reminded of her own loss, only half heard the implication that his loss now excused him from getting on with his life.

Sympathy has an unmistakable place in human relationships, but its "use," if that does not sound too inhuman, is limited in helping. In a sense, when I sympathize with someone, I become his or her accomplice. If I sympathize with my client as she tells me how awful her husband is, I take sides without knowing what the complete story is. Expressing sympathy can reinforce self-pity, which has a way of driving out problem-managing action.

Interrupting. I am reluctant to add "interrupting," as some do, to this list of shadow-side obstacles to effective listening. Certainly, when helpers interrupt their clients, by definition, they stop listening. And interrupters often say things that they have been rehearsing, which means that they have been only partially listening. My reluctance, however, comes from the conviction that the helping conversation should be a dialogue. There are benign and malignant forms of interrupting. The helper who cuts the client off in mid thought because he has something important to say is using a malignant form. But the case is different when a helper "interrupts" a monologue with some gentle gesture and a comment such as, "You've made several points. I want to make sure that I've understood them." If interrupting promotes the kind of dialogue that serves the problem-management process, then it is useful. Still, care must be taken to factor in cultural differences in storytelling.

One possible reason counselors fall prey to these kinds of shadow-side listening is the unexamined assumption that listening with an open mind is the same as approving what the client is saying. This is not the case, of course. Rather, listening with an open mind helps you learn and understand. Whatever the reason for shadow-side listening, the outcome can be devastating because of a truth philosophers learned long ago—a small error in the beginning can lead to huge errors down the road. If the foundation of a building is out of kilter, it is hard to notice with the naked eye. But by the time construction reaches the ninth floor, it begins to look like the leaning tower of Pisa. Tuning in to clients and listening both actively and with an open mind are foundation counseling skills. Ignore them and dialogue is impossible.

Myths About Nonverbal Behavior

Richmond and McCroskey (2000) spell out the shadow side of nonverbal behavior in terms of commonly held myths (pp. 2–3):

1. *Nonverbal communication is nonsense. All communication involves language. Therefore, all communication is verbal.* This myth is disappearing. It does not stand up under the scrutiny of common sense.

2. *Nonverbal behavior accounts for most of the communication in human interaction.* Early studies tried to "prove" this, but they were biased. Studies were aimed at dispelling myth number 1 and overstepped their boundaries.

3. *You can read a person like a book.* Some people, even some professionals, would like to think so. You can read nonverbal behavior, verbal behavior, and context and still be wrong.

4. *If a person does not look you in the eye while talking to you, he or she is not telling the truth.* Tell this to liars! The same nonverbal behavior can mean many different things.

5. *Although nonverbal behavior differs from person to person, most nonverbal behaviors are natural to all people.* Cross-cultural studies give the lie to this. But it isn't true even within the same culture.

6. *Nonverbal behavior stimulates the same meaning in different situations.* Too often the context is the key. Yet some professionals buy the myth and base interpretive systems on it.

THE SHADOW SIDE OF COMMUNICATION SKILLS—PART 1

Interpersonal communication competence is critical for effective everyday living. It is the principal enabling skill for just about everything we do. Yet, it is, in my view, "forgotten" by society. In that respect it suffers a fate shared by a number of essential "life skills" such as problem solving, opportunity development, parenting, and managing (knowing how to make some system work), to name a few. Let me make my point. In lecturing, I have often asked audiences to answer two questions. The first question goes something like this:

> Given the importance of effective human relationships in just about every area of life, how important is it for your kids to develop a solid set of interpersonal communication skills? On a scale from 1 to 100, how high would you rate the importance?

Inevitably, the ratings are at the high end, always near 100. The second question goes something like this:

> Given the importance of these skills, where do your kids pick them up? How does society make sure that they acquire them? In what forums do they learn them?

Then the hemming and hawing begin. "Well, I guess they get some of them at home. That is, if they find good role models at home." Or, "Life itself is the best teacher of these skills. They learn them on the run." The members of the audience mill around like that for a while, until I say,

> Let me summarize what I've been hearing. And, by the way, it's no different from what I hear every place else. Although most parents rate the importance of these sets of skills very highly, we live in a society that leaves their development to chance. Nothing is done systematically to make sure that our kids learn these skills. And, by the way, there is no assurance that they will pick them up on the run.

Children learn a bit from their parents, they might get a dash in school, perhaps a soupçon of TV helps. But, in the main, they are more often exposed to poor communicators than good ones.

Ideally, helpers-to-be would arrive at training programs already equipped with a solid set of interpersonal communication skills. Training would help them adapt these skills to the counseling process. After all, basic interpersonal communication skills are not special skills peculiar to helping. Rather, they are extensions of the kinds of skills all of us need in our everyday interpersonal interactions. However, because trainees don't ordinarily arrive so equipped, they need time to come up to speed in communication competence. This, in a strange way, creates its own problem. Some helper-training programs focus almost exclusively on interpersonal communication skills. As a result, trainees know how to communicate but not necessarily how to help. Furthermore, most adults feel that they are "pretty good" at these skills. When they compare themselves to others, they often see themselves as "above average." This may be true, but the benchmark ("average") is woefully low. Hopefully, the positive psychology movement will help make learning basic skills in living, especially interpersonal communication skills, as common as learning basic math. We're nowhere near that ideal yet.

COMMUNICATING EMPATHY: WORKING HARD AT UNDERSTANDING CLIENTS

THE SHADOW SIDE OF SHARING EMPATHIC HIGHLIGHTS
  No response
  Distracting questions
  Clichés
  Interpretations
  Advice
  Parroting
  Sympathy and agreement
  Faking it
THE IMPORTANCE OF EMPATHIC RELATIONSHIPS

Helpers listen to clients both in order to understand them and their concerns and to respond to them in constructive ways. The logic of listening includes, as we have seen, tuning in to clients, listening actively, processing what is heard contextually, and identifying the key ideas, messages, or points of view the client is trying to communicate—all at the service of understanding clients. Listening, then, is a very active process that serves understanding. But helpers don't just listen; they also respond to clients in a variety of ways. They respond by sharing their understanding, checking to make sure that they've got things right, asking questions, probing for clarity, summarizing the issues being discussed, and challenging clients in a variety of ways.

In this chapter the focus is on communicating empathic understanding to clients and checking to see if that understanding is accurate. When helpers both understand and accurately communicate their understanding to clients, they help their clients understand themselves, their problem situations, and their unused opportunities more fully—a process that helps them move to problem-managing action.

THE THREE DIMENSIONS OF RESPONDING SKILLS: PERCEPTIVENESS, KNOW-HOW, AND ASSERTIVENESS

The communication skills involved in responding to clients have three dimensions: perceptiveness, know-how, and assertiveness.

Perceptiveness. "Feeling empathy" for others is not helpful if the helper's perceptions are not accurate. Ickes defined "empathic accuracy" as "the ability to accurately infer the specific content of another person's thought and feelings" (1993, p. 588). According to Ickes, this ability is a component of success in many walks of life.

> Empathically accurate perceivers are those who are consistently good at "reading" other people's thoughts and feelings. All else being equal, they are likely to be the most tactful advisors, the most diplomatic officials, the most effective negotiators, the most electable politicians, the most productive salespersons, the most successful teachers, and the most insightful therapists. (Ickes, 1997, p. 2)

The assumption is, of course, that such people are not only accurate perceivers but they can also weave their perceptions into their dialogues with their constituents, customers, students, and clients. Helpers do this by sharing empathic highlights with their clients.

Your responding skills are only as good as the accuracy of the perceptions on which they are based. Consider the difference between these two examples.

> Beth is counseling Ivan in a community mental health center. Ivan is scared to talk about an "ethical blunder" that he made at work. Beth senses his discomfort but thinks that he is angry rather than scared. She says, "Ivan, I'm wondering what's making you so angry right now." Because Ivan does not feel angry, he says nothing. In fact, he's startled by what she says and feels even more insecure. Beth takes his silence as a confirmation of his "anger." She tries to get him to talk about it.

Beth's perception is wrong and therefore disrupts the helping process. She misreads Ivan's emotional state and tries to engage in a dialogue based on her flawed perception. Contrast this to what happens in the following example.

> Mario, a manager, is counseling Enrique, a relatively new member of his team. During the past week, Enrique has made a significant contribution to a major project, but he has also made one rather costly mistake. Enrique's mind is on his blunder, not his success. Mario, sensing Enrique's discomfort, says, "Your ideas in the meeting last Monday helped us reconceptualize and reset the entire project. It was a great contribution. That kind of 'outside the box' thinking is very valuable here. (He pauses.) I'd also like to talk to you about Wednesday. Your conversation with Acme's purchasing agent on Wednesday made him quite angry. (He pauses briefly once more.) Something tells me that you might be more worried about Wednesday's mistake than delighted with Monday's contribution. I just wanted to let you know that I'm not." Enrique is greatly relieved. They go on to have a useful dialogue about what made Monday so good and what could be learned from Wednesday's blunder.

Mario's perceptiveness and his ability to defuse a tense situation lays the foundation for an upbeat dialogue.

The kind of perceptiveness needed to be a good helper comes from basic intelligence, social intelligence, experience, reflecting on your experience, developing wisdom, and, more immediately, tuning in to clients, listening carefully to what they have to say, and objectively processing what they say. Perceptiveness is part of social-emotional maturity.

Know-how. Once you are aware of what kind of response is called for, you need to be able to deliver it. For instance, if you are aware that a client is anxious and confused because this is his first visit to a helper, it does little good if you don't know how to translate your perceptions and your understanding into words. Let's return to Ivan and Beth for a moment.

> Ivan and Beth end up arguing about his "anger." Ivan finally gets up and leaves. Beth, unfortunately, takes this as a sign that she was right in the first place. The next day Ivan goes to see his minister. The minister sees quite clearly that Ivan is scared and confused. His perceptions are right. He says something like this: "Ivan, you seem to be very uncomfortable. It may be that whatever is on your mind might be difficult to talk about. But I'd be glad to listen to it, whatever it is. But I don't want to push you into anything." Ivan blurts out, "But I've done something terrible." The minister pauses and then says, "Well, let's see what kind of sense we can make of it." Ivan hesitates a bit, then leans back into his chair, takes a deep breath, and launches into his story.

The minister is not only perceptive but also knows how to address Ivan's anxiety and hesitation. The minister says to himself, "Here's a man who is almost exploding with the need to tell his story, but fear or shame or something like that is paralyzing him. How can I put him at ease, let him know that he won't get hurt here? I need to recognize his anxiety and offer an opening." He does not use these words, of course, but these are the kind of sentiments that instinctively run through his mind. This chapter is designed to help you develop know-how.

Assertiveness. Accurate perceptions and excellent know-how are meaningless if they remain locked up inside you. They need to become part of the therapeutic dialogue. For instance, if you see that self-doubt is a theme that weaves itself throughout a client's story about her frustrating search for a better relationship with her estranged brother but fail to share your hunch with her, you do not pass the assertiveness test. Consider this example.

> Nina, a young counselor in the Center for Student Development, is in the middle of the first session with Antonio, a graduate student. During the session, he mentions briefly a very helpful session he had the previous year with Carl, a middle-aged counselor on the staff. Carl has accepted an academic position at the university and is no longer involved with the Center. Nina realizes that Antonio is disappointed that he couldn't see Carl and might have some misgivings about being helped by a new counselor—a younger woman. She has faced sensitive issues like this before and would not be offended if Antonio were to choose a different counselor. During a lull in the conversation, she says something like this: "Antonio, could we take a time-out here for a moment? I think you might be a bit disappointed to find out that Carl is no longer here. Or at least I probably would be if I were in your shoes. You were just more or less assigned to me, and I'm not sure the fit is right. Maybe you can give that a bit of thought. Then, if you think I can be of help, you can schedule another meeting with me. But you're certainly free to review who is on staff and choose whomever you want."

In this case, perceptiveness, know-how, and assertiveness all come together. This is not to suggest that assertiveness is an overriding value in and of itself. To be assertive without perceptiveness and know-how is to court disaster.

SHARING EMPATHIC HIGHLIGHTS: COMMUNICATING UNDERSTANDING TO CLIENTS

Although many people may "feel empathy" for others—that is, are motivated in many different ways by the value of empathy described in Chapter 3—the truth is that few know how to put empathic understanding into words. And so sharing empathic highlights as a way of showing understanding during conversations remains an improbable event in everyday life. Perhaps that's why it is so powerful in helping settings. When clients are asked what they find helpful in counseling sessions, being understood gets top ratings. There is such an unfulfilled need to be understood.

In some ways this section is a kind of anatomy lesson. That is, we are going to take the process of sharing empathic highlights apart and look at the pieces. Further on, we'll put them back together again.

The Basic Formula

Basic empathic understanding can be expressed in the following stylized formula:

> **You feel** . . . [here name the correct emotion expressed by the client]
> **because** . . . [here indicate the correct thoughts, experiences, and behaviors that give rise to the feelings].

For instance, Leonardo is talking with a helper about his arthritis and all it's attendant ills. There is pain, of course, but more to the point, he can't get around the way he used to. At one point the helper says:

> You feel bad, not so much because of the pain, but because your ability to get around—your freedom—has been curtailed.

Leonardo replies:

> That's just it. I can manage the pain. But not being able to get around is killing me! It's like being in jail.

They go on to discuss ways in which Leonardo, with the help of family and friends, can get out of "jail"—that is, become more mobile—together with ways of coping with the time he is in "jail."

The formula—"You feel . . . because . . ."—is a beginner's tool to get used to the concept of sharing highlights. It focuses on the key points of clients' stories, points of view, intentions, proposals, and decisions together with the feelings, emotions, and moods associated with them. The formula is used in the following examples. For the moment, ignore the fact that it might sound a bit stylized. Ordinary human language will be substituted later. In the first example, a divorced mother with two young children is talking to a social worker about her ex-husband. She has been talking about the ways he has let her and their kids down. She ends by saying:

CLIENT: I could kill him! He failed to take the kids again last weekend. This is three times out of the last 6 weeks.

HELPER: You feel furious because he keeps failing to hold up his part of the bargain.

CLIENT: I just have to find some way to get him to do what he promised to do. What he told the court he would do.

His not taking the kids according to their agreement [an experience for the client] infuriates her [an emotion]. The helper captures both the emotion and the reason for it. And the client moves forward in terms of thinking about possible actions she could take.

In the next example, a woman who has been having a great deal of gastric and intestinal distress is going to have a colonoscopy. She is talking with a hospital counselor the night before the procedure.

PATIENT: God knows what they'll find when they go in. I keep asking questions, but they keep giving me vague answers.

HELPER: You feel troubled because you believe that you're being left in the dark.

PATIENT: In the dark about my body, my life! If they'd only tell me! Then I could prepare myself better.

They go on to discuss what she needs to do to get the kind of information she wants. The accuracy of the helper's response does not solve the woman's problems, but the patient does move a bit. She gets a chance to vent her concerns, she receives a bit of understanding, and says why she wants more information. This perhaps puts her in a better position to ask for a more open relationship with her doctors.

The key elements of an empathic highlight are the same as the key elements of the client's story discussed in Chapter 4—that is, the thoughts, experiences, behaviors, and feelings that make up the client's story. Here are some guidelines.

Respond Accurately to Clients' Feelings, Emotions, and Moods

The importance of feelings, emotions, and moods in our lives was discussed in Chapter 4. Helpers need to respond to clients' emotions in such a way as to move the helping process forward. This means identifying key emotions the client either expresses or discusses (helper perceptiveness) and weaving them into the dialogue (helper know-how) even when they are sensitive or part of a messy situation (helper courage or assertiveness). Do you remember the last time you as a consumer got a problem resolved with a good customer service representative? She might have said something like this to you: "I know you're angry right now because the package didn't arrive and you have every right to be. After all, we did make you a promise. Here's what we can do to make it right for you. . . ." Rather than ignoring the customer's emotions, good customer service reps face up to them as helpfully as possible.

Use the right family of emotions and the right intensity. In the basic highlight formula, "You feel . . ." should be followed by the correct family of emotions and the correct intensity.

> **Family.** The statements "You feel hurt," "You feel relieved," and "You feel enthusiastic" specify different families of emotion.

> **Intensity.** The statements "You feel annoyed," "You feel angry," and "You're furious" specify different degrees of intensity in the same family (anger).

The words *sad, mad, bad,* and *glad* refer to four of the main families of emotion, whereas *content, quite happy,* and *overjoyed* refer to different intensities within the *glad* family.

Distinguish between expressed and discussed feelings. Clients both *express* emotions they are feeling during the interview and *talk about* emotions they felt at the time of some incident. For instance, consider this interchange between a client involved in a child custody proceeding and a counselor. She is talking about her husband.

CLIENT (calmly): I get furious with him when he says things, little snide things, that suggest that I don't take good care of the kids.

HELPER: You feel especially angry when he intimates that you're not a good mother.

The client isn't angry right now. Rather, she is talking about the anger.

The following example—a woman is talking about one of her colleagues at work—deals with *expressed* rather than *discussed* feelings.

CLIENT (enthusiastically): I threw caution to the wind and confronted him about his sarcasm and it actually worked. He not only apologized but behaved himself the rest of the trip.

HELPER: You feel great because you took a chance and it paid off.

Clients don't always name their feelings and emotions. However, if they express emotion, it is part of the message and needs to be identified and understood.

Read and respond to feelings and emotions embedded in clients' nonverbal behavior. Often helpers have to read their clients' emotions—both the family and the intensity—in their nonverbal behavior. In the following example, a North American student comes to you, sits down, looks at the floor, hunches over, and speaks haltingly:

CLIENT: I don't even know where to start. (He falls silent).

HELPER: It's pretty clear that you're feeling miserable. Maybe we can talk about why.

CLIENT (after a pause): Well, let me tell you what happened. . . .

You see that he is depressed and his nonverbal behavior indicates that the feelings are quite intense. His nonverbal behavior reveals the broad family ("You feel bad") and the intensity ("You feel very bad"). Of course, you do not yet know the thoughts, experiences, and behaviors that give rise to these emotions.

Be sensitive in naming emotions. Naming and discussing feelings and emotions threaten some clients. If this is the case, it might be better to focus on thoughts, experiences, and behaviors and proceed only gradually to a discussion of feelings. The following client, an unmarried man in his mid-30s who has come to talk about "certain dissatisfactions" in his life, has shown some reluctance to express or even to talk about feelings.

CLIENT (in a pleasant, relaxed voice): You won't believe it! My mother is always trying to make a little kid out of me. And I'm 35! Last week, in front of a group of my friends, she brought out my rubber boots and an umbrella and gave me a little talk on how to dress for bad weather (laughs).

COUNSELOR A: It might be hard to admit it, but I get the feeling that down deep you were furious.

CLIENT: Well, I don't know about that. Anyway, at work. . . .

Counselor A pushes the emotion issue and is met with some resistance. The client changes the topic.

COUNSELOR B (in a somewhat lighthearted way): So she's still playing the mother role—to the hilt, it would seem.

CLIENT (with more of a bite in his voice): And the hilt includes not wanting me to grow up. But I am grown up . . . well, pretty grown up. But I don't always act grown up around her.

Counselor B, choosing to respond to the "strong mother" issue rather than the more sensitive "being kept a kid and feeling really lousy about it" issue, gives the client more room to move. This works, for the client himself moves toward the more sensitive issue—his playing the child, at least at times, when he's with his mother.

Some clients are hesitant to talk about certain emotions. One client might find it relatively easy to talk about his anger but not his hurt. The following client is talking about his disappointment at not being chosen for a special team at work.

CLIENT: I worked as hard as anyone else to get the project up and running. In fact, I was at the meeting where we came up with the idea in the first place. . . . And now they've dropped me.

COUNSELOR A: So you feel really hurt—left out of your own project.

CLIENT (hesitating): Hmm. . . . I'm really ticked off. Why shouldn't I be? . . .

Here is a client with lots of ego. He doesn't like the idea that he has been "hurt." Counselor B takes a different tack.

COUNSELOR B: So it's more than annoying to be left out of what, in many ways, is your own project.

CLIENT: How could they do that? . . . It is more than annoying. It's . . . well . . . humiliating!

Counselor B, factoring in the client's ego, sticks to the anger, allowing the client himself to name the more sensitive emotion. Contextual listening—in this case listening to the client's emotions through the context of the pride he takes in himself and his accomplishments—is part of social intelligence. However, being sensitive to clients' sensitive emotions should not rob counseling of its robustness. Too much tiptoeing around clients' "sensitivities" does not serve them well. Remember what was said earlier. Clients are not as fragile as we sometimes make them out to be.

Use different ways to share highlights about feelings and emotions. Because clients express feelings in a number of different ways, helpers can communicate an understanding of feelings in a variety of ways.

> **By single words.** You feel good. You're depressed. You feel abandoned. You're delighted. You feel trapped. You're angry.
>
> **By different kinds of phrases.** You're sitting on top of the world. You feel down in the dumps. You feel left in the lurch. Your back's up against the wall. You're really on a roll.
>
> **By what is implied in behavioral statements.** You feel like giving up (implied emotion: despair). You feel like hugging him (implied emotion: joy). Now that you see what he's been doing to you, you almost feel like throwing up (implied emotion: disgust).
>
> **By what is implied in experiences the client is discussing.** You feel you're being dumped on (implied feeling: victimized). You feel you're being stereotyped (implied feeling: resentment). You feel you're at the top of her list (implied feeling: elation). You feel you're going to get caught (implied feeling: fear). Note that the implication of each could be spelled out: You feel angry because you're being dumped on. You resent the fact that you're being stereotyped. You feel great because it seems that you're at the top of her list.

Because ultimately you must discard formulas and use your own language—words that are yours rather than words from a textbook—it helps to develop a variety of ways of communicating your understanding of clients' feelings and emotions. It keeps you from being wooden in your responses.

Consider this example: The client tells you that she has just been given the kind of job she has been looking for over the past 2 years. Here are some possible responses to her emotion.

> **Single word.** You're really happy.
>
> **A phrase.** You're on cloud nine.
>
> **Experiential statement.** You feel you finally got what you deserve.
>
> **Behavioral statement.** You feel like going out and celebrating.

With experience, you can extend your range of expression at the service of your clients. Providing variety will become second nature.

Neither overemphasize nor underemphasize feelings, emotions, and moods. Some counselors take an overly rational approach to helping and almost ignore clients' feelings. Others become too preoccupied with clients' emotions and moods. They pepper clients with questions about feelings and at times extort answers. To say that feelings, emotions, and moods are important is not to say that they are everything. The best defense against either extreme is to link feelings, emotions, and moods to the thoughts, experiences, and behaviors that give rise to them.

Respond Accurately to the Key Experiences, Thoughts, and Behaviors in Clients' Stories

Key experiences, thoughts, and behaviors give rise to clients' feelings, emotions, and moods. The "because . . ." in the empathic-highlight formula is to be followed by an indication of the experiences, thoughts, and behaviors that underlie the client's feelings. In the following example, the client, a graduate student in law school, is venting his frustration.

CLIENT (heatedly): You know why he got an A? He took my notes and disappeared. I didn't get a chance to study them. And I never even confronted him about it.

HELPER: You feel doubly angry because not only did he steal your notes, but you let him get away with it.

The response specifies both the client's experience (the theft) and his behavior (in this case, a failure to act) that give rise to his distress. His anger is directed not only at his classmate but also himself.

In the following example, a mugging victim has been talking to a social worker to help cope with his fears of going out. Before the mugging, he had given no thought to urban problems. Now he tends to see menace everywhere.

CLIENT: This gradual approach of getting back in the swing seems to be working. Last night I went out without a companion. First time. I have to admit that I was scared. But I think I've learned how to be careful. Last night was important. I feel I can begin to move around again.

HELPER: You feel comfortable with the one-step-at-a-time approach you've been taking. And it paid off last night when you regained a big chunk of your freedom.

CLIENT: That's it! I know I'm going to be free again. . . . Here's what I've been thinking of doing. . . .

The client is talking about success in implementing a program. The helper's response recognizes the client's satisfaction and also how important it is for the client to feel both safe and free. The client moves on to describe the next phase of his program.

Another client, after a few sessions spread out over 6 months, says something like this about the progress she is making in rebuilding her life after a devastating car accident. She's back at work and has been working with her husband at rebuilding their marriage.

CLIENT (talking in an animated way): I really think that things couldn't be going better. I'm doing very well at my new job, and my husband isn't just putting up with it. He thinks it's great. He and I are getting along better than ever, even sexually, and

I never expected that. We're both working at our marriage. I guess I'm just waiting for the bubble to burst.

HELPER: You feel great because things have been going better than you ever expected—and it seems almost too good to be true.

CLIENT: Well, a "bubble bursting" might be the wrong image. I think there's a difference between being cautious and waiting for disaster to strike. I'll always be cautious, but I'm finding out that I can make things come true instead of sitting around waiting for them to happen as I usually do. I guess I've got to keep making my own luck.

This client talks about her experiences, attitudes, and behaviors and expresses feelings, the flavor of which is captured in the helper's highlight. The response, capturing as it does both the client's enthusiasm and her lingering fears, is quite useful because the client makes an important distinction between reasonable caution and expecting to worst to happen. She moves on to her need to make things happen, to become more of an agent in her life.

In the following example, the client, a 45-year-old male construction worker, married, with four children between the ages of 9 and 16, has been expressing concerns about his children.

CLIENT: I don't consider myself old-fashioned, but I think kids these days suffer from overindulgence. We keep giving them things. We let them do whatever they want. I fall into the same trap myself. It's just not good for them. I don't think we're preparing them for what the world is really like. People assume that the economy will keep booming. Everyone keeps shouting, "Free lunch!" This isn't doing kids any good.

COUNSELOR: So you see the "do-what-you-want" and "free-lunch" messages as a lot of hogwash. It's going to backfire and your kids could end up getting hurt.

CLIENT: Right. . . . But I'm not in control. My kids can get one set of messages from me and then get a flood of contradictory messages outside, and from TV, and the Internet. . . . I don't want to be a tyrant. Or come across as a killjoy. That doesn't work anyway. At work I see problems and I take care of them. But this has got me stymied.

COUNSELOR: So the whole picture seems pretty gloomy right now. You're not exactly sure what to do about it. You handle problems at work. But it's a lot harder to do something about societal problems that could hurt your kids. What have you been doing so far to handle all of this?

Once the counselor communicates understanding of the client's point of view, he (the client) moves on to share his sense of helplessness. The helper realizes, however, that the client probably has tried some approaches to managing this problem situation. He is probably not as helpless as he makes himself out to be.

In the next example, the client, who is hearing impaired, has been discussing ways of becoming, in her words, "a full-fledged member of my extended family." The discussion between client and helper takes place through a combination of lipreading and signing.

CLIENT (enthusiastically): Let me tell you what I'm thinking of doing. . . . First of all, I'm going to stop fading into the background in family and friends' conversation groups. I'll be the best listener there. And I'll get my thoughts across even if I have to use props. That's how I really am . . . inside, you know, in my mind.

HELPER: Sounds exciting. You're thinking of getting right into the middle of things . . . where you think you belong. You might even try a bit of drama.

CLIENT: And I think that, well, socially, I'm pretty smart. So I'm not talking about being melodramatic or anything. I can do all this with finesse, not just barge in.

HELPER: You'll make it all natural. . . . Draw me a couple of pictures of what this would look like.

The client comes up with a proposal for a course of action that will help her take her "rightful place" in conversations with family and friends setting her agenda (Stage III: helping clients develop strategies for accomplishing goals). The helper's response recognizes her enthusiasm and sense of determination. They go on to have a dialogue about practical tactics.

When clients announce key decisions or express their resolve to do something, it's important to recognize the core of what they are saying. In the following example, a client being treated for social phobia has benefitted greatly from cognitive-behavioral therapy. For instance, in uncomfortable social situations he has learned to block self-defeating thoughts and to keep his attention focused externally—on the social situation itself and on the agenda of the people involved—instead of turning in on himself.

CLIENT (emphatically): I'm not going to turn back. I've had to fight to get where I am now. But I can see how easy it could be to slide back into my old habits. I bet a lot of people do. I see it all around me. People make resolutions and then they peter out.

HELPER: Even though it's possible for you to give up your hard-earned gains, you're not going to do it. You're just not.

CLIENT: But what can I do to make sure that I won't? I'm convinced I won't, but. . . .

HELPER: You need some ratchets. They're the things that keep roller-coaster cars from sliding back. You hear them going click, click, click on the way up.

CLIENT: Ah, right! But I need psychological ones. . . .

HELPER: And social ones. . . . What's kept you from sliding back so far?

In a positive psychology mode, the counselor focuses on past successes. They go on to discuss the kind of "ratchets" he needs to stay on track.

PRINCIPLES FOR SHARING EMPATHIC HIGHLIGHTS

Here are a number of principles that can guide you as you share highlights. Remember that these guidelines are principles, not formulas to be followed slavishly.

Use empathic highlights at every stage and in every task of the helping process. Sharing highlights is useful at every stage and in every task of the helping process. Communicating and checking understanding is always helpful. Here are some examples of helpers sharing highlights at different stages and tasks of the helping process.

Stage I: Problem clarification and opportunity identification. A teenager in his third year of high school who has just found out that he is moving with his family to a different city. A school counselor responds, "You're miserable because you have to leave all your friends. But it sounds like you may even feel a bit betrayed. You didn't see this coming at all." The counselor realizes that he has to help his client pick up the pieces and move on, but sharing his understanding helps build a foundation to do so. The teen goes on to talk in positive terms about the large city they will be moving to and the opportunities it will offer. At one point the school counselor responds, "So there's an upside to all this. Big cities are filled with things to do. You like theater and there's loads there. That's something to look forward to."

Stage II: Evaluating goal options. A woman has been discussing the trade-offs between marriage and career. At one point her helper says, "There's some ambivalence here. If you marry Jim, you might not be able to have the kind of career you'd like. Or did I hear you half say that it might be possible to put both together? Sort of get the best of both worlds." The client goes on the explore the possibilities around "getting the best of both worlds." It helps her greatly in preparing for her next conversation with Jim.

Stage III: Choosing actions to accomplish goals. A man has been discussing his desire to control his cholesterol level without taking a medicine whose side effects worry him. He says that it might work. The counselor responds, "It's a relief to know that sticking to the diet and exercise might mean that you won't have to take any medicine. . . . Hmm. . . . Let's explore the 'might' part. I'm not exactly sure what your doctor said." The helper recognizes the client's determination, but then seeks further clarification.

The action arrow: Implementation issues. A married couple has been struggling to put into practice a few strategies to improve their communication with each other. They've both called their attempts a "disaster." The counselor replies, "OK, so you're annoyed with yourselves for not accomplishing even the simple active-listening goals you set for yourselves. . . . Let's see what we can learn from the 'disaster'" (said somewhat lightheartedly). The counselor communicates understanding of their disappointment in not implementing their plan, but, in the spirit of positive psychology, focuses on what they can learn from the failure.

Sharing empathic highlights is a mode of human contact, a relationship builder, a conversational lubricant, a perception-checking intervention, and a mild form of social influence. It is always useful. Driscoll (1984), in his common-sense way, referred to highlights as "nickel-and-dime interventions that each contribute only a smidgen of therapeutic movement, but without which the course of therapeutic progress would be markedly slower" (p. 90). Because sharing highlights provides a continual trickle of understanding, it is a way of providing support for clients throughout the helping process. It is never wrong to let clients know that you are trying to understand them from their frame of reference. Of course, thoughtful listening and processing lead to highlights that are much more than "nickel-and-dime" interventions. Clients who feel they are being understood participate more effectively and more fully in the helping process. Because sharing highlights helps build trust, it paves the way for the helper to use stronger interventions, such as challenging.

Respond selectively to core client messages. It is impossible to respond with highlights to everything a client says. Therefore, as you listen to clients, make every attempt to identify and respond to what you believe are core messages—that is, the heart of what the client is saying and expressing, especially if the client speaks at any length. Sometimes this selectivity means paying particular attention to one or two messages even though the client communicates many. For instance, a young woman, in discussing her doubts about marrying her companion, says at one time or another during a session that she is tired of his sloppy habits, is not really interested in his friends, wonders about his lack of intellectual curiosity, is dismayed at his relatively low level of career aspirations, and vehemently resents the fact that he faults her for being highly ambitious.

COUNSELOR: The whole picture doesn't look very promising, but the mismatch in career expectations is especially troubling.

CLIENT: You know, I'm beginning to think that Jim and I would be pretty good friends, even *because* we're so different. But partners? Maybe that's pushing it.

In this example, the counselor's empathic highlights help the client herself to identify what is core. The counselor follows her lead. His summary highlight at the end allows her to question the direction in which she and her friend are headed. Of course, because clients are not always so obliging, helpers must continually ask themselves as they listen, "What is key? What is most important here?" and then find ways of checking it out with the client. This helps clients sort out things that are not clear in their own minds.

Responding selectively sometimes means focusing on experiences or actions or feelings rather than all three. Consider the following example of a client who is experiencing stress because of his wife's poor health and concerns at work.

CLIENT: This week I tried to get my wife to see the doctor, but she refused, even though she fainted a couple of times. The kids had no school, so they were underfoot almost constantly. I haven't been able to finish a report my boss expects from me next Monday.

HELPER: It's been a lousy, overwhelming week all the way around.

CLIENT: As bad as they come. When things are lousy both at home and at work, there's no place for me to relax. I just want to get the hell out of the house and find some place forget it all. . . . Almost run away. . . . But I can't. . . . I mean I won't.

Here the counselor chooses to emphasize the feelings of the client, because she believes that his feelings of frustration and irritation are what is uppermost in his consciousness right now. This helps him move deeper into the problem situation— and then find a bit of resolve at the bottom of the pit.

At another time or with another client, the emphasis might be quite different. In the next example, a young woman is talking about her problems with her father.

CLIENT: My dad yelled at me all the time last year about how I dress. But just last week I heard him telling someone how nice I looked. He yells at my sister about the same things he ignores when my younger brother does them. Sometimes he's really nice with my mother and other times, too much of the time, he's just awful—demanding, grouchy, sarcastic.

HELPER: The inconsistency is killing you.

CLIENT: Absolutely! It's hard for all of us to know where we stand. I hate coming home when I'm not sure which "dad" will be there. Sometimes I come late to avoid all this. But that makes him even madder.

In this response, the counselor emphasizes the client's experience of her father's inconsistency. It hits the mark and she explores the problem situation further.

Respond to the context, not just the words. A good empathic response is based not just on the client's immediate words and nonverbal behavior. It also takes into account the context of what is said, everything that "surrounds" and permeates a client's statement. This client may be in crisis. That client may be doing a more leisurely "taking stock" of where he is in life. You are listening to clients in the context of their lives. The context modifies everything the client says.

Consider this case. Jeff, a white teenager, is accused of beating a black youth whose car stalled in a white neighborhood. The beaten youth is still in a coma. When Jeff talks to a court-appointed counselor, the counselor listens to what Jeff says in light of Jeff's upbringing and environment. The context includes the racist attitudes of many people in his blue-collar neighborhood, the sporadic violence there, the fact that his father died when Jeff was in primary school, a somewhat indulgent mother with a history of alcoholism, and easy access to drugs. The following interchange takes place.

JEFF: I don't know why I did it. I just did it, me and these other guys. We'd been drinking a bit and smoking up a bit—but not too much. It was just the whole thing.

HELPER: Looking back, it's almost like it's something that happened to you rather than something you did, and yet you know, somewhat bitterly, that you actually did it.

JEFF: More than bitter! I've screwed up the rest of my life. It's not like I got up that morning saying that I was going to bash someone that day.

The counselor's response is in no way an attempt to excuse Jeff's behavior, but it does factor in some of the environmental realities. Later on he will challenge Jeff to decide whether he is to remain a victim of his environment in terms of the prejudices he has acquired, gang membership, family history, and the like or whether he is going to do something about it.

Use highlights as a mild social-influence process. Because helpers cannot respond with highlights to everything their clients say, they are always searching for core messages. They are forced into a selection process that influences the course of the therapeutic dialogue. So even sharing highlights can be part of the social-influence dimension of counseling mentioned in Chapter 3. Helpers believe that the messages they select for attention are core primarily because they are core for the client. But helpers also believe, at some level, that certain messages *should* be important for the client.

In the following example, an incest-victim-turned-incest-perpetrator is in jail awaiting trial. In a session with a counselor he is trying to exonerate himself by blaming what happened to him in the past. He has been talking so quickly that the helper finds it difficult to interrupt. Finally, the helper, who has a pretty good working relationship with the client, breaks in.

HELPER: You've used some strong language to describe yourself. Let me see if I have it right. You said something about being "structurally deformed." I believe you also used the term "automatic reactions." You describe yourself as "haunted" and "driven."

CLIENTS: Well . . . I guess it's strong language. . . . Makes me sound like a psychological freak. Which I'm not.

The helper wants the client to listen to himself. So his "let me get this straight" response is a kind of empathic highlight form of challenge. It hits the mark because the client pulls himself up short. Of course helpers need to be careful not to put words in a client's mouth.

Use highlights to stimulate movement in the helping process. Sharing highlights is an excellent tool for building the helping relationship. But it also needs to serve the goals of the entire helping process. Therefore, sharing highlights is useful to the degree that it helps the client move forward. What does "move forward" mean? That depends on the stage or task in focus. For instance, sharing highlights helps clients move forward in Stage I if it helps them explore a problem situation or an undeveloped opportunity more realistically. It helps clients move forward in Stage II to the degree that it helps them identify and explore possibilities for a better future, craft a change agenda, or discuss commitment to that agenda. Moving forward in Stage III means clarifying action strategies, choosing specific things to do, and setting up a plan. In the action phase, moving forward means identifying obstacles to action, overcoming them, and accomplishing goals.

In the following example, a young woman visits the student services center at her college to discuss an unwanted pregnancy.

CLIENT: And so here I am, 2 months pregnant. I don't want to be pregnant. I'm not married, and I don't even love the father. To tell the truth, I don't even think I like him. Oh, Lord, this is something that happens to other people, not me! I wake up thinking this whole thing is unreal. Now people are trying to push me toward abortion.

HELPER: You're still so amazed that it's almost impossible to accept that it's true. To make things worse, people are telling you what to do.

CLIENT: Amazed? I'm stupefied! Mainly, at my own stupidity for getting myself into this. I've never had such an expensive lesson in my life. But I've decided one thing. No one, no one is going to tell me what to do now. I'll make my own decisions.

After the helper's highlight, self-recrimination over her lack of self-responsibility helps the client make a stand. She says she wants to capitalize on a very expensive mistake. It often happens that sharing highlights that hit the mark puts pressure on clients to move forward. So sharing highlights, even though it is a communication of understanding, is also part of the social-influence process.

Recover from inaccurate understanding. Although helpers should strive to be accurate in the understanding they communicate, all helpers can be inaccurate at times. You may think you understand the client and what he or she has said only to find out, when you share your understanding, that you were off the mark. Therefore, sharing highlights is a perception-checking tool. If the helper's response is accurate, the client often tends to confirm its accuracy in two ways. The first is some kind of verbal or nonverbal indication that the helper is right. That is, the client nods or gives some other nonverbal cue or uses some assenting word or phrase such as "that's right" or "exactly." This happens in the following example, in which a client who has been arrested for selling drugs is talking to his probation officer.

HELPER: So your neighborhood makes it easy to do things that can get you into trouble.

CLIENT: You bet it does! For instance, everyone's selling drugs. You not only end up using them, but you begin to think about pushing them. It's just too easy.

The second and more substantive way in which clients acknowledge the accuracy of the helper's response is by moving forward in the helping process—for instance, by clarifying the problem situation or possibilities for a better future more fully. In the preceding example, the client not only acknowledges the accuracy of the

Box 5-1   Suggestions for Sharing Empathic Highlights

How effectively do you do the following?

- Remember that empathy is a value, a way of being, that should permeate all communication skills.
- Tune in carefully, both physically and psychologically, and listen actively to the client's point of view.
- Make every effort to set your judgments and biases aside for the moment and walk in the shoes of the client.
- As the client speaks, listen especially for core messages.
- Pay attention to both verbal and nonverbal messages and their context.
- Respond with highlights fairly frequently, but briefly, to the client's core messages.
- Be flexible and tentative enough that the client does not feel pinned down.
- Use highlights to keep the client focused on important issues.
- Move gradually toward the exploration of sensitive topics and feelings.
- After sharing a highlight, attend carefully to cues that either confirm or deny the accuracy of your response.
- Determine whether your highlights are helping the client remain focused and stimulate the clarification of key issues.
- Note signs of client stress or resistance; try to judge whether these arise because you are inaccurate or because you are too accurate in your responses.
- Keep in mind that the communication skill of sharing empathic highlights, however important, is just one tool to help clients see themselves and their problem situations more clearly with a view to managing them more effectively.
- Take special care when the client's personal culture differs considerably from your own.

helper's empathy verbally—"you bet it does"—but, more important, also outlines the problem situation in greater detail. If the helper again responds by sharing a highlight, this leads to the next cycle. The next cycle may mean further clarification of the problem or opportunity or it may mean moving on to goal setting or some kind of problem-managing action.

On the other hand, when a response is inaccurate, the client often lets the counselor know in different ways. He or she may stop dead, fumble around, go off on a different tangent, tell the counselor "That's not exactly what I meant," or even try to get the helper back on track. Helpers need to be sensitive to all these cues. In the following example, Ben, a man who lost his wife and daughter in a train crash, has been talking about the changes that have taken place since the accident.

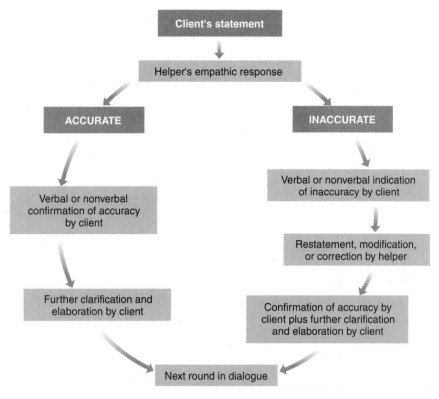

FIGURE 5-1
The Movement Caused by Accurate and Inaccurate Highlights

HELPER: So you don't want to do a lot of the things you used to do before the accident. For instance, you don't want to socialize much anymore.

BEN (pausing a long time): Well, I'm not sure that it's a question of wanting to or not. I mean that it takes much more energy to do a lot of things. It takes so much energy for me just to phone others to get together. It takes so much energy sometimes being with others that I just don't try.

HELPER: It's like a movie of a man in slow motion—it's so hard to do almost anything.

BEN: I'm in low gear, grinding away. And I don't know how to get out of it.

Ben says that it is not a question of motivation but of energy. The difference is important to him. By picking up on it, the helper gets the interview back on track. Ben wants to regain his old energy but he doesn't know how. His "lack of energy" is most likely some form of depression. And there are a number of ways to help clients deal with depression. This provides an opening for moving the helping process forward.

If you are intent on understanding your clients, they will not be put off by occasional inaccuracies on your part. Figure 5-1 indicates two different paths—one when helpers hit the mark in sharing highlights the first time, the other when they are inaccurate and then recover.

Use empathic highlights as a way of bridging diversity gaps. This principle is a corollary of the preceding two. Highlights based on effective tuning in and listening constitute one of the most important tools you have in interacting with clients who differ from you in significant ways. Sharing highlights is one way of telling clients that you are a learner, especially if the client differs from you in significant ways. Scott and Borodovsky (1990) referred to empathic listening as "cultural role taking." They could have said "diversity role taking." In the following example, a younger white male counselor is talking with an elderly African American woman who has recently lost her husband. She is in the hospital with a broken leg.

CLIENT: I hear they try to get you out of these places as quick as possible. But I seem to be lying around here doing nothing. Jimmy [her late husband] wouldn't even recognize me.

HELPER: It's pretty depressing to have this happen so soon after losing your husband.

CLIENT: Oh, I'm not depressed. I just want to get out of here and get back to doing things at home. Jimmy's gone, but there's plenty of people around there to help me take care of myself.

HELPER: Getting back into the swing of things is the best medicine for you.

CLIENT: Now you got it right. What I need right now is to know when I can go home and what I need to do for my leg once I get there. I've got to get things in order. That's what I do best.

The helper makes assumptions that might be true for him and some people in his culture, but they miss the mark with the client. She's taking her problems in stride and counting on her social system and a return to everyday household life to keep her going. The helper's second response hits the mark and she, in Stage II fashion, outlines some of the things she wants.

TACTICS FOR COMMUNICATING HIGHLIGHTS

The principles just outlined provide strategies for sharing empathic highlights. Here are a few hints—tactics, if you will—to help you improve the quality of your responses.

Give yourself time to think. Beginners sometimes jump in too quickly with an empathic response when the client pauses. "Too quickly" means that they do not give themselves enough time to reflect on what the client has just said in order to identify the core message being communicated. Watch some experts on tape. They often pause and allow themselves to assimilate what the client is saying.

Use short responses. I find that the helping process goes best when I engage the client in a dialogue rather than give speeches or allow the client to ramble. In a dialogue the helper's responses can be relatively frequent, but lean and trim. In trying to be accurate, the novice helper is often long-winded, especially if he or she lets the client go on and on before responding. Again, the question "What is the core of what this person is saying to me?" can help you make your responses short, concrete, and accurate.

Gear your response to the client, but remain yourself. If a client speaks animatedly, telling you how he finally got his partner to listen to his point of view about a new venture, and you reply accurately but in a flat, dull voice, your response is not fully empathic. This does not mean that you should mimic your clients, go overboard, or not be yourself. It means that part of being with the client is sharing in a reasonable way in his or her emotional tone. Consider this example:

12-YEAR-OLD CLIENT: My teacher started picking on me from the first day of class. I don't fool around more than anyone else in class, but she gets me anytime I do. I think she's picking on me because she doesn't like me. She doesn't yell at Bill Smith, and he acts funnier than I do.

COUNSELOR A: This is a bit perplexing. You wonder why she singles you out for so much discipline.

Counselor A's language is stilted, not in tune with the way a 12-year-old speaks. Here's a different approach.

COUNSELOR B: You're mad because the way she picks on you seems unfair.

On the other hand, helpers should not adopt a language that is not their own just to be on the client's wavelength. A older counselor using "hip" language or slang with a young client sounds ludicrous.

The Shadow Side of Sharing Empathic Highlights

Some helpers are poor communicators without even realizing it. Many responses that novice or inept helpers make are really poor substitutes for sharing accurate empathic highlights. Consider the following example, which includes a range of such responses. Robin is a young woman who has just started a career in law. This is her second visit to a counselor in private practice. In the first session she said she wanted to "talk through" some issues relating to the "transition" from school to business life. She appeared quite self-confident. In this session, after talking about a number of transition issues, she begins speaking in a rather strained voice and avoids eye contact with the counselor.

ROBIN: Something else is bothering me a bit. . . . Maybe it shouldn't. After all, I've got the kind of career that a lot of women would die for. Well—I'm glad that none of my feminist colleagues are around—I don't like the way I look. I'm neither fat nor thin, but I don't really like the shape of my body. And I'm uncomfortable with some of my facial features. Maybe this is a strange time of life to start thinking about this. I'll be thirty-three next year. . . . I bet I seem like an affluent, self-centered yuppie. . . .

Robin pauses and looks at a piece of art on the wall. What would you do or say? The following are some possibilities that are better avoided.

No response. It can be a mistake to say nothing, though cultures differ widely in how they deal with silence (Sue & Sue, 1990). In North American culture, generally speaking, if the client says something significant, respond to it, however briefly. Otherwise, the client may think that what he or she has just said doesn't merit a response. Don't leave Robin sitting there stewing in her own juices. A skilled helper would realize that a person's nonacceptance of his or her body could generalize to other aspects of life and therefore should not be treated as just a "vanity" problem.

Distracting questions. Some helpers, like many people in everyday life, cannot stop themselves from asking questions. Instead of responding with an empathic highlight, a counselor might ask something like, "Is this something new now that you've started working?" This response ignores what Robin has said and the feelings she has expressed and focuses rather on the helper's mistaken agenda to get more information.

Clichés. A counselor might say, "The workplace is competitive. It's not uncommon for issues like this to come up." This is cliché talk. It turns the helper into an insensitive instructor and probably sounds dismissive to the client. Clichés are hollow. The helper is saying, in effect, "You don't really have a problem at all, at least not a serious one." Clichés are a very poor substitute for understanding.

Interpretations. For some helpers, interpretive responses based on their theories of helping seem more important that expressing understanding. Such a counselor might say something like, "Robin, my bet is that your body-image concerns are probably just a symptom. I've got a hunch that you're not really accepting yourself at a deeper level. That's the real problem." The counselor fails to respond to the client's feelings and also distorts the content of the client's communication. The response implies that what is really important is hidden from the client.

Advice. In everyday life, giving unsolicited advice is extremely common. It happens in counseling, too. For instance, a counselor might say to Robin, "Hey, don't let this worry you. You'll be so involved with work issues that these concerns will disappear." Advice giving at this stage is out of order and, to make things worse, the advice given has a cliché flavor to it. Furthermore, advice giving robs clients of self-responsibility.

Parroting. Sharing a highlight does not mean merely repeating what the client has said. Such parroting is a parody of sharing empathic highlights. Review what Robin said about herself at the beginning of this section. Then evaluate the following response.

COUNSELOR: So, Robin, even though you have a great job, one that many people would envy, it's your feelings about your body that bother you. The feminist in you recoils a bit from this news. But there are things you don't like—your body shape, some facial features. You're wondering why this is hitting you now. You also seem to be ashamed of these thoughts. "Maybe I'm just self-centered," is what you're saying to yourself.

Most of this is accurate, but it sounds awful. Mere repetition or restatement or paraphrasing carries no sense of real understanding of, no sense of being with, the client. Real understanding, because it passes through you, should convey some part of you. Parroting doesn't. To avoid parroting, tap into the processing you've been doing as you listened, come at what the client has said from a slightly different angle, use your own words, change the order, refer to an expressed but unnamed emotion—in a word, do whatever you can to let the client know that you are working at understanding.

Sympathy and agreement. Being empathic is not the same as agreeing with the client or being sympathetic. An expression of sympathy has much more in common with pity, compassion, commiseration, and condolence than with empathic understanding. Although these are fully human traits, they are not particularly useful in counseling. Sympathy denotes agreement, whereas empathy denotes understanding and acceptance of the person of the client. At its worst, sympathy is a form of collusion with the client. Note the difference between Counselor A's response to Robin and Counselor B's response.

COUNSELOR A: This is not an easy thing to struggle with. It's even harder to talk about. It's even worse for someone who is as self-confident as you usually are.

ROBIN: I guess so.

Note that Robin does not respond very enthusiastically to collusion-talk. She is interested in managing her problem. The helping process does not move forward. Let's see a different approach.

COUNSELOR B: You've got some misgivings about how you look, yet you wonder whether you're justified even talking about it.

ROBIN: I know. It's like I'm ashamed of my being ashamed. What's worse, I get so pre-occupied with my body that I stop thinking of myself as a person. It blinds me to the fact that I more or less like the person I am.

Counselor B's response gives Robin the opportunity to deal with her immediate anx-iety and then to explore her problem situation more fully. Adriean Mancillas (2005, p. 19) suggests that clichéd forms of sympathy such as "You must be so strong" and "That must have been really hard for you" tend to "invalidate" the clients' experi-ence instead of helping them deal with it.

Faking it. Clients are sometimes confused, distracted, and in a highly emotional state. All these conditions affect the clarity of what they are saying about them-selves. Helpers may fail to pick up what the client is saying because of the client's confusion or because clients are not stating their messages clearly. Or the helpers themselves have become distracted in one way or another. In any case, it's a mistake to feign understanding. Genuine helpers admit that they are lost and then work to get back on track again. A statement like "I think I've lost you. Could we go over that once more?" indicates that you think it important to stay with the client. It is a sign of respect. Admitting that you're lost is infinitely preferable to such clichés as "uh-huh," "um," and "I understand." On the other hand, if you often catch yourself saying that you don't understand, then you'd better find out what is standing in the way. In any case, faking it is never a substitute for competence.

If you catch yourself making any of these mistakes, then find a way to recover. For instance, Figure 5-1 suggests a way of recovering from a failure to understand the client accurately. But helpers make all sorts of mistakes. Carolyn Dillon (2003) categorizes common mistakes and demonstrates how helpers can learn from them. She describes the "signals" clients send to helpers indicating a mis-take is being or has been made. Effective helpers recognize these signals and act on them.

THE IMPORTANCE OF EMPATHIC RELATIONSHIPS

In day-to-day conversations, sharing empathic highlights is a tool of civility. Making an effort to get in touch with your conversational partner's frame of refer-ence sends a message of respect. Therefore, sharing highlights plays an important part in building relationships. However, the communication skills as practiced in helping settings don't automatically transfer to the ordinary social settings of every-day life. In everyday life, understanding does not necessarily have to be put into words. Given enough time, people establish empathic relationships with one another in which understanding is communicated in a variety of rich and subtle ways without necessarily being put into words. A simple glance across a room as one spouse sees the other trapped in a conversation with a person he or she does not want to be with can communicate worlds of understanding. The glance says, "I know you feel caught. I know you don't want to hurt the other person's feelings. I can feel the struggles going on inside you. But I also know that you'd like me to rescue you, as soon as I can do so tactfully."

People with empathic relationships often express empathy in actions. An arm around the shoulders of someone who has just suffered a defeat expresses both empathy and support. I was in the home of a poor family when the father came bursting through the front door shouting, "I got the job!" His wife, without saying a word, went to the refrigerator, got a bottle of beer with a makeshift label on which "Champagne" had been written and offered it to her husband. Beer never tasted so good.

On the other hand, some people enter caringly into the world of their relatives, friends, and colleagues and are certainly "with" them but don't know how to communicate understanding through words. When a wife complains, "I don't know whether he really understands," she is not necessarily saying that her relationship with her husband is not mutually empathic. She is more likely saying that she would appreciate it if he were to put his understanding into words more often. In general, the more frequent use of empathic highlights in everyday life, especially when relationships are not going as well as they might, is often highly desirable.

The Art of Probing and Summarizing

THE SHADOW SIDE OF COMMUNICATION SKILLS—PART 2

Communication skills as necessary but not sufficient
The helping relationship versus helping technologies
Developing proficiency in communication skills

THE ART OF PROBING

In most of the examples used in the discussion of sharing empathic highlights, clients have demonstrated a willingness to explore themselves and their behavior relatively freely. Obviously, this is not always the case. Although it is essential that helpers respond with highlights when their clients do reveal themselves, it is also necessary at times to encourage, prompt, and help clients to explore their concerns when they fail to do so spontaneously. Therefore, the ability to use prompts and probes well is another important communication skill. If sharing highlights is the lubricant of dialogue, then probes provide often-needed nudges.

Prompts and probes are verbal and sometimes nonverbal tactics for helping clients talk more freely and concretely about any issue at any stage of the helping process. For instance, counselors can use probes to help clients identify and explore opportunities they have been overlooking, to clear up blind spots, to translate dreams into realistic goals, to come up with plans for accomplishing goals, and for working through obstacles to action. Probes, judiciously used, provide focus and direction for the entire helping process. Let's start with prompts.

Verbal and Nonverbal Prompts

Prompts are brief verbal or nonverbal interventions designed to let clients know that you are with them to encourage clients to talk further.

Nonverbal prompts. Counselors' various nonverbal behaviors can have the force of probes. For example, a client who has been talking about how difficult it is to make a peace overture to a neighbor she is at odds with says, "I just can't do it!" The helper says nothing but, rather, simply leans forward attentively and waits. The client pauses and then says, "Well, you know what I mean. It would be very hard for me to take the first step. It would be like giving in. You know, weakness." They go on to explore how such an overture, properly done, could be a sign of strength rather than weakness. Such things as bodily movements, gestures, nods, eye movement, and the like can be used as nonverbal prompts.

Vocal and verbal prompts. You can use such responses as "um," "uh-huh," "sure," "yes," "I see," "ah," "okay," and "oh" as prompts, provided you use them intentionally and they are not simply a sign that your attention is flagging, that you don't know what else to do, or that you are on automatic pilot. In the following example, the client, a 33-year-old married woman, is struggling with perfectionism both at work and at home.

CLIENT (hesitatingly): I don't know whether I can "kick the habit," you know, just let some trivial things go at work and at home. I know I've made a contract with myself. I'm not sure that I can keep it.

HELPER: "Oh." [The helper utters this briefly and then remains silent.]

CLIENT (pauses then laughs): Here I am deep into perfectionism and I hear myself saying that I can't do something! How ironic. Of course I can. I mean it's not going to be easy, at least at first.

The helper's "Oh" prompts the client to reconsider what she has just said. Prompts should never be the main course. They are part of the therapeutic dialogue only as a condiment.

Different Forms of Probes

Used judiciously, probes help clients name, take notice of, explore, clarify, or further define any issue at any stage of the helping process. They are designed to provide clarity and to move things forward. Probes take different forms.

Statements. One form of probe is a statement indicating the need for further clarity. For instance, a helper, talking to a client who is having problems with his 25-year-old daughter who is still living at home, says, "It's still not clear to me whether you want to challenge her to leave the nest or not." The client replies, "Well, I want to but I don't know how to do it without alienating her. I don't want it to sound like I don't care about her and that I'm just trying to get rid of her." Probes in the form of statements often take the form of the helper's confessing that he or she is in the dark in some way. "I'm not sure I understand how you intend. . . ." "I guess I'm still confused about. . . ." This kind of request puts the responsibility on clients without accusing them of failing to cough up the truth.

Requests. Probes can take the form of direct requests for further information or more clarity. A counselor, talking to a woman living with her husband and her mother-in-law, says, "Tell me what you mean when you say that three's a crowd at home." She answers, "I get along fine with my husband, I get along fine with my mother-in-law. But the chemistry among the *three* of us is very unsettling." This is helpful new information. Obviously requests should not sound like commands. "Come on, just tell me what you are thinking." Tone of voice and other paralinguistic and nonverbal cues help to soften requests.

Questions. Direct questions are perhaps the most common type of probe. "How do you react when he flies off the handle?" "What keeps you from making a decision?" "Now that the indirect approach to letting him know what your needs are is not working, what might Plan B look like?" Consider this case. A client has come for help in controlling her anger. With the help of a counselor she comes up with a solid program. In the next session, the client gives signs of backtracking. The counselor says, "You seemed enthusiastic about the program last week. But now, unless I'm mistaken, I hear a bit of hesitancy in your voice. What's going on?" The client responds, "Well, after taking a second look at the program, I'm afraid it will make me look like a wimp. My fellow workers could get the wrong idea and begin pushing me around." The counselor says, "So there's something about yourself you don't want to lose. What might that be?" The client hesitates for a moment and then says, "Spunk!" The counselor replies, "Well, let's see how you can keep your spunk and still get rid of the outbursts that get you in trouble." They go on to discuss the difference between assertiveness and aggression.

Single words or phrases that are in effect questions or requests. Sometimes single words or simple phrases are, in effect, probes. A client talking about a difficult relationship with her sister at one juncture says, "I really hate her." The helper responds simply and unemotionally, "Hate." The client responds, "Well, I know that hate is too strong a term. What I mean is that things are getting worse and worse." This kind of clarity helps. Another client, troubled with irrational fears, says, "I've had it. I just can't go on like this. No matter what, I'm going to move forward." The counselor replies, "Move forward to . . . ?" The client says, "Well . . . to not indulging myself with my fears. That's what they are, a form of self-indulgence. From our talks I've learned that it's a bad habit. A very bad habit." They go on to discuss ways of controlling such thoughts.

Whatever form probes take, they are often, directly or indirectly, questions of some sort. Therefore, a few words about the use of questions is in order.

Using Questions Effectively

Helpers, especially novices and inept counselors, tend to ask too many questions. When in doubt about what to say or do, they ask questions that add no value. It is as if gathering information were the goal of the helping interview. Using questions judiciously can be an important part of your interactions with clients. Here are two guidelines.

Do not ask too many questions. When clients are asked too many questions, they feel grilled, and that does little for the helping relationship. Furthermore, many clients instinctively know when questions are just filler, used because the helper does not have anything better to say. I have caught myself asking questions the answers to which I didn't even want to know. Let's assume that the helper is working with Rolly, an inmate in a state facility for young offenders. He is doing time for burglary and drug use. Because he is difficult to work with and blames everything on his dysfunctional family, the counselor out of frustration ends up asking a whole series of questions:

"When did you first feel caught in the messiness of your family?"

"What did you do to try to get away from their influence?"

"What could you do different?"

"What kind of friends did you have?"

These questions are no more than a random search for information, the value of which is not clear. Rolly is an expert in evading questions like this. He has been grilled by professionals. When he tires of being questioned, he just clams up. Helping that turns into question-and-answer sessions tends to go nowhere.

Ask open-ended questions. As a general rule, ask open-ended questions—that is, questions that require more than a simple yes or no or similar one-word answer. Not, "Now that you've decided to take early retirement, do you have any plans?" but, "Now that you've decided to take early retirement, how do you see the future? What plans do you have?" Counselors who ask closed questions find themselves asking more and more questions. One closed question begets another. Of course, if a specific piece of information is needed, then a closed question may be used. A career counselor might ask, "How many jobs have you had in the past 2 years?" The information is relevant to helping the client draw up a resume and a job-search strategy. Of course, occasionally, a sharp closed question can have the right impact. For instance, a client has been outlining what he was going to do to get back at his "ungrateful" son. The counselor asks, "Is getting back at him what you really want?"

Rhetorical questions like this are a form of challenge. On the other hand, open-ended questions in moderation can help clients fill in what is missing at every stage of the helping process.

Guidelines for Using Probes

Here, then, are some suggestions that can guide you in the use of all probes, whatever form they may take.

Use probes to help clients engage as fully as possible in the therapeutic dialogue. As noted earlier, many clients do not have the all the communication skills needed to engage in the problem-managing and opportunity-developing dialogue. Probes are the principal tools needed to help all clients engage in the give-and-take of the helping dialogue. The following exchange takes place between a counselor at a church parish center and a parishioner who has been struggling to tell her story about her attempts to get her insurance company to respond to the claim she filed after a car accident.

CLIENT: They just won't do anything. I call and get the cold shoulder. They ignore me and I don't like it!

HELPER: You're angry with the way you're being treated. And you want to get to the bottom of it. . . . I'd like a brief overview of what you've done so far.

CLIENT: Well, they sent me forms that I didn't understand very well. I did the best I could. I think they were trying to show that it was my fault. I even kept copies. I've got them with me.

HELPER: You're not sure you can trust them. . . . Let's see what forms look like. . . .

The forms turn out to be standard claims forms. The fact that this is the client's first encounter with an insurance company and that she has poor communication skills gives the counselor some insight into what the phone conversations between her and the insurance company must be like. By sharing highlights and using probes, he gets her to see that her experience might well be normal. The outcome is that the client gets help from a fellow parishioner who has filled in insurance forms a number of times.

Use probes to help clients achieve concreteness and clarity. Probes can help clients turn what is abstract and vague into something concrete and clear—something you can get your hands on and work with. In the next example, a man is talking about an intimate relationship that has turned sour.

CLIENT: She treats me badly, and I don't like it!

HELPER: Tell me what she actually does.

CLIENT: She talks about me behind my back. I know she does. Others tell me what she says. She also cancels dates when something more interesting comes up.

HELPER: That's pretty demeaning. . . . How have you been reacting to all this?

CLIENT: Well, I think she knows that I have an idea of what's going on. But we haven't talked about it much.

In this example, the helper's probe leads to a clearer statement of the client's experience and behavior. By sharing highlights and using probes, the helper discovers that the client puts up with a great deal because he is afraid of losing her. He goes on to help the client deal with the psychological "economics" of such a one-sided relationship.

In the next example, a man who is dissatisfied with living a somewhat impover-ished social life is telling his story. A simple probe leads to a significant revelation.

CLIENT: I do funny things that make me feel good.

HELPER: What kinds of things?

CLIENT: Well, I daydream about being a hero, a kind of tragic hero. In my daydreams I save the lives of people whom I like but who don't seem to know I exist. And then they come running to me but I turn my back on them. I choose to be alone! I come up with all sorts of variations of this theme.

HELPER: So in your daydreams you play a character who wants to be liked or loved but who gets some kind of satisfaction from rejecting those who haven't loved him back. I'm not sure I've got that right.

CLIENT: Well . . . yeah . . . I sort of contradict myself. . . . I do want to be loved but I guess I don't do very much to get a real social life. It's all in my head.

The helper's probe leads to a clearer statement of what's going on in the client's head. Helping the client to explore his fantasy life could be a first step toward find-ing out what he really wants from relationships and what he needs to do to get it.

The next client has become the breadwinner since her husband suffered a stroke. Someone takes care of her husband during the day.

CLIENT: Since my husband had his stroke, coming home at night is rather difficult for me. I just. . . . Well, I don't know.

HELPER: It sounds like it really gets you down. . . . What's it like?

CLIENT: When I see him sitting immobile in the chair, I'm filled with pity for him and the next thing I know it's pity for myself and it's mixed with anger or even rage, but I don't know what or whom to be angry at. I don't know how to focus my anger. Good God, he's only 42 and I'm only 40!

In this case, the helper's probe leads to a fuller description of the intensity of the client's feelings and emotions, her sense of desperation.

In each of these cases, the client's story gets more specific. Of course, the goal is not to get more and more detail. Rather, it is to get the kind of detail that makes the problem or unused opportunity clear enough to see what can be done about it.

Use probes to explore and clarify clients' points of view, intentions, proposals, and decisions. Clients often fail to clarify their points of view, intentions, proposals, and decisions. For instance, a client might announce some decision she has made. But the decision itself is unclear, and the reasons behind it and the implications for the client and others are not spelled out. In the following case, the client has had a bad automo-bile accident while driving under the influence. Luckily he was the only one hurt. He is recovering physically, but his psychological recovery has been slow. The accident opened up a Pandora's box of psychological problems that were not being handled—for instance, a lack of self-responsibility. A counselor has been helping him work through some of these issues. The following exchange takes place during an early session.

CLIENT: I don't think that the laws around driving under the influence should be as tough as they are. I'm scared to death of what might happen to me if I ever had an accident again.

COUNSELOR: So you feel you're in jeopardy. . . . What makes you think that the laws are too tough?

CLIENT: Well, they bully us. One little mistake and bingo! Your freedom goes out the window. Laws should make people free.

COUNSELOR: Well, let's explore a little. Hmm, let's say all laws on driving under the influence were dropped. What then?

The counselor knows that the client is running away from taking responsibility for his actions. Using probes to get him to spell out the implications of his point of view on DUI laws is the beginning of an attempt to help the client face up to himself.

In a later session, the client talks about the legal ramifications of the accident. He has to go to court.

CLIENT: I've been thinking about this. I'm going to get me a really good lawyer and fight this thing. I talked with a friend, and he thinks he knows someone who can get me off. I need a break. It might cost me a bundle. After all, I messed up someone's property a bit, but I didn't hurt anyone.

COUNSELOR: What's the best thing that could happen in court?

CLIENT: I'd get off scot-free. Well, maybe a slap on the wrist of some kind. A warning.

COUNSELOR: And what's the worst thing that could happen?

CLIENT (a long pause): I haven't given that a lot of thought. I don't really know much about the laws or the courts or how tough they might be. That sort of stuff. But with the right lawyer. . . .

COUNSELOR: Hmm. I'm trying to put myself in your shoes. . . . I think I'd try to find out how cases like mine tend to go in court. . . . I'd like to know that before spending a lot of money on a defense lawyer. . . . What do you think?

The counselor is using probes to help the client explore the implications of a decision he's making.

The state has very tough DUI laws. In the end, because the client's blood-alcohol level was so high, his license is suspended for 6 months, he is fined heavily, and he has to spend a month in jail. All of this is very sobering. The counselor visits him in jail and they talk about the future.

CLIENT: I feel like I've been hit by a train.

COUNSELOR: You had no idea that it would be this bad.

CLIENT: Right. No idea. . . . I know you tried to warn me in your own way, but I wasn't ready to listen. . . . Now I have to begin to put my life back together. Though I don't feel like it.

COUNSELOR: But now that you've had the wake-up call, a horrible wake-up call, what does the future hold . . . even if you don't feel up to looking at it?

CLIENT: I've been thinking. One thing I want to do is to make some sort of apology to my family. They're hurting as bad as I am. I feel so awkward. I know how to act in cocky mode. Humble mode I'm not used to. Do I write a long letter? Do I wait and just apologize through my actions? Do I take each one of them aside? I don't know, but I've just got to do it.

COUNSELOR: Somehow you have to make things right with them. Just how, well that's another matter. Maybe we could start by finding out what you want to accomplish through an apology, however it's done.

Here we find a much more sober and cooperative client. He proposes, roughly, a course of action. The counselor supports his need to move beyond past stupidities and present misery. It's about the future, not the past. The counselor's last statement is a probe aimed at giving substance and order to the client's proposal. It asks the client, What do you want to accomplish?

Use probes to help clients fill in missing pieces of the picture. Probes further the therapeutic dialogue by helping clients identify missing pieces of the puzzle—thoughts, experiences, behaviors, and feelings that would help both clients and helpers get a better fix on the problem situation, possibilities for a better future, or drawing up a plan of action. In the following example, the client is at odds with his wife over his mother-in-law's upcoming visit.

HELPER: I realize now that you often get angry when your mother-in-law stays for more than a day. But I'm still not sure what she does that makes you angry.

CLIENT: First of all, she throws our household schedule out and puts in her own. Then she provides a steady stream of advice on how to raise the kids. My wife sees this as an "inconvenience." For me it's a total family disruption. When she leaves, there's a lot of emotional cleaning up to be done.

Just what the client's mother-in-law does to get him going has been missing. Once the behavior has been spelled out in some detail, it is easier to help him come up with some remedies. Still missing, however, is what *he* does in the face of his mother-in-law's behavior. The helper continues:

HELPER: So when she takes over everything gets turned upside down. . . . How do you react in the face of all this turmoil?

CLIENT: Well . . . well . . . I guess I go silent. Or I just get out of there, go somewhere, and fume. After she's gone, I take it out on my wife, who still doesn't see why I'm making such a fuss.

So now it's clear that the client does little to change things. It is also obvious that he is a little taken aback by being asked how he handles the situation.

In the next example, a divorced woman is talking about the turmoil that takes place when her ex-husband visits the children. It has some similarities with the case we've just seen.

HELPER: The Sundays your husband exercises his visiting rights with the children end in his taking verbal potshots at you, and you get these headaches. I've got a fairly clear picture of what he does when he comes over and how it gets to you, but it might help if you could describe what you do.

CLIENT: Well, I do nothing.

HELPER: So last Sunday he just began letting you have it for no particular reason. Or just to make you feel bad.

CLIENT: Well . . . not exactly. I asked him about increasing the amount of the child-support payments. And I asked him why he's dragging his feet about getting a better job. He's so stupid. He can't even take a bit of sound advice.

Through probes the counselor helps the client fill in a missing part of the picture—her own behavior. She keeps describing herself as total victim and her ex-husband as total aggressor. That doesn't seem to be the full story. Her behavior contributes to the drama.

Use probes to help clients get a balanced view of problem situations and opportunities. In their eagerness to discuss an issue or make a point, clients often describe one side of a picture or one viewpoint. Probes can be used to help them fill out the picture. In the following example, the client, a manager who has been saddled with a bright, highly ambitious and aggressive young woman who plays politics to further her own interests, has been agonizing over his plight.

COUNSELOR: I've been wondering whether you see any upside to this. Any hidden opportunities.

CLIENT: I'm not sure what you mean. It's just a disaster.

COUNSELOR: Well, you strike me as a pretty bright guy. I'm wondering if there are any lessons for you hidden in all this.

CLIENT (pausing): Oh, well, you know I tend to ignore politics around here, but now it's in my face. Where there are people, there are politics, I suppose. I think she's being political to serve her own career. But I don't want to play her game. There must be some other kind of game or something that would let me keep my integrity. The days of avoiding all of this are probably over.

The problem situation has a flip side. It is an opportunity for rethinking and learning. As such, problems are incentives for constructive change. The client can learn something through all this. It's an opportunity to come to grips with the male-female dynamics of the workplace and a chance to explore "positive" political skills.

Use probes to help clients move into more beneficial stages of the helping process. Probes can be used to help clients engage in dialogue about any part of the helping process—telling their stories more fully, attacking blind spots, setting goals, formulating action strategies, discussing obstacles to action, and reviewing actions taken. Many clients do not easily move into whatever stage of the helping process might be most useful for them. Probes can help them do so. In the following example, the counselor uses a probe to help a middle-aged couple, Sean and Fiona, who have been complaining about each other, to move on to Stages II and III. Besides complaining, they have talked vaguely about "reinventing" their marriage, a term used in some kind of marriage encounter group they attended. For instance, they have hinted at doing more things in common.

COUNSELOR: What kinds of things do you like doing together? What are some possibilities?

FIONA: I can think of something, though it might sound stupid to you (she says with a glance toward her husband). We both like doing things for others, you know, caring about other people. Before we were married, we talked about spending some time in the Peace Corps together, though it never happened.

SEAN: I wish we had. . . . But those days are past.

COUNSELOR: Are they? The Peace Corps may not be an option, but there must be other possibilities. (Neither Fiona nor Sean says anything.) I tell you what. Here are a couple of pieces of paper. Jot down three ways of helping others. Do your own list. Forget what your spouse might be thinking.

The counselor uses probes to get Sean and Fiona to brainstorm possibilities for some kind of service to others. This moves them away from tortuous problem exploration toward opportunity development.

The next client has been talking endlessly about the affair her husband is having. Her husband knows that she knows.

COUNSELOR: You've said you're not going to do anything about it because it might hurt your son. But doing nothing is not the only possible option. Let's just name some of them. Who knows? We might find a gem.

CLIENT: Hmm. . . . I'm not sure I know.

COUNSELOR: Well, you know people in the same predicament. You've read novels, seen movies. What are some of the standard things people do? I'm not saying do them. Let's just review them.

CLIENT: Hmm. . . . Well, I knew someone in a situation like this who did an outrageous thing. She knew her teenaged daughter was aware of what was going on. So one night at dinner she just said, "Let's all talk about the affair you're having and how to handle it. It's certainly not news to any of us."

COUNSELOR: All right, that's one possible way. Let's hear some more.

This primes the pump. The counselor uses a few more probes to put a number of possibilities on the table. The focus on action brings energy to the session.

In the following example, Jill, the helper, and Justin, the client, have been discussing how Justin is letting his impairment—he has lost a leg in a car accident—stand in the way of his picking up his life again. The session has bogged down a bit.

JILL: Let's try a bit of drama. I'm going to be Justin for a while. You're going to be Jill. As my counselor, ask me some questions that you think might make a difference for me. Me, that is, Justin.

JUSTIN (pausing a long time): I'm not much of an actor, but here goes. . . . Why are you taking the coward's way out? Why are you on the verge of giving up?" (His eyes tear up.)

Jill gets Justin to formulate the probes. It's her way of asking Justin to "move forward" and take responsibility for his part of the session. Justin's "probes" turn out to be challenges, almost accusations, certainly much stronger than anything Jill, at this stage, would have tried. However painful this is for Justin, it's a breakthrough.

Use probes to challenge clients and help them challenge themselves. In the last chapter we saw that even sharing empathic highlights can act as mild form of social-influence or challenge. We also saw that effective highlights often act as probes. That is, they can be indirect requests for further information or ways of steering a client toward a different stage of the helping process. And, as you have probably noticed in the examples used in this chapter, probes can edge much closer to outright challenge. Many probes are not just requests for relevant information. They often place some kind of demand on the client to respond, reflect, review, or reevaluate. Probes can serve as a bridge between communicating understanding to clients and helping them challenge themselves. The following client, having committed himself to standing up to some of his mother's possessive ways, now shows signs of weakening in his resolve.

HELPER: The other day you talked of "having it out with her"—though that might be too strong a term. But just now you mentioned something about "being reasonable with her." Tell me how these two differ.

CLIENT (pausing): Well, I think you might be witnessing a case of cold feet. . . . She's a very strong woman.

The counselor helps the client revisit his decision to "get tough" with his mother and, if this is what he really wants, what he can do to strengthen his resolve. Using probes as mild forms of challenge is perfectly legitimate provided you know what you are doing.

The Relationship Between Sharing Highlights and Using Probes

The trouble with dealing with skills one at a time is that each skill is taken out of context. In the give-and-take of any given helping session, however, the skills must be intermingled in a natural way. In actual sessions, skilled helpers continually tune in, listen actively, and use a mix of probes and empathy to help clients clarify and come to grips with their concerns, deal with blind spots, set goals, make plans, and get things done. There is no formula for the right mix. That depends on the client, client needs, the problem situation, possible opportunities, and the stage of the helping process.

Here is a basic guideline about sharing highlights and using probes. After using a probe to which a client responds, respond with empathy to what the client has to say. Check your understanding. Be hesitant to follow one probe with another. The logic of this is straightforward. First, if a probe is effective, it will yield information that needs to be listened to and understood. Second, a shared highlight, if accurate, tends to place a demand on the client to explore further. It puts the ball back in the client's court.

In the following example, the client is a young Chinese American woman whose father died in China and whose mother is now dying in the United States. She has been talking about the traditional obedience of Chinese women and her fears of slipping into a form of passivity in her American life. She talks about her sister, who gives everything to her husband without looking for anything in return. The first counselor sticks to probes.

COUNSELOR: To what degree is this self-effacing role rooted in your culture?

CLIENT: Well, being somewhat self-effacing is certainly in my cultural genes. And yet I look around and see many of my North American counterparts adopt a very different style. A style that frankly appeals to me. But last year, when I took a trip back to China with my mother to meet my half sisters, the moment I landed I wasn't American. I was totally Chinese again.

COUNSELOR A: What did you learn there?

CLIENT: That I am Chinese!

The client says something significant about herself, but instead of responding with understanding, the helper uses another probe. This elicits only a repetition, with some annoyance, of what she had just said. Now a different approach.

COUNSELOR B: You learned just how deep your cultural roots go.

CLIENT: And if these roots are so deep, what does that mean for me here? I love my Chinese culture. I want to be Chinese and American at the same time. How to do that, well, I haven't figured that out yet. I thought I had, but I haven't.

In this case, an empathic highlight works much more effectively than another probe. Counselor B helps the client move forward.

In the next example, a single middle-aged woman working in a company that has reinvented itself after a downturn in the economy still has a job, but the pay is much less and she is doing work she does not enjoy. She does not have the computer and Internet-related skills needed for the better jobs. She feels both stressed and depressed.

CLIENT: Well, I suppose that I should be grateful for even having a job. But now I work longer hours for less pay. And I'm doing stuff I don't even like. My life is no longer mine.

Box 6-1 Guidelines in Using Probes

- Keep in mind the goals of probing. Use probes to:
  - Help clients engage as fully as possible in the therapeutic dialogue.
  - Help nonassertive or reluctant clients tell their stories and engage in other behaviors related to managing their problems and developing opportunities.
  - Help clients identify experiences, behaviors, and feelings that give focus to their stories.
  - Help clients open up new areas for discussion.
  - Help clients explore and clarify stories, feelings, points of view, decisions, and proposals.
  - Help clients be as concrete and specific as possible.
  - Help clients remain focused on relevant and important issues.
  - Help clients move on to a further stage of the helping process.
- Use probes to provide mild challenges to clients to examine the way they think, behave, and act both within helping sessions and in their daily lives.
- Make sure that probing is done in the spirit of empathy.
- Use a mix of statements, open-ended questions, prompts, and requests, not questions alone.
- Follow up a successful probe with an empathic highlight rather than another probe.
- Use whatever judicious mixture of highlights and probing is needed to help clients clarify problems, identify blind spots, develop new scenarios, search for action strategies, formulate plans, and review outcomes of action.

HELPER: So the extra pressure and stress makes you wonder just how "grateful" you should feel.

CLIENT: Precisely. . . . And the future looks pretty bleak.

HELPER: What could you change in the short term to make things more bearable?

CLIENT: Hmm. . . . Well, I know one way. We all keep complaining to one another at work. And this seems to make things even worse. I can stop playing that game. It's one way of making life a bit less miserable.

HELPER: So one way is to stop contributing to your own misery by staying away from the complaining chorus. . . . What might you start doing?

CLIENT: Well, there's no use sitting around hoping that what has happened is going to be reversed. I've been really jolted out of my complacency. I assumed with the economy humming again I'd find things easy. The economy may be humming, but jobs, good jobs, are still scarce. But I'm still young enough to acquire some more skills. And I do have some skills that I haven't needed to use before. I'm a good communicator, and I've got a lot of common sense. I work well with people. There are probably some jobs around here that require those skills.

HELPER: So, given the wake-up call, you think it might be possible to take unused skills and reposition yourself at work.

CLIENT: Repositioning. Hmm, I like that word. It makes a lot of pictures dance through my mind. . . . Yes, I need to reposition myself. For instance. . . .

This combination of highlights and probing gets things moving. Instead of focusing on the misery of the present situation, the client names a few possibilities for a better future, a Stage II activity.

You should be careful not to become either an empathic highlight "machine," grinding out one highlight after another, or an "interrogator," peppering your clients continually with needless probes. All responses to clients, including probes and challenges, are empathic if they are based on a solid understanding of the client's core messages and points of view. All responses that build on and add to the clients remarks are implicitly empathic, which cuts down on the need to share a steady stream of highlights.

THE ART OF SUMMARIZING: PROVIDING FOCUS AND DIRECTION

The communication skills of visibly tuning in, listening, sharing highlights, and probing need to be orchestrated in such a way that they help clients focus their attention on issues that make a difference. The ability to summarize and to help clients summarize the main points of a helping interchange or session is a skill that can be used to provide both focus and challenge.

When to Use Summaries

Brammer (1973) listed a number of goals that can be achieved by judicious use of summarizing—"warming up" the client, focusing scattered thoughts and feelings, bringing the discussion of a particular theme to a close, and prompting the client to explore a theme more thoroughly. There are certain times when summaries prove particularly useful: at the beginning of a new session, when the session seems to be going nowhere, and when the client needs a new perspective.

At the beginning of a new session. Using summaries at the beginning of a new session, especially when clients seem uncertain about how to begin, prevents clients from merely repeating what has already been said before. It puts clients under pressure to move on. Consider this example: Liz, a social worker, begins a session with a rather overly talkative man by summarizing of the main points from the previous session. This serves several purposes. First, it shows the client that she had listened carefully to what he had said in the last session and that she had reflected on it after the session. Second, the summary gives the client a jumping-off point for the new session. It gives him an opportunity to add to or modify what was said. Finally, it places the responsibility for moving forward on the client. The implied sentiment of the summary is: "Now where do you want to go with this?" Summaries put the ball in the client's court and give them an opportunity to exercise initiative.

During a session that is going nowhere. Helpers can use a summary to give focus to a session that seems to be going nowhere. One of the main reasons sessions go nowhere is that helpers allow clients to keep discussing the same things over and over again instead of helping them either go more deeply into their stories, focus on possibilities, and goals, or discuss strategies that will help clients get what they need and want. For instance, Marcia is a coach, consultant, and counselor who is working with the staff of a shelter for the homeless. One of the staff members is showing signs of burnout. In a second meeting with Marcia, she keeps going over the same ground, talking endlessly about stressful incidents that have taken place over the last few months. At one point Marcia provides a summary.

MARCIA: Let's see if I can pull together what you've been saying. The work here, by its very nature, is stressful. You've mentioned a whole string of "incidents" such as being hit by someone you were trying to help or the heated arguments with some of your coworkers. But I believe you've intimated that these are the kinds of things that happen in these places. Shelters are prone to them. They are part of the furniture. They're not going to stop. But they can be very punishing. At times you wish you weren't here. But if you're going to stay and if these kinds of incidents are not going to stop, maybe some questions might be, "How do I cope with them? How do I do my work and get some ongoing satisfaction from it? What changes can we make around here that might lessen the number of these incidents?"

The purpose of the summary here is to help the client move beyond "poor me" and find ways of coping with this kind of work. The challenge in places like shelters is creating a supportive work environment, developing a sense of organizational and personal purpose, promoting the kind of teamwork that fits the institution's mission, and fostering a culture of coping strategies.

When the client needs a new perspective. Often when scattered elements are brought together, the client sees the "bigger picture" more clearly. In the following example, a man who has been reluctant to go to a counselor with his wife has, in a solo session with the counselor, agreed to a couple of sessions "to please her." In the session, he talks a great deal of his behavior at home, but in a rather disjointed way.

COUNSELOR: I'd like to pull a few things together. You've encouraged your wife in her career, especially when things are difficult for her at work. You also encourage her to spend time with her friends as a way of enjoying herself and letting off steam. You also make sure that you spend time with the kids. In fact, time with them is important for you.

CLIENT: Yeah. That's right.

COUNSELOR: Also, if I have heard you correctly, you currently take care of the household finances. You are usually the one who accepts or rejects social invitations, because your schedule is tighter than hers. And now you're about to ask her to move because you can get a better job in Boston.

CLIENT: When you put it all together like that, it sounds as if I'm running her life. . . . She never tells me I'm running her life.

COUNSELOR: Maybe we could talk a little about this when the three of us get together.

CLIENT: Hmm. . . . Well, I'd . . . hmm. . . . (laughs). I'd better think about all of this before the next session.

The summary provides the client with a mild jolt. He realizes that he needs to face up to the "I am making many decisions for her, and some of them are big" theme implied in the summary.

In the following example, the client is a 52-year-old man who has been talking about a number of problems he is experiencing. He has come for help because he has been "down in the dumps" and can't seem to shake it.

HELPER: Let's take a look at what we've seen so far. You're down—not just a normal slump; this time it's hanging on. You worry about your health, but you check out all right physically, so this seems to be more a symptom than a cause of your slump. There are some unresolved issues in your life. One that you seem to be stressing a lot is the fact that your recent change in jobs has means that you don't see much of your old friends anymore. Because you're single, this leaves you, currently, with a rather bleak social life. Another issue—one you find painful and embarrassing—is your struggle to stay young. You don't like facing the fact that you're getting older. A third issue is the way you—to use your own word—"overinvest" yourself in work, so much so that when you finish a long-term project, suddenly your life is empty.

CLIENT (pauses): It's painful to hear it all that baldly, but that about sums it up. I've suspected I've got some screwed-up values, but I haven't wanted to stop long enough to take a look at it. Maybe the time has come. I'm hurting enough.

HELPER: One way of doing this is by taking a look at what a better future would look like.

CLIENT: That sounds interesting, even hopeful. How would we do that?

The counselor's summary hits home—somewhat painfully—and the client draws his own conclusion. Care should be taken not to overwhelm clients with the contents of the summary. Nor should summaries be used to "build a case" against a client. Helping is not a judicial procedure. Perhaps the foregoing summary would have been more effective if the helper had also summarized some of the client's strengths. That would have provided a more positive context.

Getting the Client to Provide the Summary

Summaries can be useful when clients don't seem to know where to go next, either in the helping session itself or in a real-world action program. In cases like this helpers can, of course, use probes to help them move on. Summaries, however, have a way of keeping the ball in the client's court. Moreover, the helper does not always have to provide the summary. Often it is better to ask the client to pull together the major points. This helps the client own the helping process, pull together the salient points, and move on. Because this is not meant to test clients, the counselor should provide clients whatever help they need to stitch the summary together.

In the following example, the client, who has lost her job and her boyfriend because of her alcohol-induced outbreaks of anger, has been talking about "not being able to stick to the program." The counselor asks her to summarize what she's been doing and the obstacles she has been running into. With the help of the counselor, she stumbles through a summary. At the end of it she says, "I guess it's clear to both of us that I haven't been doing a very good job sticking to the program. On paper, my plan looked like a snap. But it seems that I don't live on paper." The counselor and client take a couple of steps back, reviewing goals, plans, obstacles, and execution.

THE SHADOW SIDE OF COMMUNICATION SKILLS—PART 2

Up to this point we have been dealing with basic communication skills. In Chapters 7 and 8, we will consider advanced communication skills and processes dealing with challenging. But first, let's look again at the shadow side of communication skills.

Communication skills as necessary but not sufficient. Some training programs and helpers tend to overidentify the helping process with the communication skills—that is, with the tools that serve it. Being good at communication skills is not the same as being good at helping. Moreover, an overemphasis on communication skills can turn helping into a great deal of talk with very little action—and few outcomes that make a difference in clients' lives.

Communication skills are essential, of course, but they still must serve both the process and the outcomes of helping. These skills certainly help you establish a good relationship with clients. And a good relationship is the basis for the kind of social-emotional reeducation that has been outlined earlier. But you can be good at communication, good at relationship building, even good at social-emotional reeducation and still shortchange your clients, because they need more than that. Some who overestimate the value of communication skills tend to see a skill such as sharing highlights as some kind of "magic bullet." Others overestimate the value of information gathering. This is not a broad indictment of the profession. Rather, it is a caution for beginners.

The helping relationship versus helping technologies. On the other hand, some practitioners underestimate the need for solid communication skills. There is a subtle assumption that the "technology" of their approach, such as treatment manuals, suffices. They listen and respond through their theories and constructs rather than through their humanity. They become technologists instead of helpers. They are like some medical doctors who become more and more proficient in the use of medical technology and less and less in touch with the humanity of their patients.

Some years ago I spent 10 days in a hospital (an eternity in these days of managed care). The staff were magnificent in addressing my medical needs. But the psychological needs that sprang from my anxiety about my illness were not addressed at all. Unfortunately, my anxieties were often expressed through physical symptoms. Then those symptoms were treated medically. Out of frustration I asked the young doctor who was debriefing me about the staffing conference where my case was reviewed, "When you have conferences during which patients are discussed, do you say, 'Well, we've thoroughly reviewed his medical status and needs. Now let's turn our attention to what he's going through. What can we do to help him through this experience?'" One resident said, "No, we don't have time." Don't get me wrong. These were dedicated, generous people who had my interests at heart. But they ignored many of my needs. Having healthcare psychologists involved in such staffing sessions is a step in the right direction.

Developing proficiency in communication skills. Understanding communication skills and how they fit into the helping process is one thing. Becoming proficient in their use is another. Some trainees think that these "soft" skills should be learned easily and fail to put in the kind of hard work and practice that makes them "fluent" in them. Doing the exercises in communication-skills manuals and practicing these skills in training groups can help, but that isn't enough to make these skills second nature. Tuning in, listening, processing, sharing highlights, and probing that are trotted out, as it were, for helping encounters are likely to have a hollow ring to them. These skills must become part of your everyday communication style.

After providing some initial training in communication skills, I tell students, "Now, go out into your real lives and get good at these skills. I can't do that for you." In the beginning, it may be difficult to practice all these skills in everyday life, not because they are so difficult, but because they are relatively rare in conversations. Take sharing highlights. Listen to the conversations around you. If you were to use an unobtrusive counter, pressing a button every time you heard someone share an empathic highlight, you might go days without pressing the button. But you can make sharing highlights a reality in your own life. And those who interact with you will notice the difference. They probably will not call it empathy or sharing highlights. Rather, they will say such things as "She really listens to me" or "He takes me seriously."

On the other hand, you will hear many probes in everyday conversations. People are much more comfortable asking questions than providing understanding. However, many of these probes tend to be aimless. Worse, many will be disguised criticisms. "Why on earth did you do *that?*" Learning how to integrate purposeful probes with highlights demands practice in everyday life. Life is your lab. Every conversation is an opportunity.

These skills have a place in all the human transactions of life, including business transactions. When businesses are asked what competencies they want to see in job applicants, especially for managerial positions, communication and relationship-building skills are inevitably at or near the top of the list. I once ran a training program on these skills for a CPA firm. Although the director of training believed in their value in the business world, many of the account executives did not. They resisted the whole process. I got a call one day from one of them. He had been one of the more notable resisters. He said, "I owe you this call." "Really?" I replied with an edge of doubt in my voice. "Really," he said. He went on to tell me how he had recently called on a potential client, a man whose company was dissatisfied with its current audit firm and looking for a new one. During the interview, the account executive said to himself, "Since we don't have the slightest chance of getting this account, why don't I amuse myself by trying these communication skills?" In his phone call to me he went on to say, "This morning I got a call from that client. He gave us the account, but in doing so he said, 'You're not getting the account because you were the low bidder. You were not. You're getting the account because we thought that you were the only one that really understood our needs.' So, almost literally, I owe you this call." I forgot to ask him for a share of the fee.

This road to proficiency applies to the basic skills we have being reviewing in the past few chapters and to the advanced skills of challenging outlined and illustrated in the next two chapters.

HELPING CLIENTS
CHALLENGE THEMSELVES

CHALLENGE: THE BASIC CONCEPT

In some areas of life, failure to challenge can have devastating consequences. For instance, when nurses and technicians fail to challenge mistakes surgeons are making in the operating room, they fail to prevent needless damage to patients, hospitals, and the surgeons themselves. Therefore, many hospitals are making concerted efforts to transform the culture of the operating room (Carter, 2006; Landro, 2005) so that no one there need agree with the statement, "In the O.R.s here, it is difficult to speak up if I perceive a problem with patient care" (Carter, 2006, p. 5). In one survey, 60% of respondents agreed with this statement. Too much is at stake not to challenge. Effective helpers must go beyond understanding clients (tuning in, listening, process-ing, sharing highlights) and help them clarify their concerns (probing, summarizing). Good counselors also help their clients test reality (challenging). Because helping at its best is a constructive social-influence process, some form of challenge is central to helping. This does not signal a movement from the pro-client stance associated with the communication skills reviewed so far to some kind of "tough" or client-as-enemy stance. The reality is simpler. All effective helping is some kind of mixture of support and challenge. Although challenge without support is harsh and unjustified, support without challenge can end up being empty and counterproductive.

Martin (1994) put it well when he suggested that the helping dialogue may add the most value when it is perceived by clients as relevant, helpful, interested, sup-portive, and *"somehow inconsistent (discordant) with their current theories of themselves and their circumstances"* (pp. 53–54, emphasis added). The same point is made by Trevino (1996) in the context of cross-cultural counseling.

> Certain patterns of congruency and discrepancy . . . between client and coun-selor facilitate change. There is a significant body of research suggesting that congruency between counselor and client enhances the therapeutic relation-ship, whereas discrepancy between the two facilitates change. In a review of the literature on this topic, Claiborn (1982, p. 446) concluded that the presentation of discrepant points of view contributes to positive outcomes by changing "the way the client construes problems and considers solutions." (p. 203)

Challenge adds that discordant note. Notice that the term *challenge* rather than the harder-edged *confrontation* is used here. Most people see both confronting and being confronted as unpleasant experiences. But, at least in principle, they more readily buy the softer but still edgy option of challenge. It has a constructive ring to it. Finn (2005) stresses compassion and firmness, pointing out that at times "we must say difficult things to clients in plain nonjudgmental language, which forces us to develop courage and wisdom" (p. 29). That said, there is also room, as we shall see in Chapter 8, for some form of confrontation.

The Goals of Challenging

The overall goal of challenging is to help clients do some reality testing and invest what they learn from this in their futures.

> *Help clients challenge themselves to change ways of thinking, expressing emotions, and both reacting and acting that keep them mired in problem situations and prevent them from identifying and developing opportunities.*

A parallel goal in the spirit of positive psychology is more upbeat. It deals with new perspectives and translating these new perspectives into new ways of acting. It goes something like this:

> *Become partners with your clients in helping them challenge themselves to find possibilities in their problems, to discover unused strengths and resources, both internal and external, to invest these resources in the problems and opportunities of their lives, to spell out possibilities for a better future, to find ways of making that future a reality, and to commit themselves to the actions needed to make it all happen.*

As idealistic as this might sound, it contains the spirit of the counselor's role as "catalyst for a better future."

The Targets of Challenge

Counselors should help clients challenge whatever stands in the way of understanding and managing problem situations or identifying and developing life-enhancing opportunities. The following are some of the main targets of challenge.

- self-defeating mind-sets
- self-limiting internal behavior
- self-defeating expressions of feelings and emotions
- dysfunctional external behavior
- distorted understanding of the world
- discrepancies in thinking and acting
- unused strengths and resources
- the predictable dishonesties of everyday life
- inadequate participation in the helping process

Here are a few examples in each of these categories.

Self-defeating mind-sets. Mind-sets here refer to more or less permanent states of mind. They include such things as assumptions, attitudes, beliefs, values, bias, convictions, inclinations, norms, outlook, points of view, unexamined perceptions of self/others/the world, preconceptions, and prejudices. Mind-sets, whether productive or problematic, tend to drive external behavior—or at least leak out into external behavior. The principle: Invite clients to transform outmoded, self-limiting mind-sets and perspectives into self-enhancing and liberating new perspectives that drive problem-managing and opportunity-developing action. Here are examples dealing with two kinds of mind-sets—prejudices and self-limiting beliefs and assumptions.

Prejudices. Candace is having a great deal of trouble with a colleague at work who happens to be Jewish. As she grew up, Candace picked up the idea that "Jews are treacherous businesspeople." The counselor invites her to rethink this prejudicial stereotype. Her colleague may or may not be treacherous, but if he is, it's not because he's Jewish. Separating the individual from the prejudicial stereotype helps Candace think more clearly. But it is now clear that Candace has two problems—her troubled relationship with her colleague at work and her prejudice. She has discovered that her problem with her colleague is not his Jewishness, but she may also need to discover that in troubled relationships often both parties contribute to the mess. She realizes that she has to go back and examine her behavior and the degree to which it has been affected by her prejudice.

Self-limiting beliefs and assumptions. Albert Ellis (Ellis, 2004; Ellis & MacLaren, 2004) has developed a rational-emotional-behavioral approach (REBT) to helping. He claims that one of the most useful interventions helpers can make is to challenge clients' irrational and self-defeating beliefs. Clients (and the rest of us) have a way of talking themselves into these dysfunctional beliefs. Some of the common beliefs that Ellis has found to get in the way of effective living are these:

- **Being liked and loved.** I must always be loved and approved by the significant people in my life.
- **Being competent.** I must always, in all situations, demonstrate competence, and I must be both talented and competent in some important area of life.
- **Having one's own way.** I must have my way, and my plans must always work out.
- **Being hurt.** People who do anything wrong, especially those who harm me, are evil and should be blamed and punished.
- **Being danger-free.** If anything or any situation is dangerous in any way, I must be anxious and upset about it. I should not have to face dangerous situations.
- **Being problemless.** Things should not go wrong in life, and if by chance they do, there should be quick and easy solutions.
- **Being a victim.** Other people and outside forces are responsible for any misery I experience. No one should ever take advantage of me.
- **Avoiding.** It is easier to avoid facing life's difficulties than to develop self-discipline; making demands of myself should not be necessary.
- **Tyranny of the past.** What I did in the past, and especially what happened to me in the past, determines how I act and feel today.
- **Passivity.** I can be happy by being passive, by being uncommitted, and by just enjoying myself.

I am sure that you could add to the list. Ellis suggests that when these kinds of belief are violated in a person's life, he or she tends to see the experience as terrible, awful, even catastrophic. "People pick on me. I hate it. It shouldn't happen. Isn't it awful!" Such "catastrophizing," Ellis says, gets clients nowhere. It is unfortunate to be picked on, he says, but it's not the end of the world. Moreover, clients can often do something about the issues over which they catastrophize.

Take Allison, a widow in her early 60s, who is seeing a counselor because of "anxiety attacks." A very stressful marriage ended suddenly the previous year when her husband died of a heart attack. At the time she felt angry, guilty, and relieved. She felt guilty because he died at a time when they were having particularly serious disagreements. She wondered whether the arguments had precipitated the attack. But she was also angry because "he left me holding the bag," that is, there was now no way to resolve the conflict. Still she was "profoundly relieved that all of this is over." The relationship together with all its stress was over, she was sure that he was "in a better place," and she was free to redesign her life.

Her current anxiety is due to the fact that she has just had a cancerous intestinal polyp removed. There was no metastasis, so the prognosis is actually quite good. She will have to have tests from time to time to monitor the state of her colon, but she is in no immediate danger. Instead of rejoicing over the good news, she has begun catastrophizing about the things that could go wrong. "These are supposed to be the 'golden years' and now I'm facing the possibility of a deadly disease! And I hate those tests!" Allison is working from the assumption that she should be problemless. "These things shouldn't be happening to me!" The counselor needs to help her challenge her dysfunctional beliefs.

Blind spots are not an affliction of people who are less-gifted intellectually. Sternberg (2002, 2003) explores self-beliefs that make otherwise smart people do stupid things. He has identified four fallacies in the thinking of such people.

1. "*The egocentrism fallacy.* They think it's all about them. In planning their actions, they take into account their own interests, but no one else's.

2. *The omniscience fallacy.* They may indeed know a lot about something. However, they start to think they know everything about everything.

3. *The omnipotence fallacy.* They think that they are all-powerful—that they can do whatever they want.

4. *The invulnerability fallacy.* They think that they can get away with whatever they do—that they will not be caught, or that even if they are, they will be able to get themselves out of any fix." (2003, p. 5)

The fact that in politics, the executive suite, the entertainment industry, and sports these blind spots are displayed on the wide screen does not mean that the rest of us are exempt. In fact, they are not even restricted to smart people.

Self-limiting internal behavior. Some forms of thinking are actually behaviors. They are things we can choose to do or not do. Internally we daydream, pray, ruminate on things, believe, identify problems, review opportunities, make decisions, formulate plans, make judgments, question motives, approve of self and others, disapprove of self and others, wonder, value, imagine, ponder, create standards, fashion norms, mull things over, ignore, forgive, rehearse—we *do* all sorts of things internally. These are internal or cognitive behaviors, not just things that happen to us. The helping principle is clear. Invite clients to replace self-limiting and self-defeating internal behaviors with more creative ones. The ways in which internal behavior can be self-limiting are legion. Consider a few examples.

John daydreams a lot, seeing himself as some kind of hero whom others admire. Thinking about unrealistic success in his social life has taken the place of working for actual success. The new perspective: Daydreaming is not all that bad. It's how you use it. With the help of a counselor, John does not stop daydreaming, but he switches its focus. He daydreams about what a fuller social life might look like. This provides him with some practical strategies for expanding and enriching his interactions with others. He begins to try these out.

Nadia, when given an assignment on a project, immediately begins to try to think of reasons why the project won't work. Then her internal behavior spills out into external behavior as she goes around telling everyone that the project should be changed or shelved. This annoys her colleagues. With the help of a supervisor, she sees how self-limiting her instinct to take things apart is. She lacks intellectual balance. So her supervisor helps her get into the habit of first trying to see what value the project or program will add to the company and what she might do to improve it. She finds that after doing some of this more positive internal work, she engages her colleagues more constructively about the project. She finds better ways of critiquing projects. They listen to her, and she adds real value.

For some clients, developing new perspectives and changing their internal behavior can be enormously helpful. It certainly works for Bella, a woman whose husband died 2 years ago. She is suffering from depression, not incapacitating, but still miserable. At one point she says,

BELLA: You know, I stopped wearing black a year ago. But. . . . (She pauses for a long time.)

THERAPIST: But you're still wearing black inside?

In a flash, Bella had it. Not magic, but now she had the metaphor she needed. She knew that she could and should stop wearing black "inside." With just a couple of more sessions with the therapist, Bella begins to move beyond the grieving that had come to be the hallmark of her life.

Self-defeating expressions of feelings and emotions. Managing our emotions and the ways we express them is part of social-emotional intelligence. Some of our emotions are bottled up, some go on inside, and others are quite visible. Sheila becomes depressed when her boss fails to notice the good work she is doing. She feels taken for granted. But at work she puts on a good face. On the other hand, Ira rants and raves about his "stupid" boss to whoever will listen. Everyone, except his boss, knows where he stands.

Clients need to be helped to face up to needless denial and bottling up of emotions, letting emotions run riot internally, and self-defeating forms of emotional expression. Arthur flies off the handle whenever anyone suggests that a less acerbic interpersonal style would benefit him and his friends, relatives, and colleagues at work. Arthur needs help in seeing the world as others see it. Cynthia is reluctant to express her delight and praise her son and daughter for their success at school because she doesn't want to "spoil" them. She needs to learn that praising and spoiling are not necessarily linked. Counselors can help clients identify the *triggers* for inappropriate expression of emotion. A counselor helps Ken realize that long periods of surfing the Internet tend to trigger episodes of depression. Ken needs to get out and interact with people to maintain emotional balance.

Dysfunctional external behavior. External behavior is the stuff people could see if they were looking. For some clients, their external behavior constitutes trouble. When Achilles is with women at work, he engages in behavior that others, including the courts, see as sexual harassment. He thinks he is just being "friendly." Consider Jake. He is not an alcoholic, but when he has a couple of drinks, he tends to get mean and argumentative. He thinks that he is merely helping others "get the point." So both his thinking and his acting need challenge. Some clients engage in behaviors that keep them mired in their problems. Clarence, a self-doubting and overly cautious person, is very deferential around his manager. He does not realize that his manager interprets his deference as a "lack of ambition." Clarence's behavior keeps him mired in a job he hates. But the manager says nothing because she would rather have an "obedient" rather than an "aggressive" employee.

Not doing something is also a form of behavior. Clients often fail to make choices and engage in behaviors that would help them cope with problems or develop opportunities. When Clarence is offered an opportunity to update his skills, he turns it down. He also refuses a promotion, saying to himself, "I don't want to get in over my head." His self-defeating external behaviors are based on a self-defeating thoughts about himself. His manager sees him bypass these opportunities, but says nothing to him. She says to herself, "Anyway, I've got a hard-working drone. That's something these days. I'll leave well enough alone."

Ryan is having trouble relating to his college classmates. He is aggressive, hogs conversations, tries to get his own way when events are being planned, and criticizes others freely. One of his friends, after a couple of drinks, gets very angry with Ryan and tells him off. "Self-centered," "arrogant," and "pushy," are the kinds of words she uses. Ryan goes into a funk. Later in the week he talks things through with the dorm prefect, an older student for whom he has a great deal of respect. Ryan begins to see how self-defeating his interpersonal style is. He goes on to work with one of the counselors in the student services center to do something about it. They discuss ways of being proactive and assertive rather than aggressive. His "edge" has too much of an edge about it. He takes a course in interpersonal communication, belatedly learns the value of dialogue, and finds plenty of incentives to invest what he learns in his interactions with his classmates, in his part-time job, and at home.

Distorted understanding of the world. Clients' failure to see the world as it really is can keep them mired in problem situations and prevent them from identifying and developing opportunities. For instance, parents fail to notice signs indicating that their teenage son has started to use drugs. If at times we are blind to others and their needs—often those closest to us—we are also blind to their attitudes toward us and the impact of their behavior on us. Take Sandra. She interprets her husband's being less insistent when it comes to having sex as a sign that "he is finally coming to his senses." So she is shocked when she learns, by accident, that he is having an affair. All of us have our areas of ignorance and naiveté. Clients' failure to fully understand the environment in which they live does not mean that they are stupid. Rather, like all of us, clients have blind spots.

Discrepancies. Various kinds of discrepancies plague our lives. For instance, we don't always do what we say we're going to do. Just review last year's "New Year's Resolutions." Discrepancies keep clients mired in their problem situations. Discrepancies can include the following:

- what clients think or feel versus what they say
- what they say versus what they do
- their views of themselves versus the views that others have of them
- what they are versus what they claim they want to be
- their stated goals versus what they actually accomplish
- their expressed values versus their actual behavior

The list goes on. For instance, a helper might help the following clients challenge the discrepancies in their lives.

- Tom sees himself as witty, but his friends see him as biting.
- Minerva says that physical fitness is important, but she overeats and underexercises.
- George says he loves his wife and family, but he is seeing another woman and stays away from home a great deal.
- Clarissa, unemployed for several months, wants a job, but she doesn't want to participate in a retraining program.

Let's use the example of Clarissa to illustrate how the discrepancy between talking and acting can be challenged. Clarissa has just told the counselor that she has decided against joining the retraining program.

COUNSELOR: At one time you thought that the retraining program would be just the kind of thing you've been looking for.

CLARISSA: Well. . . . I don't know if it's the kind of thing I'd like to do. . . . The work would be so different from my last job. . . . And it's a long program.

COUNSELOR: So you feel the fit isn't good.

CLARISSA: Yes, that's right.

COUNSELOR: Clarissa, you seemed so enthusiastic when you first talked about the program. . . . (gently): What's going on?

CLARISSA (pauses): You know, I've gotten a bit lazy. . . . I don't like being out of work, but I've gotten used to it.

The counselor sees a discrepancy between what Clarissa is saying and what she is doing. She is actually letting herself slip into a "culture of unemployment." Now that the discrepancy is out in the open, they can work together on how she wants to shape her future.

Unused strengths and resources. Challenge should focus not just on clients' problems, but on the "possible self" that every client is. Part of positive psychology is helping clients get in touch with unexploited opportunities and unused or underused strengths and resources (Aspinwall & Staudinger, 2003; Tedeschi & Kilmer, 2005). Some of these resources are client-based—for instance, talents and abilities not being used—and some are external—for example, failure to identify and use social support in managing problems or developing opportunities. Effective helpers are instinctively asking themselves questions like this: "What kind of unused strengths does this client have? How can I help this client unleash and marshal both internal and external resources?" We all have resources we fail to use; they need to be mined.

Strengths are buried even in dysfunctional behavior. For instance, Driscoll (1984) has pointed out that helpers can show clients that even their "irrationalities" can be a source of strength. Instead of forcing clients to see how stupidly they are thinking and acting, helpers can challenge them to find the logic embedded even in seemingly dysfunctional ideas and behaviors. Then clients can use that logic as a resource to manage problem situations instead of perpetuating them. A psychiatrist friend of mine helped a client see the "beauty," as it were, of a very carefully constructed self-defense system. The client, through a series of mental gymnastics and external behaviors, was cocooning himself from real life. My friend helped the client see how inventive he had been and how powerful the system that he had created was. He went on to help him redirect that power into more life-enhancing channels.

The predictable dishonesties of everyday life. The "predictable dishonesties of life" refers to the distortions, evasions, games, tricks, excuse making, and smoke screens that keep clients mired in their problem situations. All of us have ways of defending ourselves from ourselves, from others, and from the world. We all have our little dishonesties. But they are two-edged swords. Although lies, whether white or not, may help me cope with difficulties—especially unexpected difficulties—in my interactions with others, they come with a price tag, especially if they become a preferred coping strategy. Blaming others for my misfortunes helps me save face, but it disrupts interpersonal relationships and prevents me from developing a healthy sense of self-responsibility. The purpose of helping clients challenge themselves with respect to the dishonesties of everyday life, whether they take place in the helping sessions or are more widespread patterns of behavior, is not to strip clients of their defenses, which in some cases could be dangerous, but to help them cope with their inner and outer worlds more creatively.

Distortions. Some clients would rather not see the world as it is—it is too painful or demanding—and therefore distort it in various ways. The distortions are self-serving. For instance:

- At work Arnie is afraid of his supervisor and therefore sees her as aloof, whereas in reality she is a caring person. He is working out of past fears rather than current realities.
- Edna sees her counselor in some kind of divine role and therefore makes unwarranted demands on him.
- Nancy sees her getting her own way with her friends as an indication of whether they really like her or not.

Let's take a look at Nancy, who is married to Milan. They are experiencing some bumps in their marriage.

> Nancy and Milan come from different cultures. They fought a great deal in the early years of their marriage, but then things settled down. Now, squabbles have broken out about the best way to bring up their children. Milan is not convinced that counseling is a good idea, so the counselor is talking to Nancy alone. She has forbidden her 12-year-old son to bicycle to school because she doesn't want "his picture to end up on a milk carton." Milan thought that she was being extremely overprotective. One day he stalks out of the house, yelling back at her, "Why don't you just keep him locked in his room?"

Nancy and her counselor have a session not long after the above incident. Nancy is defending her approach to her son.

NANCY: Milan's just too permissive. Now that Jan is entering his teenage years, he needs more guidance, not less. Let's face it, the world we live in is dangerous.

COUNSELOR: So from your point of view, this is not the time for letting your guard down. . . . Of course, I'm also making the assumption that Milan is not indifferent to Jan's welfare.

NANCY: Of course not! Good grief, he cares as much as I do. We just disagree on how to do it. "Safe, not sorry" is my philosophy.

Hopefully, this gets rid of an implied distortion: "I'm interested in my son's welfare, but his father isn't." They continue their dialogue.

COUNSELOR: Let's widen the discussion a bit. What other issues do you and Milan disagree on?

NANCY: Well, we used to disagree a lot. But we've put that behind us, it would seem. He leaves a lot of the home decisions to me.

COUNSELOR: I'm not sure whether you both decided that you should make the decisions at home or if it just happened that way.

NANCY (slowly): I suppose it just happened that way. . . . I don't really know.

COUNSELOR: I'm curious because he seems to be annoyed that you're the one making the decisions about how to bring up your son. . . .

NANCY (pausing): Like he wants to reassert himself. Take over again.

Another distortion. Perhaps Nancy feels the counselor is getting too close to a sensitive issue that she thought was resolved long ago.

COUNSELOR: You got a bit annoyed when I asked whether Milan was as committed to the kids as you. . . . Because he cares as much as you do about his son, I'm wondering what the disagreement is really about.

NANCY: Like he's drawing a line in the sand, taking a stand on this one? Or what?

COUNSELOR (caringly): I don't want to guess what's going through Milan's mind. . . . Maybe we could we try once more to get him to come with you.

The counselor has a hunch that the problem is as much about power and getting one's own way as it is about bringing up children. Nancy does seem to have trouble with her own "little dishonesties."

Games, tricks, and smoke screens. If clients are comfortable with their delusions and profit by them, they will obviously try to keep them. If they are rewarded for playing games, inside the counseling sessions or outside, they will continue a game approach to life (see Berne, 1964). Consider some examples:

- Kennard plays the "Yes, but . . . " game. He gets his therapist to recommend some things he might do to control his anger. He then points out why each recommendation will not work. When the therapist calls this game, Kennard says, "Well, I didn't think you guys were supposed to tell clients what to do." A more savvy helper might have sniffed out Kennard's tendency to play games much earlier.

- Dora makes herself appear helpless and needy when she is with her friends, but when they come to her aid, she is angry with them for treating her like a child. When she tries this in an early session, her counselor invites her to examine this "helpless and needy" routine.

The number of games we can play to avoid the work involved in squarely facing the tasks of life is endless. Clients who are fearful of changing will attempt to lay down smoke screens to hide from the helper the ways in which they fail to face up to life. Such clients use communication in order not to communicate. Therefore, helpers do well if they establish an atmosphere that discourages clients from playing games. An attitude of "nonsense is challenged here" should pervade the helping sessions.

Excuses. Snyder, Higgins, and Stucky (1983) examined excuse-making behavior in depth. Excuse making, of course, is universal, part of the fabric of everyday life (see also Halleck, 1988; Higginson, 1999; Snyder & Higgins, 1988; Yun, 1998). Like games and distortions, it has its positive uses in life. Even if it were possible, there is no real reason for setting up a world without myths. On the other hand, excuse making contributes a great deal to avoiding the problems of life. Clients routinely provide excuses for why they did something "bad," why they didn't do something "good," and why they can't do something they need to do.

For example, Roberto tells the helper that he has engaged in benign attempts to sabotage his wife's career "for her own good" because she would "get hurt" in the Anglo world. The counselor helps him explore the alternative hypothesis that "he is not ready" for the changes in style that his wife's career and behavior were demanding from him.

> Before Roberto and Maria got married, they talked a great deal about the cultural difficulties they might face. He tended to adhere to traditional Latino cultural, whereas she was much more open to what he saw as "Anglo" attitudes and behavior. For instance, she was especially concerned about cultural norms relating to the role of women in society. When asked about their differences, Roberto said he would enjoy being married to someone with a "pioneer" spirit. Maria said that she thought that they had "worked things out." That was then. Now she has put herself through college, gotten a job, developed it into a career, and assumed the role of both mother and co-breadwinner. She makes more money than Roberto. His woes include thinking that he is losing face in the community, feeling belittled by his wife's success, and being forced into an overly "democratic" marriage.

If Roberto is going to manage the conflict between himself and his wife better, he needs to challenge himself to review and make changes in some of the ways he thinks.

This only skims the surface of the games, evasions, tricks, distortions, excuses, rationalizations, and subterfuges resorted to by clients (together with the rest of the population). Skilled helpers are caring and empathic, but they do not let themselves be conned. That helps no one.

Inadequate participation in the helping process. Ideally, clients become full partners with their counselors in the helping process. They are coconspirators in the search for a better life. But, often enough, this does not happen for two basic reasons. First, clients may lack the skills, especially the communication skills, needed to participate fully. If this is the case, you can use your communication skills to help them overcome obstacles to participation. Second, clients are often reluctant to participate fully in the helping process because they are reluctant to change. Reluctance refers to the ambiguity all of us feel when we know that managing our lives better is going to cost us. Like the rest of us, clients are not sure whether they want to pay that price. Incentives for not changing often drive out or stand in the way of incentives for changing.

In helping sessions, clients manifest reluctance in many, often covert, ways. They may talk about only safe or low-priority issues, seem unsure of what they want, benignly sabotage the helping process by being overly cooperative, set unrealistic goals and then use their unreality as an excuse for not moving forward, don't work very hard at changing their behavior, and are slow to take responsibility for themselves. They may blame others or the social settings and systems of their lives for their troubles and play games with helpers. Reluctance admits of degrees; clients come "armored" against change to a greater or lesser degree.

The reasons for reluctance are many and are built into the human condition. They affect all of us. Clients have anxieties about the intensity of the helping encounter, the trustworthiness of the counselor, the compatibility of the relationship, the disorganization that delving into problem situations can cause, the pain that can come from revealing their inner lives and their secrets, and change in general. Some clients think that change is impossible, so why try? Others are dissatisfied with the pace of change. For some, counseling is too fast, for others, too slow. Some clients are looking for short-term relief. Some come with the idea that counseling is magic and are put off when change proves to be hard work. Future rewards are not compelling.

In a counseling group, I once dealt with a man in his 60s who complained about constant anxiety. He described how he had been treated brutally by his father until he finally ran away from home. His message was implicit: "No one who grows up with scars like mine can be expected to take charge of his life and live responsibly." He had been using his mistreatment as a youth as an excuse to act irresponsibly at work (he had an extremely poor job record), in his relationship with himself (he drank excessively), and in his marriage (he had been uncooperative and unfaithful and yet expected his wife to support him). The idea that he could change, that he could take responsibility for himself even at his age, frightened him, and he wanted to run away from the group. But because his anxiety was so painful, he stayed. And he changed. Chapter 9 discusses reluctance and resistance in greater depth.

In practice, the aforementioned targets for challenge are often mixed together. Minerva believes that the world is filled with dishonest people (a self-limiting mindset). Whenever she meets someone new, she views that person's behavior through this lens and thinks that he or she is guilty until proved innocent (internal behavior in need of reform). Because she is always on guard, she comes across as cold and indifferent (inadequate management of feelings and emotions). Therefore, when she meets someone new, she is defensive and often questions that person's intentions and actions (external behavior needing change). She also realizes that when she meets someone new, she expects some kind of initial trust on the part of the other person even though she doesn't give it herself (a discrepancy). She fails to see that even her closest friends are becoming uncomfortable around her (failure to see the world as it really is). In the counseling sessions, Minerva remains guarded and is slow to share what she really thinks (inadequate participation).

BLIND SPOTS

Up to this point the discussion has focused on *what* needs to be challenged, the content, as it were. However, it is also necessary to consider the client's degree of awareness of self-limiting thinking, emotional expression, and behavior. Like the rest of us, clients don't seem to realize, or realize fully, how they are doing themselves in. That is, they have blind spots. Blind spots are part of the human condition. They are things we fail to see or choose to ignore that keep us from identifying and managing problem situations or identifying and developing opportunities.

Degrees of Awareness
Some blind spots appear to be unintentional, whereas others tend to be self-inflicted. Either way, they stand in the way of change. Blind spots come in a variety of flavors.

Simple unawareness. Some things clients are simply not aware of. Becoming aware of them helps them know themselves better and both cope with problems and develop opportunities. Serge was surprised when a fellow member of a self-help group called him "talented." He was brought up in a family that prized modesty and "humility." He never thought about himself as talented, creative, or resourceful and had little idea how this lack of awareness had narrowed his life. His mind-set stood in the way of change. Katya, on the other hand, was always upbeat. Because she always tried to look at the positive side of things, she "exuded sunshine," as one of her friends put it. What she did not realize was that sometimes her exuberance was inappropriate because it stood in the way of facing problems squarely. One of her colleagues at work thought that Katya's upbeat nature was great but needed to tempered by a dose of reality.

Katya had two friends, Mia and Casper. Mia had her own problems. For instance, she thought that her emotional outbursts were part of her punchy style. She did not realize that sometimes her colleagues at work wanted to strangle her. Casper didn't know that he had an acerbic communication style. If asked, he would describe himself as "assertive" and "logical." The problem is that his "assertiveness" made people shy away from him. And others around him saw his "logic" as stubbornness. People avoided or ignored him. No wonder he was dissatisfied with his social life. Simple unawareness is itself an elastic term. There are degrees. At one end is simple ignorance—"He doesn't have a clue." At the other end the ignorance is not so simple. Mia said of Casper, "He probably has some idea that he's rubbing people the wrong way; he's not totally in the dark."

Failure to think things through. This is a very common human experience. We explore problems, examine opportunities, search for possibilities, or formulate plans of action in an incomplete and haphazard way. Then we go on to base decisions on our flawed reasoning. For instance, Kim and Lea are both women who have lost a son in Iraq. They have long been both neighbors and friends, but recently have had some bitter encounters. Kim believes that the war was a mistake and that her son has died because of the "politicians," whereas Lea believes that the war is essential for national security and that her son has died fighting for a very just cause. Their beliefs are so charged with emotion that neither has any feeling for the other's point of view. They are both so set in their convictions that they are willing to destroy their relationship, something that neither son would have wanted. In this case, strong convictions made stronger by emotion keep them from thinking things through. In other cases, habit or laziness or thoughtlessness is the villain. Counselors who challenge and help clients think key issues through provide an invaluable service.

Self-deception. Where does simple lack of awareness or failure to think things through end and self-deception start? There are things clients would rather not know, because, if they knew them, they would be challenged to change their behavior in some way. So they'd rather stay in the dark. Goleman, who now writes extensively about "emotional intelligence" (1995, 1998), early on wrote a book called *Vital Lies, Simple Truths* (1985) on the psychology of self-deception and its pervasiveness in human life. Self-deception and the kind of social-emotional maturity outlined in his book *Emotional Intelligence* are incompatible.

Yet, as Eduardo Giannetti (1997) points out in *Lies We Live By: The Art of Self-Deception*, it is ubiquitous. "How," he marvels, "do we carry out such feats as believing in what we don't believe in, lying to ourselves and believing the lie . . . ?" (p. viii). Stan thought that he could get away with flirting with other women even though he was engaged. "It's just natural for a man," was his excuse. His fiancée saw his behavior as insulting. When she called off the wedding, Stan painfully realized the seriousness of his self-deception.

Choosing to stay in the dark. Choosing to stay in the dark is a common human experience. It is as if someone were to say, "I could find out, but I don't want to, at least not yet." Lots of people, when they have physical symptoms such as pain in their guts, avoid thinking about it. Finding out whether the pain indicates something serious could be uncomfortable. Finding out that the pain is life-threatening might be terrifying. Like the rest of us, clients often enough choose to stay in the dark. Ilia, recently released from jail, knows that she should have a clear understanding of the conditions of her probation, but chooses not to. When asked about the conditions, she says, "I don't know. No one really explained them to me." She knows that if she gets caught violating any of the conditions, she could go back to jail. But she puts that out of her mind. When clients are being vague or evasive with their helpers, they may well not want to know. And it is likely that they don't want you to know either.

Knowing, not caring, failing to see consequences. Clients sometimes know that their thinking, forms of emotional expression, and acting are getting them into trouble or keeping them there, but they don't seem to care. Or they know what they should do to get out of trouble, but they don't lift a finger. We can use the term *blind spot*, at least in an extended sense, to describe this kind of behavior because clients don't seem to fully understand or appreciate the degree to which they are choosing their own misery. Or they do not see the implications and consequences of not caring. For instance, Tanel says he knows nagging his wife to get a job even though there are two young children at home annoys his wife, but he keeps on anyway. "I can't help it." This creates a great deal of tension, but he continues to focus on his wife's reluctance to get a job rather than the negative consequences of his nagging. He has no idea of the pool of resentment that is building in his wife.

One way of helping clients challenge both internal and external actions is to help them explore the consequences of their actions. Let's return to Roberto. He has made some "mild" attempts at sabotaging his wife's career. He refers to his actions as "delaying tactics."

HELPER: It might be helpful to see where all of this is leading.

ROBERTO: What do you mean?

HELPER: I mean let's review what impact your "delaying tactics" have had on Maria and your marriage. And then let's review where these tactics are most likely ultimately to lead.

ROBERTO: Well, I can tell you one thing. She's become even more stubborn.

Through their discussion, Roberto discovers that his sabotage is working against rather than for him. He is endangering the marriage by keeping himself in the dark.

So, as you can see, the term *blind spot* as used here is somewhat elastic. We are unaware, we deceive ourselves, we don't want to know, we ignore, we don't care, or we know, but not fully; that is, we do not fully understand the implications or the consequences of what we know. But it's a good term. It has great face validity. As soon as you say "blind spot," people generally know what you mean.

Box 7-1   Questions to Uncover Blind Spots

These are the kinds of questions you can help clients ask themselves in order to develop new perspectives, change internal behavior, and change external behavior.

- What problems am I avoiding?
- What opportunities am I ignoring?
- What's really going on?
- What am I overlooking?
- What do I refuse to see?
- What don't I want to do?
- What unverified assumptions am I making?
- What am I failing to factor in?
- How am I being dishonest with myself?
- What's underneath the rocks?
- If others were honest with me, what would they tell me?

Helping clients deal with blind spots is one of the most important things you can do as a helper. For instance, if Lester has a prejudice and doesn't advert to it, he has a blind spot. If he has a prejudice but adverts to it only vaguely (though he probably does not refer to his attitude as a prejudice), then he is keeping himself in the dark. If he knows he is prejudiced and, when asked, says, "Everyone I know is like that!" then he doesn't care and fails to explore the human meaning of prejudice. He is prejudiced, knows it, fosters it, and lets it spill over into the way he deals with people. Lester has the full dysfunctional package.

Contrast this with Bernice. Initially she is unaware that she is prejudiced, becomes aware of her prejudice, tries to get rid of it as part of her internal mental furniture, refuses to act on it, and even learns something about herself and the world as she does all this. She deals with her prejudice creatively. She has turned a problem into an opportunity. Helping clients deal with dysfunctional blind spots can prevent damage, limit damage already done, and turn problems into opportunities. Box 7-1 outlines the kinds of questions you can help clients ask themselves in order to surface blind spots and develop new perspectives.

From Blind Spots to New Perspectives

Challenge focuses on the kind of understanding that leads to constructive action together with the constructive action itself. We do our clients a disservice if all that we do is help them identify and explore self-limiting blind spots. The positive psychology part of challenging is helping clients transform blind spots into new perspectives and translate these new perspectives into more constructive patterns of both internal and external behavior.

There are many upbeat names for this process of transforming blind spots into new perspectives: seeing things more clearly, getting the picture, getting insights, developing new perspectives, spelling out implications, transforming perceptions, developing new frames of reference, looking for meaning, shifting perceptions, seeing the bigger picture, developing different angles, seeing things in context, context breaking, rethinking, getting a more objective view, interpreting, overcoming blind spots, second-level learning, double-loop learning (Argyris, 1999), thinking creatively, reconceptualizing, discovering, having an "ah-ha" experience, developing a new outlook, questioning assumptions, getting rid of distortions, relabeling, and making connections. You get the idea. Some terms used to describe this process are framebreaking, framebending, and reframing. All of these imply some kind of cognitive restructuring that is needed in order to identify and manage both problems and opportunities. Developing new perspectives, although painful at times, tends to be ultimately prized by clients.

In the following example, the client, Leslie, a fairly religious 83-year-old woman, is a resident of a nursing home. She is talking to one of the aides.

LESLIE: I've become so lazy and self-centered. I can sit around for hours and just reminisce . . . letting myself think of all the good things of the past—you know, the old country and all that. Sometimes a whole morning can go by.

AIDE: I'm not sure what's so self-indulgent about that.

LESLIE: Well, it's in the past and all about myself. . . . I don't know if it's right.

AIDE: The way you talk about it, the reminiscing sounds almost like a kind of meditation for you.

LESLIE: You mean like a prayer?

AIDE: Well, yes. . . . like a prayer.

The client sees reminiscing as laziness. The aide helps her develop a new, more positive perspective. Reminiscing as a kind of meditation on life or even a prayer—Leslie is comfortable with that. With that new slant, Leslie feels that she can "indulge" herself.

Effective helpers assume that clients have the resources to see themselves and the world in which they live in a less distorted way and to act on what they see. Another way of putting it is that skilled counselors help clients move from what the Alcoholics Anonymous movement calls "stinkin' thinkin'" to healthy thinking. And from "stinkin'" emoting to constructive emotional expression. And from dysfunctional actions to healthy behavior. Consider Carla. Facing menopause, she is lumbered with the outmoded view of menopause as a "deficiency disease." Without minimizing Carla's discomfort and stress, a counselor helps her see menopause as a natural developmental stage of life. Although it indicates the ending of one phase, it also opens up new life-stage possibilities. Looking forward to those possibilities rather than looking back at what she's lost helps Carla a great deal.

It's important to note that identifying blind spots is not always the same as developing a new perspective. When Sandra, mentioned earlier, discovered that her husband was having an affair, she discovered a fact. She was blind to how serious her husband was about achieving a more satisfactory sexual life and deceived herself by telling herself that he was finally "coming to his senses." A new perspective for her was a step further: "I now realize more fully that all aspects of marriage are a two-way street and that we have to come up with an arrangement about sexual behavior that is satisfactory to both of us." This might sound pedestrian, but it is a new perspective for her, which, if acted on, could help save the marriage.

LINKING CHALLENGE TO ACTION

More and more theoreticians and practitioners are stressing the need to link insight to problem-managing and opportunity-developing action. Wachtel (1989, p. 18) put it succinctly: "There is good reason to think that the really crucial insights are those closely linked to new actions in daily life and, moreover, that insights are as much a product of new experience as their cause." Although new perspectives are important, they are not magic. Overstressing insight and self-understanding—that is, new perspectives—can actually stand in the way of action instead of paving the way for it.

Linking insight to action. Unfortunately, the search for insight can too easily become a goal in itself. Much more attention has been devoted to developing insight than to linking insight to action. I do not mean to imply that achieving useful insights into oneself and one's world is not hard work and often painful. But the pain must be turned into gain. Constructive behavioral change leading to valued outcomes is the goal. Consider this case.

> Ned, an inveterate micromanager, comes to realize that despite a company leadership program that most of his direct reports have taken, he still keeps making all the decisions in his team. This slows down the work and makes the department less efficient than it might be. In a leadership training program he gets "360" feedback, that is feedback from a survey filled in by his boss, his direct reports, his peers, and himself. Once he gets over the shock of reading the ratings, he sits down with a coach and talks about what constructive changes are needed in his managerial style—more delegating, getting and acting on suggestions from members of the team, and providing constructive feedback for them. They work together on drawing up an implementation plan.

Effective counselors, as they help their clients develop insights into themselves and their thinking, the ways they express or suppress emotion, and their behavior, maintain a constructive action-centered approach. Clients should be able to say, "I now see much more clearly how I am putting myself in jeopardy, but *I can do something about it*" and then actually move to action.

For instance, Joanna, a woman in her late fifties, has her annual physical exam. The doctor discovers that she has a higher than average risk for stroke. She communicates this information to Joanna.

JOANNA: What puts me at risk?

DOCTOR: A few things. The biggest factor in your smoking. Next is your cholesterol level.

JOANNA: So, if I stop smoking and get my diet under control. . . .

DOCTOR: It won't make you live forever, but it will make you live better.

JOANNA: I don't want to live forever. And I don't want to be too prudent. That would take all the fun out of life. . . . But now that I am a person of a certain age, as they say, I have something to think about. I think that I can rise to the challenge of stopping smoking. And I bet that I can come up with some delicious cholesterol-lowering meals. By the way, what about exercise? Isn't that part of the package?

Joanna has been ignoring the ways in which she is putting her health at risk. That her risk level is higher than average is the basic part of the new perspective. Now that she is approaching her 60s, this realization moves to center stage. Joanna herself adds the "I can do something about it" part.

An invitation to stop, start, continue. Challenge is basically simple. It deals with clients' thinking, their ways of managing their emotions, and their behavior. It is about "stop, start, and keep going." Because self-challenge is the ideal, helpers' challenge at its best is an *invitation* to clients to do the following:

- **Stop.** Challenge is an invitation to clients to identify and *stop* engaging in self-defeating thinking, self-indulgent emotions, and unconstructive behavior—that is, the kinds of things that keep them mired in their problem situations or that keep them from identifying and developing opportunities. For instance, Sakae, an Asian American, thinks that he is genetically better than European Americans. He never says this outright, but he communicates this attitude to others in a variety of often-not-too-subtle ways. This mars both work and social relationships, yet he wonders why he feels isolated and depressed. There is a catch, however. Simply trying to stop an established behavior can be notoriously difficult.

- **Start.** More positively, challenge is an invitation to clients to identify and *start* engaging in self-enhancing kinds of thinking, emotional expression. For instance, Grace—young, intelligent, savvy, and aggressive—was fired from three different jobs, good jobs, for "insubordination." A counselor asked her why she, with all that talent permeated by an engaging, punchy style, was working for other people. Were there not other ways she could channel her energy? Taking a cue from this counseling session, Grace decided to start her own business, which quickly blossomed. Substitution of a productive behavior for an unproductive one, especially when the substituted behavior rules out the problem behavior, is a form of starting that is often a more powerful intervention than just trying to stop a behavior. For instance, programs that teach a person how to shop for and prepare nutritious, balanced meals have a much better chance of being sustained, especially when they are paired with a sensible exercise plan, than one that merely calls for the elimination of snacks and junk food. Anyway, it's hard eating a bag of potato chips while jogging.

- **Keep going or increase.** Finally, challenge is an invitation to identify, *continue*, and even *increase* activities that manage problems or develop opportunities, especially when clients don't know what they are doing well or give signs of giving up. For instance, Melissa, a middle-aged woman in an arduous rehabilitation program, feels depressed because she feels she isn't making any progress. The counselor encourages her to keep to her exercises. He suggests that at this stage, she needs to define success as simply engaging in the exercises even though she sees no dramatic improvement so that she can say to herself, "There, I've done the exercises because they are heading me in the right direction. I've succeeded, and I feel good about myself."

Let's return to Roberto, seen earlier nagging his wife. Here are some possibilities for him in terms of starting, stopping, and continuing internal behaviors.

- *Stop* telling himself that his wife is the one with the problem, stop seeing her as the offending party when conflicts arise, and stop telling himself there is no hope for the relationship. He needs to become a co-owner of both the problems and the opportunities of the marriage.

- *Start* thinking of his wife as an equal in the relationship, start understanding her point of view, and start imagining what an improved relationship with her might look like. He needs a new communication style with his wife.

- *Continue* to take stock of the ways he contributes to their difficulties and increase the number of times he tells himself to let her live her life as fully as he wants to live his own. He needs to identify what he does best in the marriage—he is not a dud as a communicator—and reinforce these behaviors.

That is, Roberto has to get his head straight, mobilize his resources, and turn his reconstructed thinking into action. Of course, given that any relationship is a two-way street and that both parties need to change, Maria has to develop some new perspectives and change some of the ways she thinks and acts.

Roberto also needs to start, continue, increase, and stop certain external behaviors to do his part in developing a better relationship with his wife. Here are some possibilities.

- *Stop* criticizing her in front of others, stop creating crises at home and blaming her, and stop making fun of her business friends.

- *Start* activities that will help him develop his own career, including sharing his feelings with her instead of just expressing them in negative ways, engaging in mutual decision making, and taking more initiative in household chores and child care.

- *Continue* visiting her parents with her and increasing the number of times he goes to business-related functions with her.

With the counselor's help, Roberto begins to challenge himself to develop and implement a set of possibilities that would help him keep his relationship with Maria on an even keel while they work through the issue of Maria's career.

Is challenge enough to stimulate problem-managing action? For some, yes. A few well-placed challenges might be all that some clients need to move to constructive action. Once challenged, they are off to the races. But others may well have the resources to manage their lives better, but not the will. They know what they need to do but are not doing it. A few nudges in the right direction help them overcome their inertia. I have had many one-session encounters that included a bit of listening and thoughtful processing, some sharing of highlights, and a brief challenge that produced a new perspective that sent the client off on some useful course of action. On the other hand, if helping clients challenge themselves to develop new perspectives leads to one profound insight after another but no behavioral change, then once more we are whistling in the wind.

THE SHADOW SIDE OF CHALLENGING

By its very nature, challenge has a strong shadow side. The very fact that blind spots are involved puts it squarely in the shadows. We often dislike being challenged, even when it is done well. If helpers are very effective in challenge, they still might well sense some reluctance in their clients' responses. If they are poor at it, they are more likely to experience resistance. Inviting clients to challenge themselves and helping them do it to themselves is your best bet. In this chapter, we look at the shadow side in terms of the client's attitudes and behavior. In the next, we consider the shadow side of challenging in terms of helper attitudes and behavior.

Even when challenge is a response to a client's plea to be helped to live more effectively, it can precipitate some degree of disorganization in the client. Professional counselors refer to this experience under different names: "crisis," "disorganization," "a sense of inadequacy," "disequilibrium," and the like. Counseling-precipitated crises can be beneficial for the client. Whether they are or not depends, to a great extent, on the skill of the helper.

But even when an invitation to self-challenge or a direct challenge is accurate and delivered caringly, some clients still dodge and weave. Cognitive-dissonance theory (Draycott & Dabbs, 1998; Festinger, 1957) gives us some insight into the dynamics of this. Because dissonance (discomfort, crisis, disequilibrium) is an uncomfortable state, the client will often try to get rid of it. Outlined below are five of the more typical ways clients experiencing dissonance attempt to rid themselves of their discomfort. Let's examine them briefly as they apply to being challenged.

Discredit challengers. The client might confront the helper whose challenges are getting too close for comfort. Some attempt is made to point out that the helper is no better than anyone else. In the following example, the client has been discussing her marital problems and has been invited by the helper to take a look at her own behavior rather than giving example after example of what her spouse does wrong.

CLIENT: It's easy for you to sit there and suggest that I be more responsible in my marriage. You've never had to experience the misery in which I live. You've never experienced his brutality. You probably have one of those nice middle-class marriages.

Counterattack is a common strategy for coping with challenge. You may even have to field some sarcastic barbs: "Thanks for coming down out of your ivory tower to tell me how to shape up." Effective counselors are not surprised by clients' counterchallenges. Nor do they rise to the bait. Respect and empathy always pervade their attempts to help clients face up to difficult issues.

Persuade challengers to change their views. In this approach, clients reason with their helpers, urging them to see what they (helpers) have said as misinterpretations and to revise their views. In the following example, the client has been talking about the way she blows up at her husband whenever he makes "some stupid mistake." The counselor has invited her to explore the consequences of this pattern of behavior.

CLIENT: I'm not so sure that my anger at home isn't called for. I think that it's a way in which I'm asserting my own identity. If I were to lie down and let others do what they want, I would become a doormat at home. And, as you have helped me see, assertiveness should be part of my style. I think you see me as a fairly reasonable person. I don't get angry here with you because there is no reason to.

Sometimes a client like this will lead an unwary counselor into an argument about who's really right. In this situation, the counselor might help the client explore the difference between assertiveness and aggression. Getting angry may not be the problem but rather how she channels her anger.

Devalue the issue. This is a form of rationalization. A client who is being invited to challenge himself about his sarcasm points out that he is rarely sarcastic, that "poking fun at others" is just that, good-natured fun, that everyone does it, and that it is a very minor part of his life not worth spending time on. The client has a right to devalue a topic if it really isn't important. The counselor has to be sensitive enough to discover which issues are important and which are not. How would you handle the devaluing this client engages in?

Seek support elsewhere for the views being challenged. Some clients, once challenged, go out and gather testimonials supporting their views. In extreme cases, clients leave one counselor and go to another because they feel they aren't being understood. They try to find helpers who will agree with them. More commonly, the client remains with the same counselor and offers evidence from others contesting the helper's point of view.

CLIENT: I asked my wife about my so-called sarcasm. She said she doesn't mind it at all. And she thinks that my friends see it as humor and as a part of my style.

This is an indirect way of telling the counselor she is wrong. The counselor might well be wrong, but if the client's sarcasm is really dysfunctional in his interpersonal life, the counselor should find some way of pressing the issue.

Cooperate in the helping session, but then do nothing about it outside. The client can agree with the counselor as a way of dismissing an issue. However, the purpose of challenging is not to get the client's agreement but to develop new perspectives that lead to constructive action. Consider this client.

CLIENT: To tell you the truth, I am pretty lazy and manipulative. And I'm not even very clever about it.

In truth, this client is lazy and manipulative. She has given little evidence so far that she will do anything about it. If it were up to you break through this screen of "honesty," what would you do?

Of course, none of this makes clients evil—or at least no more "evil" than the rest of us. Challenge is a form of "tough love." Supporting clients is not the same as coddling them. And effective challenge is not a form of bullying.

CHALLENGING SKILLS AND THE WISDOM TO USE THEM WELL

SPECIFIC CHALLENGING SKILLS

1. Sharing Advanced Highlights: Capturing the Message Behind the Message

    Help clients make the implied explicit

    Help clients identify themes in their stories

    Help clients make connections that may be missing

    Share educated hunches based on empathic understanding

2. Information Sharing

3. Helper Self-Disclosure: Using It Carefully

    Make sure that your disclosures are appropriate

    Be careful of your timing

    Keep your disclosure selective and focused

    Don't disclose too frequently

    Do not burden the client

    Remain flexible

4. Immediacy: Discussing the Helper-Client Relationship

    Types of immediacy in helping and principles for using them

    Situations calling for immediacy

5. Making Suggestions and Giving Recommendations

6. Confrontation

7. Encouragement

FROM SMART TO WISE: GUIDELINES FOR EFFECTIVE CHALLENGING

    Keep the goals of challenging in mind

    Encourage self-challenge

    Earn the right to challenge

    Be tentative but not apologetic in the way you challenge clients

    Challenge unused strengths more than weaknesses

    Build on the client's successes

    Be specific in your challenges

Respect clients' values

Deal honestly, caringly, and creatively with client defensiveness

LINKING CHALLENGE TO ACTION

CHALLENGE AND THE SHADOW SIDE OF HELPERS

The "MUM effect"

Excuses for not challenging

Helpers' blind spots

SPECIFIC CHALLENGING SKILLS

There are any number of ways in which helpers can challenge clients to develop new perspectives, change their internal behavior, and change their external behavior. The following are discussed and illustrated in this chapter: (1) identifying and sharing the message behind the message, (2) sharing information, (3) helper self-disclosure, (4) immediacy, that is, discussing the helper-client relationship, (5) suggestions and recommendations, and (6) confrontation, and (7) encouragement. It has already been noted that both probing and summarizing—and even sharing empathic highlights—can challenge clients to rethink their attitudes and behavior.

1. Sharing Advanced Highlights: Capturing the Message Behind the Message

One way of challenging is to share with the client your understanding of the message behind the message. For instance, Gordon gets angry when he talks about his interactions with his ex-wife, but as he talks, the helper hears not just anger but also hurt. It may be that Gordon can talk relatively easily about and express his anger but is reluctant to talk about his feelings of hurt. When you share basic empathic highlights—provided, of course, that you are accurate—clients recognize themselves almost immediately: "Yes, that's what I meant." However, because responding with empathy to messages that are more covert (let's call them advanced highlights) digs a bit deeper, clients might not immediately recognize themselves in the helper's response. Or they might experience a bit of disequilibrium. That's what makes sharing advanced highlights a form of challenge. For instance, the helper says something like this to Gordon: "It's pretty obvious that you really get steamed when she acts like that. . . . But I thought I sensed, mixed in with the anger, a bit of hurt." At that, Gordon looks down and pauses. He finally says, "She can still get to me. She certainly can." This appreciably broadens or deepens the discussion of the problem situation.

Here are some questions helpers can ask themselves to probe a bit deeper as they listen to clients.

- What is this person only half saying?
- What is this person hinting at?
- What is this person saying in a confused way?
- What covert message is behind the explicit message?

Note that advanced empathic listening and processing focuses on what the client is actually saying or at least expressing, however tentatively or confusedly. It is not an interpretation of what the client is saying. Sharing advanced highlights is not an attempt to "psych the client out."

In the hands of skilled helpers, capturing and sharing the message behind the message focuses not just on the problematic dimensions of clients' behavior, but also on unused opportunities and resources. Effective helpers listen for the resources that are buried deeply in clients and often have been forgotten by them. Consider the following example.

> The client, a soldier who has been thinking seriously about making the army his career, has been talking to a chaplain about his failing to be promoted. He has seen service in Bosnia, Kosovo, and Iraq and has performed very well. As he talks, it becomes fairly evident that part of the problem is that he is so quiet and unassuming that it is easy for his superiors to ignore him.

> SOLDIER: I don't know what's going on. I work hard, but I keep getting passed over when promotion time comes along. I think I work as hard as anyone else, and I work efficiently, but all of my efforts seem to go down the drain. I'm not as flashy as some others, but I'm just as substantial.

> CHAPLAIN A: You feel it's quite unfair to do the kind of work that merits a promotion and still not get it.

> SOLDIER: Yeah. . . . I suppose there's nothing I can do but wait it out. (A long silence ensues.)

Chaplain A tries to understand the client from the client's frame of reference. He deals with the client's feelings and the experience underlying those feelings. In responding, he shares a basic empathic highlight. But the client merely retreats more into himself. Here's a different approach.

> CHAPLAIN B: It's depressing to put out so much effort and still get passed by. . . . Tell me more about this "not as flashy bit." I'm curious. What in your style might make it easy for others not to notice you, even when you're doing a good job?

> SOLDIER: You mean I'm so unassuming that I could get lost in the shuffle? Or maybe it's the guys who make more noise, the squeaky wheels, who get noticed. . . . I guess I've never really thought of selling myself. It's not my style.

From the context, from the discussion of the problem situation, from the client's manner and tone of voice, Chaplain B picks up a theme that the client states in passing in the phrase "not as flashy." They go on to discuss how he might "market himself" in a way that is consistent with his values. Advanced highlights can take a number of forms. Here are some of them.

Help clients make the implied explicit. The most basic form of an advanced highlight involves helping clients give fuller expression to what they are implying rather than saying directly. In the following example, the client has been discussing ways of getting back in touch with his wife after a recent divorce, but when he speaks about doing so, he expresses very little enthusiasm.

> CLIENT (somewhat hesitatingly): I could wait to hear from her. But I suppose there's nothing wrong with calling her up and asking her how she's getting along.

> COUNSELOR A: It seems that there's nothing wrong with taking the initiative to contact her. After all, you'd like to find out if she's doing okay.

> CLIENT (somewhat drearily): Yeah, I suppose I could.

Counselor A's response might have been fine at an earlier stage of the helping process, but it misses the mark here, and the client does not move on.

COUNSELOR B: You've been talking about getting in touch with her, but, unless I'm mistaken, I don't hear a great deal of enthusiasm in your voice.

CLIENT: To be honest, I don't really want to talk to her. But I feel guilty—guilty about the divorce, guilty about her going out on her own. Frankly, all I'm doing is taking care of her all over again. And that's one of the reasons we got divorced. I had a need to take care of her, and she let me do it. That was the story of our marriage. I don't want to do that anymore.

COUNSELOR B: What would a better way of going about all this be?

CLIENT: I need to get on with my life and let her get on with hers. Neither of us is help-less. (His voice brightens.) For instance, I've been thinking of quitting my job and starting a business with a friend of mine—helping small businesses use the Internet to improve their businesses, you know, websites, marketing, all of that.

Counselor B bases her response not only on the client's immediately preceding remark but also on the entire context of his story. Her response hits the mark, and the client moves forward. As with basic highlights, there is no such thing as a good advanced highlight in itself. Does the response help the client clarify the issue more fully so that he or she might begin to see the need to act differently?

Help clients identify themes in their stories. When clients tell their stories, cer-tain themes emerge. Thematic material might refer to feelings (such as hurt, depres-sion, anxiety), thoughts (continually ruminating about a past mistake), behavior (controlling others, avoiding intimacy, blaming others, overwork), experiences (being a victim; being seduced, punished, ignored, picked on), or some combination of these. Once you see a self-defeating theme or pattern emerging from your discus-sions, you can share your perception and help the client check it out.

In the following example, a counseling-psychology trainee is talking with his supervisor. The trainee has four clients. In the past week, he has seen each of them for the third time. This dialogue takes place in the middle of a supervisory session.

SUPERVISOR: You've had a third session with each of four clients this past week. Even though you're at different stages with each because each started in a different place, you have a feeling, if I understand what you've been saying, that you're going around in circles with a couple of them.

TRAINEE: Yes, I'm grinding my wheels. I don't have a sense of movement.

SUPERVISOR: Any thoughts on what's going on?

TRAINEE: Well, they seem willing enough. And I think I've been very good at listening and sharing highlights. It keeps them talking.

SUPERVISOR: But this doesn't seem to be enough to get them moving forward. I tell you what. Let's listen to one of the tapes.

They listen to a segment of one of the sessions. The trainee turns off the recorder.

TRAINEE: Oh, now I see what I'm doing! It's all sharing highlights with a few uh-huhs. And I thought I was being pushy. But this is as far from pushy as you can get.

SUPERVISOR: So what's missing?

TRAINEE: There are very few probes and nothing close to summaries or mild challenges. Certainly some probes would have given much more focus and direction to the session.

SUPERVISOR: Let me role-play the client as well as I can and see how you might redo the session.

They then spend about 15 minutes in a role-playing session. The trainee mingles probes with highlights, and the result is quite different.

SUPERVISOR: How close did you get to challenging, even mild challenging?

TRAINEE: I didn't get there at all. . . . You know, I think that I see probes as challenges. . . . The thread through all of this is "playing it safe." I think I'm playing it safe because I don't want to damage the client. I'm afraid to push.

The theme that the supervisor helps the trainee surface is a fear of "being pushy," which explains his "playing it safe" behavior.

Help clients make connections that may be missing. Clients often tell their stories in terms of experiences, thoughts, behaviors, and emotions in a hit-or-miss way. The counselor's job, then, is to help them make the kinds of connections that provide insights or perspectives that enable them to move forward in the helping process.

- Her counselor helps Chu Hua see that she is having difficulty developing strategies for her chosen goals because she is only half-heartedly committed to her goals. They revisit the goals she has set for herself.

- A managerial coach helps Finnbar relate the trouble he is having with a strong woman supervisor to sexist attitudes that leak out into his behavior.

- A therapist helps Joanna see the link between her ingratiating style and her inability to influence her colleagues at work.

- A supervisor helps Dieter see that the persistent anxiety he feels when working with clients are related to the perfectionistic standards he has set for himself.

The following client, John, has a full-time job and is finishing the final two courses for his college degree. His father has recently had a stroke and is incapacitated. John talks about being progressively more anxious and tired in recent weeks. He visits his father regularly. He meets frequently with his mother, his two sisters, and his two brothers to discuss how to manage the family crisis. Under stress, fault lines in family relationships appear. John does his best to be the peacemaker. He has deadlines for turning in papers for current courses.

JOHN: I don't know why I'm so tired all the time. And edgy. I'm supposed to be the calm one. I wonder if it's something physical. You know, what's happened to Dad and all that. I never even think about my health.

COUNSELOR: A lot has happened in the past few weeks. Work. School. Your dad's stroke. Juggling schedules.

JOHN (interrupting): But that's what I'm good at. Working hard. Juggling schedules. I do that all the time. And I don't get tired and edgy.

COUNSELOR: Add in your dad's illness. . . .

JOHN: You know, I could handle that, too. If I were the only one, you know, just me and mom, I bet I could do it.

COUNSELOR: All right, so besides you dad's illness, what's so new?

JOHN (slowly): Well, I hate to say it. It's the squabbling. We usually get on pretty well. We all like getting together. But the meetings about Dad, they can be awful. I keep thinking about them at work. And the other evening when I was trying to write a paper for school, I was still ticked off at my older sister.

COUNSELOR: So the family stuff is getting at you no matter what you're doing.

JOHN: I'm just not used to all that. I thought we'd rally together. You know, get support from one another. Sometimes it's just the opposite.

John handles the normal stress of everyday life quite well. But the "family stuff" is acting like a multiplier. They go on to discuss what the family dynamics are like and what John can do to cope with them.

Share educated hunches based on empathic understanding. As you listen to clients, thoughtfully process what they say, and put it all into context, you will naturally begin to form hunches about the message behind the message or the story behind the story. You can share the hunches that you feel might add value. The more mature and socially competent you become and the more experience you have helping others, the more "educated" your hunches become. Here are some examples.

Hunches can help clients see the bigger picture. In the following example, the counselor is talking with a client who is having trouble with his perfectionism. He also mentions problems with his brother-in-law, whom his wife enjoys having over. He and his brother-in-law argue, and sometimes the arguments have an edge to them. At one point the client describes him as "a guy who can never get anything right." Later the counselor says, "We started out by talking about perfectionism in terms of the inordinate demands you place on yourself. I wonder whether it could be 'spreading' a bit. You should be perfect. But so should everyone else." They go on to discuss the ways his perfectionism is interfering with his social life.

Hunches can help clients see more clearly what they are expressing indirectly or merely implying. In this next example, the counselor is talking to a client who feels that a friend has let her down: "I think I might also be hearing you say that you are more than disappointed—perhaps even betrayed." Because the client has been making every effort to avoid her friend, "betrayal" rings truer than "let down." She gets in touch with the depth of her feelings.

Hunches can help clients draw logical conclusions from what they are saying. A manager is having a discussion with one of his team members who has expressed in a rather tentative way some reservations about one of the team's projects. At one point he says, "If I stitch together everything that you've said about the project, it sounds as if you are saying that it was ill-advised in the first place and probably should be shut down. I know that might sound drastic and you've never put it in those words. But if that's how you feel, we should discuss it in more detail."

Hunches can help clients open up areas they are only hinting at. In this case, a school counselor is talking to a senior in high school: "You've brought up sexual matters a number of times, but you haven't pursued them. My guess is that sex is a pretty important area for you but perhaps pretty sensitive, too."

Hunches can help clients see things they may be overlooking. A counselor is talking to a client who probably has only 6 months to live. The man is unmarried and has never made a will. He has some money, but has expressed indifference to money matters. "I'm financially lazy" is his theme. He adds, "I'm ready to die." Later in the session, the counselor says, "I wonder if your financial laziness has spread a bit. For instance, you live alone and, if I'm not mistaken, you haven't given anyone power of attorney in health matters either. That could mean that how you die will be in the hands of the doctors." This helps the client begin to rethink how he wants to die. They even discuss finances. He may not be a slave to money, but whatever money he has could go to a good cause.

Hunches can help clients take fuller ownership of partially owned experiences, behaviors, feelings, points of view, and decisions. For example, a counselor is talking to a client who is experiencing a lot of pain in a physical rehabilitation program following an automobile accident. She keeps focusing on how difficult the program is. At one point the counselor says, "You sound as if you have already decided to quit. Or I might be overstating the case. . . ." This helps the client enormously. She has been thinking of quitting but she has been afraid to discuss it. They go on to discuss her wanting to give up and her dread of giving up. When the counselor finds out that she has never even mentioned the pain to the members of the rehabilitation staff, they discuss strategies for coping with the pain, including direct conversations with the staff about the pain.

Like all responses, hunches should be based on your understanding of your clients. If your clients were to ask you where your hunches come from—"What makes you think that?"—you should be able to identify the experiential and behavioral clues on which they are based. Of course, sharing advanced empathic highlights is not license to draw inferences from clients' history, experiences, or behavior at will. Nor is it license to load clients with interpretations that are more deeply rooted in your favorite psychological theories than in the realities of the client's world. Sharing advanced empathic highlights constructively takes emotional intelligence and social competence.

2. Information Sharing

Sometimes clients are unable to explore their problems fully, set goals, and proceed to action because they lack information of one kind or another. Information can help clients at any stage of the helping process. For instance, in Stage I, it helps many clients to know that they are not the first to try to cope with a particular problem. In Stage II, information can help them further clarify possibilities and set goals. In the implementation stage, information on commonly experienced obstacles can help clients cope and persevere.

The skill or strategy of information sharing is included under challenging skills because it helps clients develop new perspectives on their problems or shows them how to act. It includes both giving information and correcting misinformation. In some cases, the information can prove to be quite confirming and supportive. For instance, a parent who feels responsible following the death of a newborn baby may experience some relief through an understanding of the earmarks of the sudden infant death syndrome. This information does not "solve" the problem, but the parent's new perspective can help him or her handle self-blame.

In some cases, the new perspectives clients gain from information sharing can be both comforting and painful. Consider the following example.

> Adrian was a college student of modest intellectual means. He made it through school because he worked very hard. In his senior year, he learned that a number of his friends were going on to graduate school. He, too, applied to a number of graduate programs in psychology. He came to see a counselor in the student services center

after being rejected by all the schools to which he had applied. In the interview, it soon became clear to the counselor that Adrian thought that many, perhaps even most, college students went on to graduate school. After all, most of his closest friends had been accepted in one graduate school or another. The counselor shared with him the statistics of what could be called the educational pyramid—the decreasing percentage of students attending school at higher levels. Adrian did not realize that just finishing college made him part of an elite group. Nor was he completely aware of the extremely competitive nature of the graduate psychology programs to which he had applied. He found much of this relieving but then found himself suddenly faced with what to do now that he was finishing school. Up to this point he had not thought much about it. He felt disconcerted by the sudden need to look at the world of work.

Giving information is especially useful when lack of accurate information either is one of the principal causes of a problem situation or is making an existing problem worse.

In some medical settings, doctors team up with counselors to give clients messages that are hard to hear and to provide them with information needed to make difficult decisions. For instance, Doug, a 67-year-old accountant, has been given a series of diagnostic tests for possible prostate cancer. He finds out that he does have cancer, but now he faces the formidable task of choosing what to do about it. The doctor sits down with him and lays out the alternatives. Because there are many different options, including doing nothing, the doctor also describes the pluses and minuses of each. Later Doug has a discussion with a counselor who helps Doug cope with the news, process the information, and begin the process of making a decision.

There are some cautions helpers should observe in giving information. When information is challenging, or even shocking, be tactful and help the client handle the disequilibrium that comes with the news. Do not overwhelm the client with information. Make sure that the information you provide is clear and relevant to the client's problem situation. Don't let the client go away with a misunderstanding of the information. Be supportive; help the client process the information. Finally, be sure not to confuse information giving with advice giving. Professional guidance is not to be confused with telling clients what to do. Neither the doctor nor the counselor tells Doug which treatment to choose. But Doug needs help with the burden of choosing.

3. Helper Self-Disclosure: Using It Carefully

A third skill of challenging involves the ability of helpers to constructively share some of their own experiences, behaviors, and feelings with clients (Edwards & Murdoch, 1994; Farber, 2003a, 2003b; Knox, Hess, Petersen, & Hill, 1997; Mathews, 1988; Peterson, 2002; J. C. Simon, 1988; Sticker & Fisher, 1990; Watkins, 1990; M. F. Weiner, 1983). In one sense counselors cannot help but disclose themselves: "The counselor communicates his or her characteristics to the client in every look, movement, emotional response, and sound, as well as with every word" (Strong & Claiborne, 1982, p. 173). This is the kind of indirect disclosure that goes on all the time. Effective helpers, as they tune in, listen, process, and respond, try to track and manage the impressions they are making on clients.

Here, however, it is a question of direct self-disclosure. Research into direct helper self-disclosure has led to mixed and even contradictory conclusions. Some researchers have discovered that helper self-disclosure can frighten clients or make them see helpers as less well adjusted. Or, instead of helping, helper self-disclosure might place another burden on clients. Other studies have suggested that helper self-disclosure is appreciated by clients. Some clients see self-disclosing helpers as "down-to-earth" and "honest."

Direct self-disclosure on the part of helpers can serve as a form of modeling. Self-help groups such as Alcoholics Anonymous use such modeling extensively. Some would say that helper self-disclosure is most appropriate in such settings. This helps new members get an idea of what to talk about and find the courage to do so. It is the group's way of saying, "You can talk here without being judged and getting hurt."

> Beth is a counselor in a drug rehabilitation program. She herself was a substance abuser for a number of years but, with the help of the help of the agency where she is now a counselor, she is clean and sober. It is clear to all people with addictions in the program that the counselors there were once substance abusers themselves and are not only rehabilitated but also intensely interested in helping others both rid themselves of drugs and develop a kind of lifestyle that helps them stay drug-free. Beth freely shares her experience, both of being a drug user and of her rather agonizing journey to freedom, whenever she thinks that doing so can help a client.

Other things being equal, counselors who have struggled with addictions themselves often make excellent helpers in programs like this. They know from the inside the games clients afflicted with addictions play. Sharing their experience is central to their style of counseling and is accepted by their clients. It helps clients develop both new perspectives and new possibilities for action. Such self-disclosure is challenging. It puts pressure on clients to talk about themselves more openly or in a more focused way.

Helper self-disclosure is challenging for at least two reasons. First, it is a form of intimacy and, for some clients, intimacy is not easy to handle. Therefore, helpers need to know precisely why they are divulging information about themselves. Second, the message to the client is, indirectly, a challenging "You can do it, too," because revelations on the part of helpers, even when they deal with past failures, often center on problem situations they have overcome or opportunities they have seized. However, done well, such disclosures can be very encouraging for clients.

In the following example, the helper, Rick has had a number of sessions with Tim, a client who has had a rather tumultuous adolescence. For instance, he fell into the "wrong crowd" and got into trouble with the police a few times. His parents were shocked, and his relationship with them became very strained. Rick believes it will be helpful to share some of his own experiences.

RICK: You know, Tim, I've had experiences like yours. It might be helpful to compare notes. In my junior year in high school I was expelled for stealing. I thought that it was the end of the world. My Catholic family took it as the ultimate disgrace. We even moved to a different neighborhood in the city.

TIM: What did it do to you?

Rick briefly tells his story, one that includes setbacks not unlike Tim's. But Rick, with the help of a very wise and understanding uncle, was able to put the past behind him. He does not overdramatize the events. In fact, his story makes it clear that developmental crises are normal. How we interpret and manage them is the critical issue.

Current research does not give us definitive answers about helper self-disclosure, but it does offer some commonsense guidelines. Because clients sometimes misinterpret helpers' self-disclosures and their intent, caution is in order. Here are some guidelines.

Make sure that your disclosures are appropriate. Sharing yourself is appropriate if it helps clients achieve treatment goals. Don't disclose more than is necessary. Helper self-disclosure that is exhibitionistic is obviously inappropriate. Jeffrey (2004) provides us some outrageous examples of inappropriate self-disclosure. One helper discussed her own problems with the father-daughter bond, her guilt at disappointing a parent, personal religious issues, and her love life. The client switched to a therapist who said nothing about herself. Helper self-disclosure should be a natural part of the helping process, not a gambit. Rick's self-disclosure helps give Tim a different view of the "bad things" that have happened. Rick's developmental perspective gives Tim a different lens, a new way of looking at his problem. But Rick also makes it clear that he is not trying to make excuses for Tim's behavior. Of course, disclosures should be culturally appropriate (Burkard et al., 2006; Kim et al., 2003).

Be careful of your timing. Timing is critical. Common sense tells us that premature or poorly timed helper self-disclosure can distract clients or turn them off. Rick's disclosures did not take place in the first meeting. He waited for a few sessions. However, once he saw a natural opening, he thought that sharing some of his own experiences would help.

Keep your disclosure selective and focused. Don't distract clients with rambling stories about yourself. In the following example, the helper is talking to a first-year grad student in a clinical-psychology program. The client is discouraged and depressed by the amount of work he has to do. The counselor wants to help him by sharing his own experience of graduate school.

COUNSELOR: Listening to you brings me right back to my own days in graduate school. I don't think that I was ever busier in my life. I also believe that the most depressing moments of my life took place then. On any number of occasions, I wanted to throw in the towel. I remember once toward the end of my third year when. . . .

It may be that selective bits of this counselor's experience in graduate school would be useful, but he wanders off into a kind of reminiscing that meets his needs rather than the client's. In contrast, Rick's disclosure was selective and focused.

Don't disclose too frequently. Helper self-disclosure is inappropriate if it occurs too frequently. When helpers disclose themselves too frequently, clients may see them as self-centered, phony, or immature. Or they may suspect that they have hidden motives. If Rick had continued to share his experiences whenever he saw a parallel with Tim's, Tim might have wondered who was helper and who was client.

Do not burden the client. Do not burden an already overburdened client. One novice helper thought that he would help make a client who was sharing some sexual problems more comfortable by sharing some of his own sexual experiences. After all, he saw his own sexual development as not too different from the client's. However, the client reacted by saying: "Hey, don't tell me your problems. I'm having a hard enough time dealing with my own." This novice counselor shared too much of himself too soon. He was caught up in his own willingness to disclose rather than its potential usefulness to the client. In more extreme cases intimate disclosure might appear to be seductive.

Remain flexible. Take each client separately. Adapt your disclosures to differences in clients and situations. In some cultures such disclosure would be totally unwarranted. When asked directly, clients say that they want helpers to disclose themselves (see Handpick, 1988), but this does not mean that every client in every situation wants it or would benefit from it. Even though Rick's disclosure to Tim was natural, it was Rick's explicit decision.

4. Immediacy: Discussing the Helper-Client Relationship

Many clients have trouble with interpersonal relationships either as a primary or a secondary concern. Helping sessions, therefore, constitute a kind of interpersonal lab, as key features of clients' interpersonal styles show up in client-helper interactions. Both have much to learn. Some of the difficulties clients have in their day-to-day relationships are reflected in their relationships with their helpers. For instance, when they are compliant with authority figures in their everyday lives, they may be compliant with their helpers. Or they may move to the opposite pole and become aggressive and angry. Therefore, helpers can learn something about a client's interpersonal style (and, of course, their own) by being sensitive to the give-and-take within the helping sessions. Even more important, clients can learn a great deal about their own style. If counseling takes place in a group, the opportunity to learn is even greater. Robert Carkhuff (Carkhuff 1969a, 1969b; Carkhuff & Anthony, 1979) called the package of skills enabling helpers and clients to explore their relationship at the service of greater productivity "immediacy." Immediacy is a useful tool for monitoring and managing the working alliance.

Types of immediacy in helping and principles for using them. Two kinds of immediacy are reviewed here: first, immediacy that focuses on the overall relationship—"How are you and I doing?" and second, immediacy that focuses on some particular event in a session—"What's going on between you and me right now?"

Relationship-focused: Review your ongoing relationship with the client if this adds value to the helping process. General relationship immediacy refers to your ability to discuss with a client where you stand in your overall relationship with him or her and vice versa. The focus is not on a particular incident but on the way the relationship itself has developed and how it is helping or standing in the way of progress. In the following example, the helper is a 44-year-old woman working as a counselor for a large company. She is talking to a 36-year-old man she has been seeing once every other week for about 2 months. One of his principal problems is his relationship with his supervisor, who is also a woman.

COUNSELOR: We seem to have developed a good relationship here. I feel we respect each other. I have been able to make demands on you, and you have made demands on me. There has been a great deal of give-and-take in our relationship. You've gotten angry with me, and I've gotten impatient with you at times, but we've worked it out. If you see it more or less the same way, I'm wondering what our relationship has that might be missing in your relationship with your supervisor.

CLIENT: Yes, we do pretty well. My boss? That's another matter. For one thing, you listen to me, and I don't think she does. On the other hand, I listen pretty carefully to you, but I don't think I listen to her at all, and she probably knows it. I think she's dumb, and I guess I "say" that to her in a number of ways even without using the words. She knows how I feel.

The review of the relationship helps the client focus more specifically on his relationship with his supervisor.

Here is another example. Norman, a 38-year-old trainer in a counselor-training program, is talking to Weijun, 25, one of the trainees.

NORMAN: Weijun, let's take a time out and talk a bit about our relationship. I'm a bit bothered about some of the things that are going on between you and me. When you talk to me, I get the feeling that you are being very careful. You talk slowly. And you seem to be choosing your words, sometimes to the point that what you are saying sounds almost prepared. You have never challenged me on anything in the group. When you talk most intimately about yourself, you seem to avoid looking at me. I find myself giving you less feedback than I give others. I've even found myself putting off talking to you about all this. Perhaps some of this is my own imagining, but I want to check it out with you.

WEIJUN: I've been afraid to talk about all this, so I keep putting it off, too. I'm glad that you've brought it up. A lot of it has to do with how I relate to people in authority, even though you don't come across as an "authority figure." You don't act the way an authority figure is supposed to act.

In this case, cultural differences play a role. For Weijun, a naturalized American citizen born in China, giving direct feedback to someone in authority is not natural. However, authority is not the issue here. He has some misgivings about Norman's style and the way he conducts the class. He tells Norman that he thinks that Norman's interventions in the training group are too "unorganized." He also thinks Norman plays favorites. He has not wanted to bring these things up because he fears that his position in the program will be jeopardized. But now that Norman has made the overture, he accepts the challenge. Of course, an immediacy interaction can be initiated by the client, though many clients for obvious reasons would hesitate to do so. Who wants to take on the leader?

Event-focused immediacy: Address relationship issues as they come up. Here-and-now immediacy refers to your ability to discuss with clients what is happening between the two of you at any given moment. It is not the state of the relationship that is being considered, but rather some specific interaction or incident. In the following example, the helper, a 43-year-old woman, is a counselor in a neighborhood human services center. Agnes, a 49-year-old woman who was recently widowed, has been talking about her loneliness. Agnes seems to have withdrawn quite a bit, and the interaction has bogged down.

COUNSELOR: I'd like to stop a moment and take a look at what's happening right now between you and me.

AGNES: I'm not sure what you mean.

COUNSELOR: Well, our conversation today started out quite lively, and now it seems rather subdued. I've noticed that the muscles in my shoulders have become tense. I sometimes tense up that way when I feel that I might have said something wrong. It could be just me, but I sense that things are a bit strained between us right now.

AGNES (hesitatingly): Well, a little. . . .

Agnes goes on to say how she resented one of the helper's remarks early in the session. She thought that the counselor had intimated that she was lazy. Agnes knows that she isn't lazy. They discuss the incident, clear it up, and move on.

Both kinds of immediacy are at work in the following example. Carl Rogers, the dean of client-centered therapy (see Landreth, 1984, p. 323) talks about one of his clients:

> I am quite certain, even before I stopped carrying individual cases, I was doing more and more of what I would call confrontation. . . . For example, I recall a client with whom I began to realize I felt bored every time he came in. I had a hard time staying awake during the hour, and that was not like me at all. Because it was a persisting feeling, I realized I would have to share it with him. . . . So with a good deal of difficulty and some embarrassment, I said to him, "I don't understand it myself, but when you start talking on and on about your problems in what seems to me a flat tone of voice, I find myself getting very bored." This was quite a jolt to him and he looked very unhappy. Then he began to talk about the way he talked and gradually he came to understand one of the reasons for the way he presented himself verbally. He said, "You know, I think the reason I talk in such an uninteresting way is because I don't think I have ever expected anyone to really hear me." . . . We got along much better after that because I could remind him that I heard the same flatness in his voice I used to hear.

Rogers's self-involving statement, genuine but quite challenging, helped the client move forward. But there is another point of view. Someone once said, "Boredom is a self-indictment." In my opinion, Rogers was bored because he restricted himself to sharing empathic highlights with clients. On principle, he did not ordinarily use probing, summaries, and challenging lest he rob them of responsibility. He was a master at understanding clients and sharing highlights. Without doubt he helped many clients. This story also points to the direction in which Rogers was moving toward the end of his career—adding the "spice" of probing and challenging to his interactions with clients.

Situations calling for immediacy. Part of skilled helping—and, more generally, social-emotional intelligence—is knowing when to use any given communication skill. Immediacy can be useful in the following situations:

- When a session is directionless and it seems that no progress is being made.
- When there is tension between helper and client.
- When trust seems to be an issue.
- When diversity, some kind of "social distance," or widely differing interpersonal styles between client and helper seem to be getting in the way.
- When dependency seems to be interfering with the helping process: "You don't seem willing to explore an issue until I give you permission to do so or urge you to do so. And I seem to have let myself slip into the role of permission giver and urger."

Make sure that you know why you are using either form of immediacy. Talking about your relationship with a client is not an end in itself.

5. Making Suggestions and Giving Recommendations

This section begins with a few imperatives. Don't tell clients what to do. Don't try to take over their lives. Let clients make their own decisions. All these imperatives flow from the values of respect and empowerment. Does this mean, however, that suggestions and recommendations are forbidden? Of course not. It was mentioned earlier that there is a natural tension between helpers' desire to have their clients manage their lives better and respecting their freedom. If helpers build strong, respectful relationships with their clients, then stronger interventions can make sense. In this context, suggestions and recommendations can stimulate clients to move to problem-managing action. Helpers move from counseling mode to guidance role. Research has shown that clients will generally go along with recommendations from helpers when the recommendations are clearly related to the problem situation, challenge clients' strengths, and are not too difficult. Effective helpers can provide suggestions, recommendations, and even directives without robbing clients of their autonomy or their integrity.

Here is a classic example of this from Cummings's (1979, 2000) work. Substance abusers came to him because they were hurting in many ways. He used every communication skill available to listen to and understand their plight.

> During the first half of the first session the therapist must listen very intently. Then, somewhere in mid-session, using all the rigorous training, therapeutic acumen, and the third, fourth, fifth, and sixth ears, the therapist discerns some unresolved wish, some long-gone dream that is still residing deep in that human being, and then the therapist pulls it out and ignites the client with a desire to somehow look at that dream again. This is not easy, because if the right nerve is not touched, the therapist loses the client. (1979, p. 1123)

So Cummings shared both basic and advanced highlights to let clients know that he understood their plight, their longings, and also their games. They came knowing how to play every game in the book. But Cummings knew all the games, too. Toward the end of the first session he told them they could have a second session—which they invariably wanted—only when they were "clean." The time of the second session depended on the withdrawal period for the kind of substance they were abusing. They screamed, shouted "foul," tried to play games, but he remained adamant. The directive "Get clean, then return" was part of the therapeutic process. And most did return. Clean.

Suggestions, advice, and directives need not always be taken literally. They can act as stimuli to get clients to come up with their own package. One client said something like this to her helper: "You told me to let my teenage son have his say instead of constantly interrupting and arguing with him. What I did was make a contract with him. I told him that I would listen carefully to what he said and even summarize it and give it back to him. But he had to do the same for me. That has produced some useful monologues. But we avoid our usual shouting matches. My hope is to find a way to turn it into dialogue." In everyday life, people feel free to give one another advice. It goes on all the time. But helpers must proceed with caution. Suggestions, advice, and directives are not for novices. It takes a great deal of experience with clients and a great deal of savvy to know when they might work.

6. Confrontation

What about clients who keep dragging their feet? Some clients who don't want to change or don't want to pay the price of changing simply terminate the helping relationship. However, those who stay stretch across a continuum from mildly to extremely reluctant and resistant. Or they may be collaborative on some issues but reluctant on others. For instance, Hester is quite willing to work on career development but very reluctant to work on improving relationships, even though relationship building is an important part of the career package. "That's my private world," she says of her relationships.

If *inviting* clients to challenge themselves is at one end of the continuum, what's at the other? Where does respecting clients' right to be themselves stop and placing demands on them to live more fully begin? Because this is a values issue, different helpers give different answers. As a consequence, helpers differ, both theoretically and personally, in their willingness to confront. For instance, Lowenstein (1993) used what he called "traumatic confrontation" (one wonders about the name) to challenge youths to face up to their dysfunctional behavior. He gives the example of confronting a 12-year-old boy who had become involved in criminal activity after the disappearance of his father. At first the boy denied everything, but then decided to face up to the situation.

Helpers confront to "make the case" for more effective living. Confrontation does not involve "do this or else" ultimatums. More often it is a way of making sure that clients understand what it means not to change—that is, making sure they understand the consequences of persisting in dysfunctional patterns of behavior or the cost of failing to seize opportunities. Confrontation, like strong medicine, is actually another way of caring for the client. But, like strong medicine, it needs to be used sparingly and carefully. Confrontation should be empathic and respectful, empower the client, and lead to action. It should not be used by helpers to vent their frustrations on reluctant and resistant clients.

7. Encouragement

This section ends on a positive psychology note. You may not have noticed, but in the last two chapters some form of the word "encouragement" has been used only a couple of times. If the whole purpose of challenging is to help client move forward, and if encouragement (sugar) works as well as challenge (vinegar), then why don't we hear more about encouragement? The sugar-vinegar analogy is not exactly right, however, because many clients find challenge both refreshing and stimulating. Challenge certainly does not preclude encouragement. Encouragement itself is a mild form of challenge. Furthermore, encouragement is a form of support and research shows that support is one of the main ingredients in successful therapy (Beutler, 2000, p. 1004).

Miller and Rollnick (1991, 2002, 2004; Rollnick & Miller, 1995) introduced an approach to helping called "motivational interviewing." A simple Internet search on "motivational interviewing" reveals an extensive body of literature, including theory, research (Burke, Dunn, Atkins, & Phelps, 2004), and case studies. Their original work focused on helping clients deal with addictive behavior, but their methodology over the years has been adapted to much wider range of human problems. Much of the literature highlights the main elements of a problem-management approach. "Motivational interviewing is a directive, person-centered clinical method for helping clients resolve ambivalence and move ahead with change. It can be applied as a preparation for treatment, a freestanding brief intervention, an enduring clinical style, or a fallback approach when motivational obstacles are encountered" (Miller & Rollnick, 2004, p. 299).

As such, it can be used at any stage or for any task in the problem-management framework. The values of respect, empathy, self-empowerment, and self-healing are emphasized.

Although it is admittedly a directive approach, the spirit of encouragement rather than confrontation pervades the approach. Typically, clients (for instance, pregnant women who smoke or use alcohol) receive personal feedback on their problem area (such as how smoking has been affecting their lungs and the harm smoking and drinking can cause the fetus). There are discussions of personal responsibility and advice on ways of managing the problem situation is offered. Clients are encouraged to find the motives, incentives, or levers of change that make sense to them and to use the change options that they find fit best. Intrinsic motives, that is, motives that clients have internalized for themselves ("I want to be free"), rather than extrinsic motives ("I'll get in trouble if I don't change") are emphasized. Clients are also given help on identifying obstacles to change and ways of overcoming them. Empathy, both as a value and as a form of communication (empathic highlights), is used extensively.

Common sense suggests that realistic encouragement be included among any set of helping skills. Like most of the skills we have been discussing, encouragement can be used at any stage of the helping process. Clients can be encouraged to identify and talk about their problems and unused opportunities, to review possibilities for a better future, to set goals, to engage in actions that will help them achieve their goals, and to overcome the inevitable obstacles. Effective encouragement is not patronizing. It is not the same as sympathy, nor does it rob the client of autonomy. It respects the client's self-healing abilities. It is a fully human nudge in the right direction.

FROM SMART TO WISE: GUIDELINES FOR EFFECTIVE CHALLENGING

Your challenges might be on the mark (smart) and still be ineffective or even hurtful. All challenges should be permeated by the spirit of the client-helper relationship values discussed in Chapter 2; that is, they should be based on understanding, caring (not power games or put-downs), genuine (not tricks or games), and designed to increase the client's self-responsibility (not expressions of helper control). They should also serve the stages and tasks of the helping process. Empathy should permeate every kind of challenge. Clearly, challenging well is not a skill that comes automatically. It needs to be learned and then practiced with care. The following principles constitute some basic guidelines for making challenging not just accurate but wise.

Keep the goals of challenging in mind. Challenge must be integrated into the entire helping process. Keep in mind that the goal is to help clients develop the kinds of alternative perspectives, internal behavior, and external actions needed to get on with the stages and tasks of the helping process. Are the insights relevant to real problems and opportunities rather than merely being dramatic in themselves? To what degree do the new perspectives developed lead to problem-managing and opportunity-developing action?

Encourage self-challenge. Invite clients to challenge themselves, and give them ample opportunity to do so. You can provide clients with probes and structures that help them engage in self-challenge. In the following excerpt, the counselor is talking to a man who has discussed at length his son's ingratitude. There has been something cathartic about his complaints, but it is time to move on.

COUNSELOR: People often have blind spots in their relationships with others, especially in close relationships. Picture your son sitting with some counselor. He is talking about his relationship with you. What's he saying?

CLIENT: Well, I don't know . . . I guess I don't think about that very much. . . . Hmm. . . . He'd probably say . . . well, that he loves me. . . . (pauses). And then he might say that since his mother died, I have never really let him be himself. I've done too much to influence the direction of his life rather than let him fashion it the way he wanted. . . . Hmm. He'd say that he loves me but he has always resented my "interference."

COUNSELOR: So both love for you and resentment for all that control.

CLIENT: And he'd be right. I'm still doing it. Not with him. He won't let me. But with lots of others. Especially in my business.

The counselor provides a structure that enables the client to challenge himself with respect to his son and then apply his what he learns to other settings. Would that all clients would respond so easily! Alternatively, the counselor might have asked this client to list three things he thinks he does right and three things he thinks he should reconsider in his relationship with his son. The point is to be inventive with the probes and structures you provide clients to help them challenge themselves.

Earn the right to challenge. Long ago Berenson and Mitchell (1974) claimed that some helpers don't have the right to challenge others because they are not doing a good job keeping their own houses in order. They made a point. Here are some of the factors that earn you the right to challenge.

- **Develop a working relationship.** Challenge only after you have spent time and effort building a relationship with your client. If your rapport is poor or you have allowed your relationship with the client to stagnate, then challenge yourself to deal with the relationship more creatively.

- **Make sure you understand the client.** Empathy drives everything. Effective challenge flows from accurate understanding. Only when you see the world through the client's eyes can you begin to see what he or she is failing to see.

- **Be open to challenge yourself.** Hesitate to challenge unless you are open to being challenged. If you are defensive in the counseling relationship, in your relationship with supervisors, or in your everyday life, your challenges might ring hollow. Model the kind of nondefensive attitudes and behavior that would like to see in your clients.

- **Work on your own life.** How important is constructive change in your own life? Berenson and Mitchell claimed that only people who are striving to live fully according to their value system have the right to challenge others, for only such persons are potential sources of human nourishment for others. You may disagree with them, but what they say should get all helpers thinking.

**Box 8-1   Evaluation Questions:
The Wisdom of Challenging**

How well do I do each of the following as I try to help my clients?

- Invite clients to challenge themselves.
- Earn the right to challenge.
- Be tactful and tentative in challenging without being insipid or apologetic.
- Be specific, developing challenges that hit the mark.
- Challenge clients' strengths rather than their weaknesses.
- Don't ask clients to do too much too quickly.
- Invite clients to clarify and act on their own values.

The Shadow Side of Challenging

How well do I do the following?

- Identify the games my clients attempt to play with me without becoming cynical in the process.
- Become comfortable with the social-influence dimension of the helping role, with the kind of "intrusiveness" that goes with helping.
- Incorporate challenge into my counseling style without becoming a confrontation specialist.
- Develop the assertiveness needed to overcome the MUM effect.
- Challenge the excuses I give myself for failing to challenge clients.
- Come to grips with my own imperfections and blind spots both as a helper and as a "private citizen."

In summary, ask yourself, "To what degree am I the kind of person from whom clients would be willing to accept challenges?"

Be tentative but not apologetic in the way you challenge clients. Tentative challenges are generally viewed more positively than strong, direct challenges. The principle is this: When you challenge clients tentatively, they are more likely to *respond* rather than *react*. The same challenge can be delivered in such a way as to invite the cooperation or arouse the resistance of the client. Deliver challenges tentatively, as hunches that are open to review and discussion rather than as accusations. Challenging is certainly not an opportunity to browbeat clients or put them in their place.

On the other hand, challenges that are delivered with too many qualifications—either verbally or through the helper's tone of voice—sound apologetic and can be easily dismissed by clients. I was once working in a career-development center. As I listened to one of the clients, it soon became evident that one reason he was getting nowhere was that he engaged in so much self-pity. When I shared this observation with him, I overqualified it. These are not my exact words, but it must have sounded something like this:

HELPER: Has it ever, at least in some small way, struck you that one possible reason for not getting ahead, at least as much as you would like, could be that at times you tend to engage in a little bit of self-pity?

I still remember his response. He paused, looked me in the eye for what seemed to be an eternity, and said, "A little bit of self-pity?" When he paused again, I said to myself, "I've been too harsh!" He continued, "I *wallow* in self-pity." We moved on to explore what he might do to move beyond self-pity to constructive change.

Challenge unused strengths more than weaknesses. Berenson and Mitchell (1974) found that successful helpers tend to challenge clients' strengths rather than their weaknesses. The positive psychology movement is just catching up to what they said long ago. What we talk about sets the tone for helping. Individuals who focus on their failures find it difficult to change their behavior. Clients who dwell too much on their shortcomings tend to belittle their achievements, to withhold rewards from themselves when they do achieve, and to live with anxiety. All of this tends to undermine performance.

Challenging strengths means pointing out to clients the assets and resources they have but fail to use. In the following example, the helper is having a one-to-one sessions with a woman who is a member of a self-help group in a rape crisis center. She is very good at helping others but is always down on herself.

COUNSELOR: Ann, in the group sessions, you provide a great deal of support for the other women. You have an amazing ability to spot a person in trouble and provide an encouraging word. And when one of the women wants to give up, you are the first to challenge her, gently and forcibly at the same time. But when Ann is dealing with Ann. . . .

ANN: I know where you're headed. . . . I know I'm a better giver than receiver. I'm much better at caring than being cared about. I'm not sure why that is. . . . Or that it even matters. I'm sure this is not lost on the other members of the group. . . . You know, I've been this way for a long time. I think I've got some bad habits when it comes to dealing with myself. I'm so fearful of being self-indulgent.

The counselor helps her place a demand on herself to use her rather substantial resources in her dealings with herself. Because she isn't self-indulgent, it's time to take a look at her resistance to being cared about.

Even adverse life experiences can be a source of strength. For instance, McMillen, Zuravin, and Rideout (1995) studied adult perceptions of benefit from child sexual abuse. Almost half the adults reported some kind of benefit, including increased knowledge of child sexual abuse, protecting other children from abuse, learning how to protect themselves from others, and developing a strong personality— without, of course, discounting the horror of the abuse. Counselors, therefore, can help clients "mine" benefits from adverse experiences, putting to practical use the age-old dictum that "good things can come from evil things." People are more resilient than we make them out to be.

Build on the client's successes. Effective helpers do not urge clients to place too many demands on themselves all at once. Rather, they help clients place reasonable demands on themselves and, in the process, help them appreciate and celebrate their successes. In the following example, the client is a boy in a detention center who is rather passive and allows the other boys to push him around. Recently, however, he has made a few half-hearted attempts to stick up for his rights in the counseling group. The counselor is talking to him alone after a group meeting.

COUNSELOR A: You're still not standing up for your own rights the way you need to. You said a couple of things in there, but you still let them push you around.

This counselor emphasizes the negative and browbeats the client. The following counselor takes a different tack.

COUNSELOR B: Here's what I've noticed. In the group meetings, you have begun to speak up. You say what you want to say, even though you don't say it really forcefully. And I get the feeling that you feel good about that. You've got some power. Now the challenge is to find ways of using it more effectively.

CLIENT: I didn't think anyone noticed. You think it was a good start?

COUNSELOR B: Certainly. . . . But what you think is more important. And it sounds like you're proud of what you did.

The second counselor emphasizes the client's success, however modest, and goes on to provide some encouragement to do even better.

Be specific in your challenges. Specific challenges hit the mark. Vague challenges get lost. Clients don't know what to do about them. Statements such as "You need to pull yourself together and get on with it" may satisfy some helper need, such as the ventilation of frustration, but they do little for clients. Specific statements, on the other hand, can hit the mark. In the following example, the client is experiencing a great deal of stress both at home and at work.

HELPER: You say that you really want to spend more time at home with the kids and you really enjoy it when you do, but you keep taking on new assignments at work, like the Eclipse project, that will add to your travel schedule. Maybe it would be helpful to talk a bit more about work-life balance.

CLIENT: Boy, there's that phrase! Work-life balance. The company talks a lot about it, but nothing much happens. I'm not sure there's anyone at work who's got the work-life balance right.

HELPER: You know what they say about career—"If you're not in charge of your career, no one is." It sounds like the same is true with work-life balance.

CLIENT: I hadn't thought about it like that. . . . But I'm afraid you're right. It's right where it belongs, I suppose, on my shoulders. I've been waiting for my family and my company to figure it out for me.

Some helpers avoid clarity and specificity because they feel that they're being too intrusive. Helping has to be intrusive to make a difference.

Respect clients' values. Challenge clients to clarify their values and to make reasonable choices based on them. Be wary of using challenging, even indirectly, to force clients to accept your values. This violates the empowerment value discussed in Chapter 3. In the following example, the client is a 21-year-old woman who has curtailed her social life, her education, and her career to take care of her elderly mother who is suffering from incipient Alzheimer's.

CLIENT: I admit that juggling work, home, and school is a real challenge. I keep feeling that I'm not doing justice to any of them.

COUNSELOR A: You have every right to have a life of your own. Why not get your mother into a nursing facility? You can still visit her regularly. Then get on with life. That's probably what she wants anyway.

Challenging clients to clarify their values is, of course, legitimate. But this counselor does little to help the client clarify her values. She makes suggestions without finding out why the client is doing what's she's doing. Counselor B takes a different approach.

COUNSELOR B: You're trying to juggle four very important areas of your life—caring for your mother, school, work, and social life. That's a tough assignment. It might help to explore what's driving you in all this. Maybe we could take a look at the values that drive your behavior. Then you could ask yourself about your priorities.

CLIENT: I've never thought about values as things that drive what I do. I thought we had values and just, well, did them. So let's take a look.

This counselor challenges her gently to find out what she really wants. Helpers can assist clients to explore the consequences of the values they hold, but that is not the same as questioning them.

Deal honestly, caringly, and creatively with client defensiveness. Do not be surprised when clients react strongly to being challenged, even when you're trying to help them respond rather than react. If they react negatively, help them work through their emotion-laden reluctance and resistance. If they seem to "clam up," try to find out what's going on inside. In the following example, the helper has just delivered a brief summary of the main points of the problem situation they have been discussing, gently pointing out the self-destructive nature of some of the client's behaviors.

HELPER: I'm not sure how all this sounds to you.

CLIENT: I thought you were on my side. Now you sound like all the others. And I'm paying you to talk like this to me!

Even though the helper was tentative in his challenge, the client still reacted defensively. Here are two different approaches (A and B) to the client's defensiveness.

A: All I've done is summarize what you have been saying about yourself. And you know you're doing yourself in. Let's look at each point we've been discussing and see if this isn't the case.

This helper takes a defensive, judicial approach. She's about to assemble the evidence. This could well lead to an argument rather than further dialogue. Helper B backs off a bit.

B: So, I'm sounding harsh and unfair to you. . . . Kind of dumping on you. . . . Let's back up.

This helper backs off without saying that her summary was wrong. She is giving the client some space. It may be that the client needs time to think about what the helper has said. Helper B tries to find a way into a constructive dialogue with the angered client.

The principles outlined above are, of course, guidelines, not absolute prescriptions. In the long run, use your common sense. Put yourself in the client's shoes. Get the client to tell what he or she needs from you, The more flexible you are, the more likely you are to add value to your clients' search for solutions.

Linking Challenge to Action

More and more theoreticians and practitioners are stressing the need to link insight to problem-managing and opportunity-developing action. Wachtel (1989, p. 18) put it succinctly: "There is good reason to think that the really crucial insights are those closely linked to new actions in daily life and, moreover, that insights are as much a product of new experience as their cause." Is challenge enough to stimulate problem-managing action? For some, yes. A few well-placed challenges might be all that some clients need to move to constructive action. Once challenged, they are off to the races.

But others may well have the resources to manage their lives better, but not the will. They know what they need to do but are not doing it. A few nudges in the right direction may help them overcome their inertia. I have had many one-session encounters that included a bit of listening and thoughtful processing, some sharing of highlights, and a brief challenge that produced a new perspective that sent the client off on some useful course of action. On the other hand, if helping clients challenge themselves to develop new perspectives leads to one profound insight after another but no action or behavioral change, then once more we are whistling in the wind. Helping becomes an intellectual game rather than a serious attempt to partner with clients in their search for a better life.

Challenge and the Shadow Side of Helpers

Effective helpers have many characteristics and engage in many behaviors that add value to clients' efforts to manage their lives better. If they were to ask their clients what was helping and what was not, they might well be surprised. It seems that most helpers don't. Let's take a brief look at helper behavior that may be part of the shadow side of helping.

The "MUM effect." Initially, some counselor trainees are quite reluctant to help clients challenge themselves. They become victims of what has been called the "MUM effect," the tendency to "keep mum about undesirable messages," to withhold bad news even when it is in the other's interest to hear it (Rosen & Tesser, 1970, 1971; Tesser & Rosen, 1972; Tesser, Rosen, & Bachelor, 1972; Tesser, Rosen, & Tesser, 1971). In ancient times, the person who bore bad news to the king was sometimes killed. That obviously led to a certain reluctance on the part of messengers to bring such news. Bad news—and, by extension, the kind of "bad news" involved in invitations to self-challenge—arouses negative feelings in the challenger, no matter how he or she thinks the receiver will react. If you are comfortable with the supportive dimensions of the helping process but uncomfortable with helping as a social-influence process, you could fall victim to the MUM effect and become less effective than you might otherwise be.

Excuses for not challenging. Reluctance to challenge is not a bad starting position. In my estimation, it is a far better approach than being too eager to challenge. However, all helping, even the most client-centered, involves social influence. It is important for you to understand your reluctance (or eagerness) to challenge—that is, to challenge yourself on the issue of challenging and on the very notion of helping as a social-influence process. When trainees examine how they feel about challenging others, here are some of the things they discover.

- I am just not used to challenging others. My interpersonal style has had a lot of the live-and-let-live in it. I have misgivings about intruding into other people's lives.
- If I challenge others, then I open myself to being challenged. I may be hurt, or I may find out things about myself that I would rather not know.
- I might find out that I like challenging others, and the floodgates will open and my negative feelings about others will flow out. I have some fears that deep down I am an angry person.
- I am afraid that I will hurt others, damage them in some way or other. I have seen others hurt by heavy-handed confrontations.
- I am afraid that I will delve too deeply into others and find that they have problems that I cannot help them handle. The helping process will get out of hand.
- If I challenge others, they will no longer like me. I want my clients to like me.

Vestiges of this kind of thinking can persist long after trainees move out into the field as helpers. People in all sorts of people-oriented occupations, including human-service workers and managers, are bedeviled by the MUM effect and come up with their own set of excuses for not giving feedback. Of course, being willing to challenge responsibly is one thing; having the skills and wisdom to do so is another. One reason managers fail to give feedback is that do not have the communication, coaching, and counseling skills needed to do it well.

Helpers' blind spots. There is an interesting body of literature on the humanity and flaws of helpers (Kottler, 2000) that can be of enormous help to both beginners—because prevention is infinitely better than cure—and to old-timers—because you *can* teach old dogs new tricks. Kottler has provided trainees and novices an upbeat view of what passion and commitment in the helping professions should look like.

Because helpers are as human as their clients, they too can have blind spots that detract from their ability to help. Some helpers have difficulty adopting a resource-collaborator role. For instance, in one study (Atkinson and his associates, 1991), counselors were almost unanimous in their preference for a "feeling" approach to counseling, whereas the majority of male clients preferred either a "thinking" or an "acting" orientation. This is not collaboration.

One of the critical responsibilities of supervisors is to help counselors identify their blind spots and learn from them. Once out of training, skilled helpers use different forums or methodologies to continue this process, especially with difficult cases. They take counsel with themselves, asking, "What am I missing here?" They take counsel with colleagues. Without becoming self-obsessed, they scrutinize and challenge themselves and the role they play in the helping relationship. Or, more simply, throughout their careers they continue to learn about themselves, their clients, and their profession. One way to identify and do something about your own blind spots as a helper is to elicit clear, honest feedback from your clients.

By seeking help and embarking on a journey of constructive change, clients are making a promise or what Hoover and DiSilvestro (2005) call a "covenant" with themselves and often with other key people in their lives. As clients talk, they signal commitments through both words and gestures (Amrhein, 2004). Helping them constructively challenge themselves to live up to this covenant or set of commitments is a way of showing concern and care, not an intrusion or imposition.

HELPING DIFFICULT CLIENTS MOVE FORWARD: RELUCTANCE, RESISTANCE, AND RESILIENCE

Part Two closes with an overview of the three Rs—reluctance, resistance, and resilience. The communication skills described and illustrated in Chapters 4–8 are the principal tools counselors use to help clients face up to their reluctance to change, deal with their resistance to being helped, and tap into the pool of resilience found, to a greater or lesser extent, in every human being. Some call clients who are reluctant, resistant, and/or slow to tap into the resilience within themselves *difficult clients*. There is also a tendency to see difficult clients as *bad* clients. But they are just clients.

Wessler, Hankin, and Stern (2001) evened the therapy playing field by suggesting that the term "difficult client" is too one-sided. The therapist is often enough part of the difficulty: "When a therapist uses the world 'difficult' to describe a client, what he or she really means is: 'I am having difficulty working with this person due to either my own emotional issues or a lack of experience working with clients like this.' In essence, using the world difficult to describe a client should be a signal to the therapist that he/she needs to grow in some way personally (and interpersonally)" (p. 5). Fair enough, but clients come for help precisely because they are not managing some part of their lives effectively. Their behavior in the helping sessions is often enough an indication or an example of poor self-management.

DIFFICULT CLIENTS: RELUCTANCE AND RESISTANCE

It is impossible to be in the business of helping people for long without encountering both reluctance and resistance (Clark, 1991; A. Ellis, 1985; Fremont & Anderson, 1986; Friedlander & Schwartz, 1985; Hanna, 2002; G. A. Harris, 1995; Kottler, 1992; Otani, 1989; Wessler, Hankin, & Stern, 2001).

In these pages a distinction is made between reluctance and resistance. *Reluctance* refers to clients' hesitancy to engage in the work demanded by the stages and steps of the helping process. Problem management and opportunity development involve a great deal of work. Therefore, there are sources of reluctance in all clients—indeed, in all human beings. For instance, part of problem management is trying new behaviors. Many clients are reluctant to do so. Unused opportunities also provide challenges. Developing unused opportunities means venturing into unknown waters. Although this is a charming idea for some, it strikes something akin to terror in others. Many clients are reluctant to talk about themselves, especially about themselves as flawed. Mahalik (1994) developed a Client Resistance Scale to measure clients' reluctance to deal with painful emotions, disclose intimate or painful material, develop a working alliance with the helper, deal with blind spots, develop new perspectives, and embrace constructive change.

Resistance, on the other hand, refers to the push-back from clients when they feel they are being coerced. Clients who think that they are being mistreated by their helpers in some way tend to resist. Clients who believe that their cultural beliefs, values, and norms—whether group or personal—are being violated by the helper can be expected to resist. For instance, because individual and cultural norms regarding self-disclosure differ widely (see Wellenkamp, 1995), clients who believe that self-disclosure is being extorted from them might well resist.

Although the behaviors through which reluctance and resistance are expressed might either seem or be the same, the distinction is still a useful one. The seeds of reluctance are in the client, whereas the stimulus for resistance is external—in the helper (Bischoff & Tracey, 1995) or the social setting surrounding the helping process. In practice, of course, a mixture of reluctance and resistance is often found in the same client. If therapy is to become more efficient, then counselors need to find ways of helping their clients deal with reluctance and resistance as expeditiously as possible.

RELUCTANCE: MISGIVINGS ABOUT CHANGE

Clients exercise reluctance in many, often covert, ways. They talk about only safe or low-priority issues, seem unsure of what they want, benignly sabotage the helping process by being overly cooperative, set unrealistic goals and then use them as an excuse for not moving forward, don't work very hard at changing their behavior, and are slow to take responsibility for themselves. They tend to blame others or the social settings and systems of their lives for their troubles and play games with helpers. Or they don't come for counseling in the first place.

Tim is reluctant to join his wife in her sessions with a counselor. He says that he'll "think about it," that he doesn't feel "any real need" to talk to a counselor, that right now the demands of his job are too pressing and that he can't "find the time" for the sessions, and so forth. Deep down he's afraid of what might happen were he to go. There are many ways clients drag their feet. Reluctance admits of degrees; clients come "armored" against change to a greater or lesser degree. The reasons for reluctance are many. They are built into the human condition. Here is a sampling.

Fear of intensity. If the counselor uses high levels of tuning in, listening, sharing empathic highlights, and probing, and if the client cooperates by exploring the feelings, experiences, behaviors, points of view, and intentions related to his or her problems in living, the helping process can be an intense one. This intensity can cause both helper and client to back off. Skilled helpers know that counseling is potentially intense. They are prepared for it and know how to support a client who is not used to such intensity.

Lack of trust. Some clients find it very difficult to trust anyone, even a most trustworthy helper. They have irrational fears of being betrayed. Even when confidentiality is an explicit part of the client-helper contract, some clients are very slow to reveal themselves. A combination of patience, encouraging, and challenging is demanded of the helper.

Fear of disorganization. Some people fear self-disclosure because they feel that they cannot face what they might find out about themselves. The client feels that the façade he or she has constructed, no matter how much energy must be expended to keep it propped up, is still less burdensome than exploring the unknown. Such clients often begin well but retreat once they start to be overwhelmed by the data produced in the problem-exploration process. Digging into one's inadequacies always leads to a certain amount of disequilibrium, disorganization, and crisis. But growth takes place at crisis points. A high degree of disorganization immobilizes the client, whereas very low disorganization is often indicative of a failure to get at the client's core concerns. A framework for helping provides clients with "channels" that act as safety devices that help them contain their fears. By challenging them to take "baby steps" that don't end in disaster, counselors help clients build confidence.

Shame. Shame is a much overlooked variable in human living (Egan, 1970; Kaufman, 1989; Lynd, 1958). It is an important part of disorganization and crisis. The root meaning of the verb *to shame* is "to uncover, to expose, to wound," a meaning that suggests the process of painful self-exploration. Shame is not just being painfully exposed to another; it is primarily an exposure of self to oneself. In shame experiences, particularly sensitive and vulnerable aspects of the self are exposed, especially to one's own eyes. Shame is often sudden—in a flash, the client sees heretofore unrecognized inadequacies without being ready for such a revelation. Shame is sometimes touched off by external incidents, such as a casual remark someone makes, but it could not be touched off by such insignificant incidents unless, deep down, one was already ashamed. A shame experience might be defined as an acute emotional awareness of a failure to be in some way. Once more, empathy and support help clients deal with whatever shame they might experience.

The cost of change. Some people are afraid to take stock of themselves because they know, however subconsciously, that if they do, they will have to change—that is, surrender comfortable but unproductive patterns of living, work more diligently, suffer the pain of loss, acquire skills needed to live more effectively, and so on. For instance, a husband and wife may realize, at some level, that if they see a counselor, they will have to reveal themselves and that once the cards are on the table, they will have to go through the agony of changing their style of relating to each other. Some clients come with the idea that counseling is magic and are put off when change proves to be hard work.

A loss of hope. Some clients think that change is impossible, so why try? Recall the man in his 60s discussed in Chapter 7, the one who was a participant in a counseling group and complained about constant anxiety. He had given up hope. How could anyone who had been treated as brutally as he was by his father have any hope? Running away from home was just the beginning. He kept running from hope the rest of his life. But after being challenged by both the helper and his fellow participants, he rediscovered hope and, with it, self-responsibility. He no longer focused on the "scars" inflicted by his father's mistreatment. He no longer focused on the self-inflicted scars of a life lived irresponsibly. He found hope in both the care he experienced in the group and in the life struggles revealed by the other participants. He found hope in community.

Each one of us needs only to look at his or her own struggles with growth, development, and maturity to add to this list of the roots of reluctance.

RESISTANCE: REACTING TO COERCION

Resistance refers to the reaction of clients who in some way feel coerced. Reluctance is often passive. Resistance can be both active and passive. It is the client's way of fighting back (see Dimond and his associates, 1978, and Driscoll, 1984). Spouses who feel forced to come to marriage counseling sessions are often resistant. They resist because they resent what they see as a power play. Tony gets angry when his wife suggests that he come with her to her counseling sessions. Knowing that she has talked this over with her mother, he feels that he is the focus of a conspiracy. They are looking for ways to coerce him to go. "I don't care what happens, but they're not going to get me," are his sentiments. Of course, some clients see coercion where it does not exist. But because people act on their perceptions, the result is still some form of active or passive fighting back.

Resistant clients, feeling abused, let everyone know that they have no need for help, show little willingness to establish a working relationship with the helper, and often enough try to con counselors. They are often resentful, make active attempts to sabotage the helping process, or terminate the process prematurely. They can be either testy or actually abusive and belligerent. Resistance to helping is, of course, a matter of degree, and not all resistant clients engage in extreme forms of resistance behaviors.

Involuntary clients (Brodsky, 2005)—sometimes called "mandated" clients—are often resisters. A high school student gets into trouble with a teacher and sees being sent to a counselor as a form of punishment. A felon receives probation on the condition of being involved in some kind of counseling process. A manager accused of sexual harassment keeps his job only if he agrees to a series of counseling sessions. Clients like these are found in schools, especially schools below college level, in correctional settings, in marriage counseling, especially if it is court-mandated, in employment agencies, in welfare agencies, in court-related settings, and in other social agencies. But any client who feels that he or she is being coerced or treated unfairly can become a resister. Clients can experience coercion in a wide variety of ways. The following kinds of clients are often resistant.

- Clients who see no reason for going to the helper in the first place.
- Clients who resent third-party referrers (parents, teachers, correctional facilities, social service agencies) and whose resentment carries over to the helper.
- Clients who don't know what helping is about and fear the unknown.
- Clients who have a history of rebelliousness.
- Clients who see the goals of the helper or the helping system as different from their own. For instance, the goal of counseling in a welfare setting may be to help clients become financially independent, whereas some clients may be satisfied with financial dependency.
- Clients who have developed negative attitudes about helping and helping agencies and who harbor suspicions about helping and helpers. They don't trust "shrinks."
- Clients who believe that going to a helper is the same as admitting weakness, failure, and inadequacy. They feel that they will lose face by going. By resisting the process, they preserve their self-esteem.
- Clients who feel that counseling is something that is being done to them. They feel that their rights are not being respected.
- Clients who feel a need for personal power and find it through resisting a powerful figure or agency. "I may be relatively powerless, but I still have the power to resist" is the subtext.
- Clients who dislike their helpers but do not discuss their dislike with them.
- Clients who differ from their helpers about the degree of change needed.
- Clients who differ greatly from their helpers—for instance, a poor kid with an older middle-class helper.

Of course, resistance can be a healthy sign. Clients are standing up for their rights and fighting back.

Many sociocultural variables—gender, prejudice, race, religion, social class, upbringing, cultural and subcultural blueprints, and the like—can play a part in resistance. For instance, a man might instinctively resist being helped by a woman and vice versa. An African American person might instinctively resist being helped by a white person and vice versa. A person with no religious affiliation might instinctively think that help coming from a minister will be "pious" or will automatically include some form of proselytizing.

THE ROLE OF PSYCHOLOGICAL DEFENSES IN RELUCTANCE AND RESISTANCE

Clients don't talk openly about their reluctance and resistance; rather, they demonstrate them. In many ways they try to disguise them, call them something else. An overview of psychological defenses gives us some insight into the ways clients defend themselves from change.

The very concept of psychological defenses, which is usually covered in abnormal psychology courses, has had a long, somewhat confusing history (Blackman, 2004; Cramer, 2000, 2005, 2006; Hentschel, Smith, Draguns, & Ehlers, 2004). Cramer (1998) differentiates between coping mechanisms and defenses. She sees coping as a conscious and intentional process, whereas defenses tend to be unconscious and unintentional. To muddy the waters, Vaillant (2000) talks about defenses as "involuntary coping mechanisms" (p. 89). Perhaps we can say that defenses play a role in coping. The "unconscious" part, with its ties to psychoanalytic theory, is a sore point for many. Reifying—that is, making a "thing" out of the unconscious—has been too much for some to swallow.

But today, cognitive psychologists generally agree that some mental processes go on outside of awareness. For instance, there is evidence that this is the case with decision making. Indeed, later in the book we will see that decision making is not as rational as it seems. Using language such as "mental processes outside of awareness" to describe defenses keeps the unconscious from sounding like some kind of black hole inside us. Although everyone knows that we defend ourselves from the onslaught of reality in a variety of ways, theoreticians, researchers, and practitioners have not always been able to agree on the precise nature of these defenses. Are they good? Are they bad? Are some good and some bad? Yes. Yes. Yes.

Early psychoanalytic theory saw defenses as ways of managing instinctual drives, but some theoreticians and researchers now see them as playing a role in the maintenance of self-esteem and, generally, keeping ourselves intact (Cooper, 1998). Social psychologists today talk about defenses as processes "by which humans deceive themselves, enhance self-esteem, and foster unrealistic self-illusions" (Cramer, 2000, p. 639; see also Baumeister, Dales, & Sommer, 1998). "Cognitive dissonance" has been a respected social psychology concept for years. It became almost a brand name, and there have been endless studies on it. But some now say that this term itself is just a nicer word for defenses (Paulhus, Fridhandler, & Hayes, 1997).

Cramer (2000) places defenses on a maturity continuum. High on the scale are such things as anticipation, altruism, humor, sublimation, and suppression, whereas delusional projection, psychotic denial, and psychotic distortion sit at the bottom. One problem with such a list is that some high-end defenses such as anticipation, altruism, and humor need not be "outside of awareness." Or, if you will, there are forms that are outside of awareness and forms that are not.

Admitting that psychologists really don't know how defenses—either mature or immature—work, Vaillant (2000) says that one possibility is that so-called mature defenses might just as well be called virtues. Going south on the maturity scale we find (without naming them all) intellectualization and isolation, then idealization, then nonpsychotic denial and rationalization, then autistic fantasy, and finally acting out and apathetic withdrawal before hitting the bottom. Vaillant (2000) discusses the positive-psychology role of mature defenses like altruism and anticipation. There is some evidence that as clients get better, the use of immature defenses recedes and the use of mature defenses increases. Cramer (2000) cautions us on this hierarchical ranking of defenses, however. She uses denial as an example, noting that everyone uses denial from time to time, "but it is certainly not always pathogenic or pathological" (p. 672). It is not right, she suggests, to say that denial is always developmentally immature because sometimes it is merely adaptive.

Although defenses defend people from, let's say, anxiety, they also get people in trouble. For instance, people with serious physical conditions such as diabetes and obesity use defenses to deny the seriousness of their condition. Defenses work to lower anxiety in the short term, but they often put people with these conditions at risk in the longer term. Defenses, then, can keep us from doing what is in our best interest. In counseling settings, defenses can keep clients from developing insights that would help them admit and deal with their problems more creatively.

The point is that some defenses, or "adaptive mental mechanisms," can help get us into and keep us mired in our troubles, whereas others can help prevent us from getting into trouble and help us cope when trouble strikes. The former need to be challenged and the latter supported. This not mean that helpers should mount an attack on every immature defense they run across. This could backfire: "By thoughtlessly challenging irritating, but partly adaptive, immature defenses, a clinician can evoke enormous anxiety and depression in a patient" (Vaillant, 1994, p. 49) and put the helping relationship itself at risk.

Immature defenses contribute a great deal to both reluctance and resistance. The problem is that clients don't show up with a list of defenses tattooed on their foreheads. Personality traits are also important, but they, too, are not always easily identifiable. Take locus of control. A client with a pronounced external locus of control might well have problems with responsibility and accountability. He or she might be looking for a great deal of guidance from a helper. The point is that clients can arrive on your doorstep with built-in tendencies to reluctance and resistance. This puts a burden on you to listen fully to your clients and to process what they say and do by what you know about human beings. Understanding psychological defenses can help. Understanding personality traits and their implications can help. But in the end, it is you and the client trying to do what is best for the client.

Finally, Clark (1998) has written a book describing defense mechanisms as they manifest themselves in counseling and suggesting ways of helping clients challenge them. Although he deals with traditional defense mechanisms—denial, displacement, identification, isolation, projection, rationalization, reaction formation, regression, repression, and undoing—he does so with a fresh voice. The book is filled with examples of dialogue in which helpers listen, process, share highlights, probe, and challenge in the spirit of empathy as they try to help their clients face up to their defenses and what these defenses might be costing them.

PRINCIPLES FOR MANAGING RELUCTANCE AND RESISTANCE

Because both reluctance and resistance are such pervasive phenomena, helping clients manage them is part and parcel of all our interactions with clients (Kottler, 1992). Here are some principles that can act as guidelines.

Avoid Unhelpful Responses to Reluctance and Resistance

Helpers, especially beginning helpers who are unaware of the pervasiveness of reluctance and resistance, are often disconcerted when they encounter uncooperative clients. Such helpers are prey to a variety of emotions—confusion, panic, irritation, hostility, guilt, hurt, rejection, depression. Distracted by these unexpected feelings, they react in any of several unhelpful ways.

- They accept their guilt and try to placate the client.
- They become impatient and hostile and manifest these feelings either verbally or nonverbally.
- They do nothing in the hope that the reluctance or the resistance will disappear.
- They lower their expectations of themselves and proceed with the helping process, but in a halfhearted way.
- They try to become warmer and more accepting, hoping to win the client over by love.
- They blame the client and end up in a power struggle with him or her.
- They allow themselves to be abused by clients, playing the role of a scapegoat.
- They lower their expectations of what can be achieved by counseling.
- They hand the direction of the helping process over to the client.
- They give up.

In short, when helpers engage "difficult" clients, they experience stress, and some give in to self-defeating "fight or flight" approaches to handling it.

The source of this stress is not just clients' behavior; it also comes from the helper's own self-defeating attitudes and assumptions about the helping process. Here are some of them.

- All clients should be self-referred and adequately committed to change before appearing at my door.
- Every client must like me and trust me.
- I am a consultant and not a social influencer; it should not be necessary to place demands on clients or even help them place demands on themselves.
- Every unwilling client can be helped.

- No unwilling client can be helped.
- I alone am responsible for what happens to this client.
- I have to succeed completely with every client.

These unrealistic beliefs are never spoken, but they can linger in the background. Effective helpers neither court reluctance and resistance nor are surprised by them.

Develop Productive Approaches to Dealing With Reluctance and Resistance

In a book like this, it is impossible to identify every possible form of reluctance and resistance, much less provide a set of strategies for managing each. Here are some principles and a general approach to managing reluctance and resistance in whatever forms they take.

Explore your own reluctance and resistance. Examine reluctance and resistance in your own life. How do you react when you feel coerced? What do you do when you feel you are being treated unfairly? How do you run away from personal growth and development? If you are in touch with the various forms of reluctance and resistance in yourself and are finding ways of overcoming them, you are more likely to help clients deal with theirs.

See some reluctance and resistance as normal. Help clients see that their reluctance and resistance are not "bad" or odd. After all, yours isn't. Beyond that, help them see the positive side of resistance. It may well indicate that they have fiber. It may be a sign of self-affirmation.

Accept and work with the client's reluctance and resistance. Teyber (2005) talks about "honoring" the client's resistance. This is a central principle. Start with the client's frame of reference. Accept both the client and his or her reluctance or resistance. Do not ignore it or be intimidated by what you find. Let clients know how you experience it and then explore it with them. Model openness to challenge. Be willing to explore your own negative feelings. The skill of direct, mutual talk (called *immediacy*), discussed in Chapter 8, is extremely important here. Help clients work through the emotions associated with reluctance and resistance. Avoid moralizing. Befriend the reluctance or the resistance instead of reacting to it with hostility or defensiveness.

See reluctance as avoidance. Reluctance is a form of avoidance that is not necessarily tied to client ill will. Therefore, you need to understand the principles and mechanisms underlying avoidance behavior, which is often discussed in texts dealing with the principles of behavior (Watson & Tharp, 2007). Some clients avoid counseling or give themselves to it only halfheartedly because they see it as lacking in suitable rewards or even as punishing. If that is the case, then counselors have to help them search for suitable incentives. Constructive change is usually more rewarding than a miserable status quo, but that might not be the client's perception, especially in the beginning. Find ways of presenting the helping process as rewarding. Talk about outcomes.

Examine the quality of your interventions. Without setting off on a guilt trip, examine your helping behavior. What are you doing that might seem unfair to the client? In what ways does the client feel coerced? For example, you may have become too directive without realizing it. Furthermore, take stock of the emotions that are welling up in you because clients lash back or drag their feet. How are these emotions "leaking out"? No use denying such feelings. Rather, own them and find ways of coming to terms with them. Do not overpersonalize what the client says and does. If you are allowing a hostile client to get under your skin, you are probably reducing your effectiveness. Of course, the client might be resistant, not because of you, but because he is under pressure from others to deal with his problems. But you take the brunt of it. Find out, if you can.

Be realistic and flexible. Remember that there are limits to what a helper can do. Know your own personal and professional limits. If your expectations for growth, development, and change exceed the client's, you can end up in an adversarial relationship. Rigid expectations of the client and of yourself become self-defeating.

Establish a "just society" with your client. Deal with the client's feelings of coercion. Provide what Smaby and Tamminen (1979) called a "two-person just society" (p. 509). A just society is based on mutual respect and shared planning. Therefore, establish as much mutuality as is consonant with helping goals. Invite participation. Help clients participate in every step of the helping process and in all the decision making. Share expectations. Discuss and get reactions to helping procedures. Explore the helping contract with your clients and get them to contribute to it.

Help the client search for incentives for moving beyond resistance. Help the client find incentives for participating in the helping process. Use client self-interest as a way of identifying these. Use brainstorming as a way of discovering possible incentives. For instance, the realization that he or she is going to remain in charge of his or her own life may be an important incentive for a client.

Do not see yourself as the only helper in your client's life. Engage significant others, such as peers and family members, in helping the client face reluctance and resistance. For instance, lawyers who belong to Alcoholics Anonymous may be able to deal with a fellow lawyer's reluctance to join a treatment program more effectively than you can.

Employ the client as a helper. If possible, find ways to get a reluctant or resistant client into situations to help others. The change of perspective can help the client come to terms with his or her own unwillingness to work. One tactic is to take the role of the client in the interview and manifest the same kind of reluctance or resistance he or she does. Have the client take the counselor role and help you overcome your unwillingness to work or cooperate. One person who did a great deal of work for Alcoholics Anonymous had a resistant alcoholic go with him on his city rounds, which included visiting hospitals, nursing homes for alcoholics, jails, flophouses, and down-and-out people on the streets. The alcoholic saw through all the lame excuses other alcoholics offered for their plight. After a week, he joined AA himself. Clients can become helpers in group counseling, too.

Reach-accept-relate strategies. Hanna, Hanna, and Keys (1999; see also Hanna, 2002, and Sommers-Flanagan & Sommers-Flanagan, 1995) drew up a list of fifty strategies—some original, many drawn from the helping literature—for counseling defiant, aggressive adolescents. Many of the strategies have wider application to both reluctant and resistant clients of all ages and can be used to put into practice the principles outlined above. The authors divide the strategies into three categories: reaching clients, accepting them, and relating to them. Here is the selection I have made (see Hanna, Hanna, & Keys, 1999, for descriptions of each strategy). Some of them are reworded.

Reach out to clients. If you do not reach out to resistant clients, you will not engage them. Brodsky (2005) notes that therapy tends to fail when therapists "passively accept problematic aspects of the client's behavior and attitudes, such as evasiveness and negativism" (p. 260). Here are some guidelines:

- Avoid sitting behind a desk during a counseling session.
- Be genuine and unpretentious.
- Show deep respect.
- Keep and use your sense of humor.
- Be able to laugh at yourself.
- Educate the client about counseling.
- Avoid being a symbol of authority.
- Avoid taking an expert stance until the relationship is fairly stable.
- Avoid asserting your credentials.
- Avoid thinking in clinical labels.
- Convey a brief therapy attitude.
- Let the client "circle in" on more sensitive issues.
- Balance insight and action.
- Admire terrifically defensive behavior.
- Address the hurt behind the anger.
- Encourage resistance; fighting back is a sign of spunk.

Accept clients. Resistant clients deserve the same kind of respect and empathy you would give to any other client. Here are suggestions to show your acceptance:

- Be clear about the boundaries of acceptable behavior in counseling sessions.
- Avoid power struggles.
- Deal nondefensively with verbal disrespect; for example, when the client is angrily disrespectful, a reply such as "I wonder who you're really mad at" would be inappropriate.
- Affirm the client's perceptions when they are accurate. "You're right. I don't find these sessions easy or fun."
- Use immediacy with issues in the counseling relationship.
- Treat shocking statements with equanimity. Reframe the message embedded in such statements.

Actively relate to clients. Do your best to establish a genuine working alliance with clients even in the face of opposition. Some tactics you might employ include the following:

- Admit when you are confused or uninformed.
- Expect crises in clients' lives.

- Tell stories of other clients in similar situations who made changes in their lives.
- Let clients know how much you are learning from your sessions with them.
- Stay in touch with similar problems you have had.
- If you think another counselor will be a better fit with a client, talk to the client about switching.
- Use sound bytes rather than paragraphs when communicating a point.
- Share only the things about yourself that you have worked through and would not mind hearing repeated.
- Do not allow the depth of caring to interfere with a clear understanding of your client and his or her problem situation.
- Encourage the client to establish a therapeutic peer culture.
- Identify true victimization, whatever its source.
- If the client is seeking attention, give it: "All right, now that you have my full attention, what do you want to do with it?"
- Make confrontation friendly and empathic.
- Reframe apathy as an attempt to avoid hurt, hassles, and difficulties.

In summary, do not avoid dealing with reluctance and resistance, but do not reinforce these processes either. Work with your client's unwillingness, and become inventive in finding ways of dealing with it. But be realistic about the human tendency to evade and avoid.

HELP CLIENTS TAP INTO THEIR RESILIENCE

Like the rest of us, clients stumble and fall as they try to assess their problems, formulate goals for a better future, and implement constructive-change programs. They can "mess up" at any stage of the helping process either in the helping sessions themselves or in their everyday lives. However, everyone has some degree of *resilience* within that enables them, after "messing up," to get up, pull themselves together, and move on once more. There has been a mini-explosion of theory and research about resilience (Alvord & Grados, 2005; Flach, 1997; Reivich & Shatté, 2002; Tedeschi & Kilmer, 2005; Tugade & Fredrickson, 2004), and the American Psychological Association has even launched a "resilience initiative" (Kersting, 2003, 2005; Newman, 2005). Newman defines resilience as "the human ability to adapt in the face of tragedy, trauma, adversity, hardship, and ongoing significant life stressors" (p. 227).

The ability to bounce back is an essential life capability. Holaday and McPhearson (1997) have compiled a list of common factors that influence resilience. Although they focused on severe-burn survivors, their discussion of resilience applies to all of us and our clients. They distinguish between *outcome* resilience and *process* resilience. Whereas resilience in general is the ability to overcome or adapt to significant stress or adversity, outcome resilience implies a return to a previous state. This is "bounce back" resilience. Dora goes through the trauma of divorce, but within a few months she bounces back. Her friends say to her, "You seem to be your old self now." She replies, "Both older and wiser." Process resilience, on the other hand, represents the continuous effort to cope that is a "normal" part of some people's lives. The burn survivors in Holaday and McPhearson's study would say such things as, "Resilience? It's my spirit and it's the reason I'm here," "[Resilience] is deep inside of you, it's already there, but you have to use it," and "To do well takes a lot of determination, courage, and struggling, but it's your choice" (p. 348).

You can encourage both kinds of resilience in clients. Take outcome resilience. Kerry finds himself in a financial mess because of a tendency to be a spendthrift and because of a few poor financial decisions. Although he makes a couple more mistakes, he works his way out of the mess. Once he reaches his goal, he puts himself on a strict budget and things stabilize. It's not difficult for him to walk the financial straight and narrow because the mess has been too painful to repeat.

There is a growing literature on "posttraumatic growth" (Calhoun & Tedeschi, 2006; Tedeschi & Calhous, 2004), which is related to outcome resilience but goes beyond it. For instance, Phillips and Daniluk (2004), in exploring the stories of seven adult women who had experienced sexual abuse as children, found that a "strong sense of resiliency and growth was a persistent theme that wove itself throughout the interviews" (p. 181). One of the women said:

> I wouldn't wish the pain of child abuse on anybody, but I realize that if I hadn't been abused or been through hell like that, I might never have known how unlimited the human spirit is around finding its way through impossible odds. I wouldn't have missed the experience of learning that for the world. (p. 182)

Process resilience is another matter. Oscar finds that controlling his anger is a constant struggle. He has to keep finding the process-resilience resources within himself needed to keep plugging away. And then there is Nadia, a middle-aged single woman suffering from chronic fatigue syndrome. She has to dig deep within herself every day to find the will to go on. Like many people suffering from this condition, she wants to do her best and even make a good impression (Albrecht & Wallace, 1998). On the days she's successful in pulling herself together, the people she meets cannot believe that she is ill. For her, running into this kind of disbelief is a two-edged sword. On the one hand, her affliction is so painful that she wonders how intelligent people could possibly not notice. On the other hand, she realizes that working hard at showing her best face to the world and often enough succeeding in doing so is a victory.

Holaday and McPhearson (1997) suggest that care factors that influence resilience include social support, cognitive skills, and psychological resources.

Social support. This includes society's overall values toward people, especially people in trouble; community support—that is, support in the neighborhood, at work, at church, and so forth; personal support through friends and other special relationships; and familial support, the "affectional ties within a family system." One burn survivor said, "My wife made me get out of the hospital bed and learn to walk again."

Cognitive skills. It seems that at least average intelligence contributes greatly to resilience. But there are many different kinds of intelligence—academic intelligence, social intelligence, street smarts, and so forth. And, as Holaday and McPhearson point out (1997, p. 350), "intelligence is also associated with the ability to use fantasy and hope." Cognitive skills also include coping style. For instance, a "belligerent style" (Zimrin, 1986), rather than a passively enduring, accepting, or yielding style, often contributes more to resilience. "I don't take what others say, it's *not* over; don't tell me I can't do something."

Clients can also cope by discussing feelings. One burn survivor said, "Sometimes I still choose to feel sorry for myself and have a bad day, and that's OK." Other useful coping strategies include avoiding self-blame and using the energy of anger to cope with the world rather than damage the self. One client said, "When I was little, I wanted the scars to go away, but now I don't care about them any more. They're part of me."

Other cognitive factors in resilience include the degree and the way clients exercise personal control in their lives and how they interpret their experiences. One client who fell off the wagon and got drunk for a couple of days said, "It's a glitch, not a pattern. I can expect a glitch now and again. Glitches can be dealt with. Patterns are damaging."

Psychological resources. Certain personality characteristics or dispositions protect people from stress and contribute to "bounce back." They include an internal locus of control, empathy, curiosity, a tendency to seek novel experiences, a high activity level, flexibility in new situations, a sense of humor, the ability to elicit positive regard from others, accurate and positive self-appraisal, personal integrity, a sense of self-protectiveness, pride in accomplishments, and a capacity for fun.

There a range of "resilience levers" in every client. Your job is to help clients discover and pull them in order to bounce back. Resilience is "deep inside you" and inside your clients. It's part of our self-healing nature.

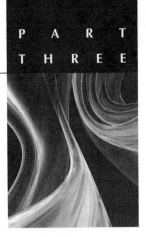

THE STAGES AND TASKS
OF THE HELPING MODEL

The communication skills reviewed in Part Two are critical tools. With them, you can help clients engage in all the stages and tasks of the helping model. But those communication skills are not the helping process itself. Part Three is a detailed exposition and illustration of the stages of the helping model. Chapter 10 explores Stage I and its three interrelated tasks. Stage I involves helping clients tell their stories in terms of problem situations and unused opportunities. In Stage I counselors help clients tell their stories, develop new perspectives, and work on issues that can make a difference in clients' lives. Chapter 11 prepares the reader for Stages II and III by pointing out the importance of goals and a solution-focused approach to managing problem situations and developing opportunities. Chapter 12 describes Stage II and its three related tasks. In Stage II counselors help clients explore possibilities for a better future, turn the best possibilities into problem-managing goals, and explore clients' commitment to these goals. If Stage II explores with clients *what* they want in terms of problem management and opportunity development, Stage III, outlined in Chapter 13, deals with *how* to get what they need and want. Here again, there are three tasks: helping clients explore strategies for achieving goals, helping them choose the best-fit package of strategies, and helping them turn this package into a viable plan.

STAGE I: HELP CLIENTS TELL THEIR STORIES

An Introduction to Stage I

Task 1: Help Clients Tell Their Stories—"What Are My Concerns?"

Guidelines for Helping Clients Explore Problem Situations and Unexploited Opportunities

Learn to work with all styles of storytelling

Help clients spot unused opportunities

Start where the client starts

Help clients clarify key issues

Help clients discuss the context of their concerns

Assess the severity of the client's problems

Help clients talk productively about the past

As clients tell their stories, search for resources, especially unused resources

Help clients see every problem as an opportunity

Action Right From the Beginning

The importance of being proactive

Using the time between sessions productively

Task 2: Help Clients Develop New Perspectives and Reframe Their Stories—"What Am I Overlooking or Avoiding?"

Help clients paint an objective picture

Invite clients to own their problems and unused opportunities

Invite clients to state their problems as solvable

Invite clients to explore their "problem-maintenance structure"

Invite clients to move on to the next needed stage of the helping process

Task 3: Help Clients Work on Issues that Make a Difference—"What Problems, if Managed, Will Help Me Most?"

Guidelines in the Search for Leverage

1. Determine whether or not helping is called for

2. If there is a crisis, first help the client manage the crisis

3. Begin with the problem that seems to be causing the client the most pain

4. *Begin with issues the client sees as important and is willing to work on*

5. *Begin with some manageable subproblem of a larger problem situation*

6. *Move as quickly as possible to a problem that, if handled, will lead to some kind of general improvement*

7. *Focus on a problem for which the benefits will outweigh the costs*

Leverage and Action

The Shadow Side of Leverage

Is Stage I Sometimes Enough?

A declaration of intent and the mobilization of resources

Coming out from under self-defeating emotions

An Introduction to Stage I

Clients come to helpers because they need help in managing their lives more effectively. Stage I illustrates three ways in which counselors can help clients understand themselves, their problem situations, and their unused opportunities with a view to managing them more effectively. There are three tasks.

- *Task 1: Help clients tell their stories.* Help clients tell their stories in terms of problem situations and unused opportunities. The questions here are, "What's going on? What are my concerns? What are the issues?"

- *Task 2: Help clients develop new perspectives and reframe their stories.* Help clients identify and clarify the critical elements of the problem situation or unused opportunity. This often means helping clients identify blind spots, develop new perspectives, and reframe the story itself. The questions here are, "What's really going on? What are my real concerns? What am I failing to see? What are the critical elements of the story?" A judicious use of challenging skills is needed here.

- *Task 3: Help clients achieve leverage by working on issues that make a difference.* Clients often have multiple problems. If that is the case, help clients work on substantive issues. The questions here are, "What are my key concerns? What should I work on? What will make a difference in my life?"

Even though these tasks are described separately here, in actual helping sessions they are intermingled. It is not a question of moving from the first to the second to the third task in any rigid way. Furthermore, these tasks are not restricted to Stage I for the following reasons. First, clients don't tell all of their stories at the beginning of the helping process. Often the full story "leaks out" over time. Second, blind spots can appear at any stage of the helping process. Blind spots affect choosing goals, setting strategies, and implementing programs. New perspectives are always welcome.

Stage I of the helping process can be seen as the assessment stage—finding out what's going wrong, what opportunities lie fallow, what resources are not being used. Client-centered assessment means helping clients understand themselves, find out "what's going on" with their lives, see what they have been ignoring, and make sense from the messiness of their lives. Assessment in this sense is not something helpers

The Skilled-Helper Model

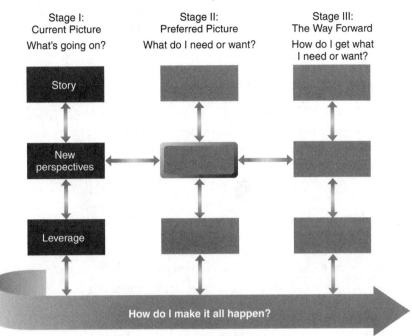

Stage I:	Stage II:	Stage III:
Current Picture	Preferred Picture	The Way Forward
What's going on?	What do I need or want?	How do I get what I need or want?

Story

New perspectives

Leverage

How do I make it all happen?

FIGURE 10-1
The Helping Model—Stage 1

do to clients. Rather, it is a kind of learning in which, ideally, both client and helper participate through their ongoing dialogue. Other forms of assessment such as psychological testing and applying psychiatric diagnostic categories may or may not be useful, but they are beyond the scope of this book. The title of each of the tasks includes a question clients can ask themselves. Figure 10-1 highlights the three tasks of Stage I.

TASK 1: HELP CLIENTS TELL THEIR STORIES—"WHAT ARE MY CONCERNS?"

The importance of helping clients tell their stories well should not be underestimated. As Pennebaker (1995b) has noted, "An important . . . feature of therapy is that it allows individuals to translate their experiences into words. The disclosure process itself, then, may be as important as any feedback the client receives from the therapist" (p. 3). Farber, Berano, and Capobianco (2004) outline the bright side of client self-disclosure.

> [M]ost clients feel that therapy is a safe place to disclose, made especially so by the goodness of the therapeutic relationship; that the disclosure process initially generates shame and anticipatory anxiety but ultimately engenders feelings of safety, pride, and authenticity; that keeping secrets inhibits the work of therapy, whereas disclosing produces a sense of relief from physical

as well as emotional tension; that disclosures in therapy facilitate subsequent disclosures to one's therapist as well as to family members and friends; and that therapists should actively pursue material that is difficult to disclose. (p. 340)

Of course, other research has shown that fear of self-disclosure is a leading factor in not seeking therapy (Vogel & Wester, 2003). So it is important for you to discern whether the client you are helping rejoices in or fears the process. One possibility for clients who fear face-to-face disclosure is writing (Sloan & Marx, 2004a, 2004b; Sloan, Marx, & Epstein, 2005). For many people, writing about their problems provides both physical and emotional relief. Researchers in counseling have not given this possibility the attention it deserves. At any rate, self-disclosure provides the grist for the mill of problem solving and opportunity development. Here, then, are some of the goals for Stage I:

- **Initial stress reduction.** Help clients "get things out on the table." This can and often does have a cathartic effect that leads to stress reduction. Some clients carry their secrets around for years. Helping them unburden themselves is part of the social-emotional reeducation process alluded to earlier.

- **Clarity.** Help clients spell out their problem situations and unexploited opportunities with the kind of concrete details—specific experiences, thoughts, behaviors, and emotions—that enables them to do something about them. Clarity opens the door to more creative options in living. Vague stories lead to vague options and actions.

- **Establishing rapport and relationship building.** Help clients tell their stories in such a way that the helping relationship develops and strengthens. The communication skills outlined in earlier chapters are basic tools for both clarity and relationship building.

- **Client action.** Right from the beginning, help clients act on what they are learning. Clients do not need "grand plans" before they can act on their own behalf.

Clients differ radically in their ability to talk about themselves and their problem situations. Reluctance to disclose oneself within counseling sessions is often a window into the client's inability to share himself or herself with others and to be reasonably assertive in the social settings of everyday life. If this is the case, then one of the goals of the entire counseling process is to help clients develop the skills, confidence, and courage they need to share themselves appropriately.

Guidelines for Helping Clients Explore Problem Situations and Unexploited Opportunities

There are a number of principles that can guide you as you help clients tell their stories. Hanna (2002) has developed a list of "precursors," which "taken together form a comprehensive picture of how people change and why they do not" (p. 6). They include, on the part of the client, a sense of necessity ("I've got to do something about this"), a willingness/readiness to experience anxiety or difficulty, awareness, confronting the problem, will/effort directed toward change, hope, and social support. Hanna calls these conditions "regulators of change" in the sense that "the more they are present in a person, the more quickly change will occur, and in some cases, the deeper the change will be in the psyche of the person" (p. 6). As clients tell their stories, you can listen and probe for these regulators and even help clients challenge themselves to develop them. Here, then, are the guidelines.

Learn to work with all styles of storytelling. There are both individual and cultural differences in clients' willingness to talk about themselves. Both affect storytelling. Some clients are highly verbal and quite willing to reveal almost everything about themselves at the first sitting. Take the case of Martina.

> Martina, 27, asks a counselor in private practice for an appointment to discuss "a number of issues." Martina is both verbal and willing to talk, and her story comes tumbling out in rich detail. Although the helper uses the skills of attending, listening, sharing highlights, and probing, she does so sparingly. Martina is too eager to tell her story.
>
> Although trained as a nurse, Martina is currently working in her uncle's business because of an offer she "could not turn down." She is doing very well financially, but she feels guilty because service to others has always been a value for her. And although she likes her current job, she also feels hemmed in by it. A year of study in Europe during college whetted her appetite for "adventure." She feels that she is nowhere near fulfilling the great expectations she has for herself.
>
> She also talks about her problems with her family. Her father is dead. She has never gotten along well with her mother, and now that she has moved out of the house, she feels that she has abandoned her younger brother, who is 12 years younger than she is and whom she loves very much. She is afraid that her mother will "smother" her brother with too much maternal care.
>
> Martina is also concerned about her relationship with a man who is 2 years younger than she. They have been involved with each other for about 3 years. He wants to get married, but she feels that she is not ready. She still has "too many things to do" and would like to keep the arrangement they have.
>
> This whole complex story—or at least a synopsis of it—comes tumbling out in a torrent of words. Martina feels free to skip from one topic to another. The way Martina tells her story is part of her enthusiastic style. At one point she stops, smiles, and says, "My, that's quite a bit, isn't it!"

As the helper listens to Martina, he learns a number of things about her. She is young, bright, and verbal and has many resources; she is eager and impatient; some of her problems are probably of her own making; she has some blind spots that could stand in the way of her grappling more creatively with her problems; she has many unexplored options, many unexploited opportunities. That said, the counselor surmises that Martina would probably her way in life, however erratically, with no counseling at all.

Contrast that example with the following one of a man who comes to a local mental health center because he feels he can no longer handle his 9-year-old boy.

> Nick is referred to the center by a doctor in a local clinic because of the trouble he is having with his son. He has been divorced for about 2 years and is living in a housing project on public assistance. After introductions and preliminary formalities have been taken care of, he just sits there and says nothing; he does not even look up. Because Nick offers almost nothing spontaneously, the counselor uses a relatively large number of probes to help him tell his story. Even when the counselor responds with empathic highlights, Nick volunteers very little. Every once in a while, tears well up in his eyes. When asked about the divorce, he says he does not want to talk about it. "Good riddance" is all he can say about his former wife. Gradually, almost

torturously, the story gets pieced together. Nick talks mostly about the "trouble" his son is getting into, how uncontrollable he seems to be getting, and how helpless he [Nick] feels.

Martina's story is full of possibilities, whereas Nick's is mainly about limitations. In both content and communication style, they are at opposite ends of the scale.

Each client is different and will approach the telling of the story in a different way. Some clients will come voluntarily; others will be sent. Some of the stories you will help clients tell will be long and detailed, others short and stark. Some will be filled with emotion; others will be told coldly, even though they are stories of terror. Some stories will be, at least at first blush, single-issue stories—"I want to get rid of these headaches"—whereas others, like Martina's, will be multiple-issue stories. Some stories will deal with the inner world of the client almost exclusively—"I hate myself," "I'm depressed," "I feel lonely"—whereas others will deal with the outer world, for instance, problems with finances, work, or relationships. Still others will deal with a combination of inner and outer concerns.

Some clients will tell the core story immediately, whereas others will tell a secondary story first to test your reactions. Some clients will make it clear that they trust you just because you are a helper, but you will read mistrust in the eyes of others, sometimes just because you are a helper.

In all these cases, your job is to establish a working relationship with your clients and help them tell their stories as a prelude to helping them manage the problems and take advantage of the opportunities buried in those stories. A story that is brought out into the open is the starting point for possible constructive change. Often the very airing of the story is a solid first step toward a better life.

When clients like Martina pour out their stories, you may let them go on or you may insist on some kind of dialogue. If the client tells the "whole" story in a more or less nonstop fashion, it will be impossible for you to share highlights relating to every core issue the client has brought up. But you can then help the client review the most salient points in some orderly way. For example, you might want to help the client review the core parts of the story by saying something like this: "You've said quite a bit. Let's see if I've understood some of the main points you've made. First of all. . . ." At this point the highlights you share will let the client know that you have been listening intently and that you are concerned about him or her. With clients like Nick, however, it's a different story. Those who lack the skills needed to tell their stories well or who are reluctant to do so constitute a different kind of challenge. Engaging in dialogue with them can be tough work. Box 10-1 provides sample questions you can help clients ask themselves to identify problem situations.

Help clients spot unused opportunities. Note that Martina's story is about both problem situations and opportunities. Many stories are a mixture of both. Early in the history of modern psychology, William James remarked that few people bring to bear more than about 10% of their human potential on the problems and challenges of living. Others since James, though changing the percentages somewhat, have said substantially the same thing, and few have challenged their statements. It is probably not an exaggeration to say that unused human potential constitutes a more serious social problem than emotional disorders, because it is more widespread. If this is the case, then most clients you meet will have unused opportunities that can play a

Box 10-1   Problem Finding

Here are some of the kinds of questions counselors can help clients ask themselves to "find" and specify problem situations. You can probably think of others.

- What are my concerns?
- What's problematic in my life?
- What issues do I need to face?
- What's troubling me?
- What would those who know me best say of me?
- What's keeping me back from being what I want to be?
- What keeps me from doing what I want to do?
- What do I need to resolve?

role in helping them manage their problem situation. Pursuing an opportunity can be a way of transcending rather than "solving" a problem.

> Bruno, 23, a graduate student in accounting, came from a strong religious background. He said that he was "obsessed" by sexual thoughts. When he gave into them he felt very guilty. He liked the fact his religion put a great deal of emphasis on helping others. He was pursuing a degree in accounting because he believed that it would help him get a decent job, but he did not find the subject matter intellectually stimulating. Working in a bank and attending school did not give him much time for socializing, but he wasn't a big socializer anyway. Nor did he have time for doing the kind of volunteer work encouraged by his church. He thought he would get to that later. In short, there was a kind of vacuum in his life.
>
> Sensing the vacuum as possible opportunity, the counselor asked Bruno how wide his church's focus on helping others extended. Bruno thought for a moment and then replied, "To the whole world." In the discussion that followed, Bruno realized that he didn't know much about the world. Developing a "sense of the world" made sense to him; it was an opportunity. The counselor doubted that "sexual obsession" was Bruno's main issue. In fact, Bruno set up a program for developing a sense of the world for himself that included reading, an occasional lecture, a course in developing-country finance, and the luxury of a very occasional conference. He made friends with some of the people he met through these activities and eventually met a like-minded young woman whom he began to date. His "obsession" disappeared.

Clients are much more likely to talk about problem situations than about unused opportunities. That's a pity because clients can manage many problems better by developing unused opportunities instead of dealing directly with their problems. Here is one example.

> Lou Anne, a single woman in her mid-20s, was arrested when she broke into a government computer. She was fined and sent to jail, a defeated and demoralized hacker. While in jail, she became friends with Marion, who one day convinced her to go to a religious service. "Just this one time," Lou Anne said. She discovered that the minister was a man with lots of street smarts and "not too pious." She liked the service more than she wanted to admit. After a few weeks, she went back to the minister "for a bit

Box 10-2 Opportunity Finding

Here are some questions counselors can help clients ask themselves to identify unused opportunities.

- What are my unused skills/resources?
- What are my natural talents?
- How could I use some of these?
- What opportunities do I let go by?
- What ambitions remain unfulfilled?
- What could I accomplish if I put my mind to it?
- What could I become good at if I tried?
- Which opportunities should I be developing?
- Which role models could I be emulating?

of advice." After he found out that she had been a hacker, he rolled his eyes and said, "Boy, could we use you." She met with him on and off, and they discussed what she'd do "on the outside."

After she was released from jail, she helped the minister set up a computer training program for disadvantaged kids. The fact that she had been a hacker appealed to the kids. They listened to her. She supported herself by working as a computer security consultant. This kept her up-to-date on the latest hacking techniques, which was still her passion.

How fortunate that the minister saw talent, however misdirected, rather than depravity. Box 10-2 outlines some of the questions you can help clients ask themselves in order to identify unused opportunities.

Start where the client starts. Clients launch into their stories at different starting points. They can start with any stage of the helping process. Join them there. Therefore, "story" is used in its widest sense. It does not mean, narrowly, "This is what happened to me, here's how I reacted, and now this is how I feel." Your job is to stay with your clients no matter where they are, not where you would like them to be. Consider the following:

- Martha starts by saying, "I thought I knew how to handle my son when he reached his teenage years. I knew he might to want to try all sorts of crazy things, so I might have to keep the reins pretty tight. And that's what I did. But now things are awful. It's not working. He's out of control." Her starting point is a *failed solution*, which has spawned a new problem. Her version of problem management went wrong somewhere.

- Thad says, "I don't know whether I want to be a doctor or a politician—or at least a political scientist. I love both, but I can't do both. I mean I have to make my mind up this coming year and choose my college courses. I hate being stuck with a decision." Thad's starting point is *choosing a goal*. He has an approach-approach conflict. He wants both goals.

- Kimberley, a human resources executive for a large company, says, "I've found out that our chief executive has been involved in some unethical and, I think, immoral behavior. He's due to retire within the next 6 months. I don't know whether it's best to bring all this to light or just monitor him till he goes. If I move on him, this could blow up into something big and hurt the reputation of the company itself. If I just monitor him till he goes, he gets away with it. I want to do what's best for the company." Her starting point is a dilemma about which *strategy* to use.

- Owen is having problems sticking to his resolve to restrain himself when one of his neighbors on his block "does something stupid." He says, "I know when I speak up [his euphemism for flying off the handle] things tend to get worse. I know I should leave it to others who are more tactful than I am. But they don't move quickly enough—or forcefully enough." His starting point is difficulty in *implementing* a course of action to which he has committed himself.

Although it is important to start where the client starts, you may have to help any given client "back up" a bit. Take Owen. He is concerned about implementing a strategy. But Owen probably needs to back up and take a look at his style of dealing with people. What gets him going in the first place?

Help clients clarify key issues. To clarify means to discuss problem situations and unused opportunities—including possibilities for the future, goals, strategies for accomplishing goals, plans, implementation issues, and feelings about all of these—as concretely as possible. Vagueness and ambiguity lead nowhere. Clarity means helping clients move from the general to the specific—specific experiences, thoughts, behaviors intentions, points of view, and decisions.

Consider this case. Janice's husband has been suffering from severe depression for over a year. One day, after Janice suffers a fainting spell, she, too, talks with a counselor. At first, feeling guilty about her husband, she is hesitant to discuss her own concerns. In the beginning she says only that her social life is "a bit restricted by my husband's illness." With the help of empathic highlights and probing on the part of the helper, her story emerges. "A bit restricted" turns, bit by bit, into the following, much fuller story. This is a summary. Janice did not say this all at once.

> John has some sort of "general fatigue syndrome" illness that no one has been able to figure out. It's like nothing I've ever seen before. I move from guilt to anger to indifference to hope to despair. I have no social life. Friends avoid us because it is so difficult being with John. I feel I shouldn't leave him, so I stay at home. He's always tired, so we have little interaction. I feel like a prisoner in my own home. Then I think of the burden he's carrying and the roller-coaster emotions start all over again. Sometimes I can't sleep, then I'm as tired as he. He is always saying how hopeless things are and, even though I'm not experiencing what he is, some kind of hopelessness creeps into my bones. I feel that a stronger woman, a more selfless woman, a smarter woman would find ways to deal with all of this. But I end up feeling that I'm not dealing with it at all. From day to day I think I cover most of this up, so that neither John nor the few people who come around see what I'm going through. I'm as alone as he is.

This is the fuller story spelled out in terms of specific experiences, thoughts, behaviors, and feelings. The actions Janice takes—staying at home, covering her feelings up—are part of the problem, not the solution. But now that the story is out in the open, there is some possibility of doing something about it.

In another case, a woman suffering from bulimia is now under psychiatric care, says that she acted "a little erratically at times" with some of her classmates in law school. The counselor, sharing highlights and using probes, helps her tell her story in much greater detail. Like Janice, she does not say all of this at once, but this is the fuller picture of "a little erratic."

> I usually think about myself as plain looking, even though when I take care of myself some say that I don't look that bad. Ever since I was a teenager, I've preferred to go it alone, because it was safer. No fuss, no muss, and especially no rejection. In law school, right from the beginning I entertained romantic fantasies about some of my classmates who I didn't think would give me a second look. I pretended to have meals with those who attracted me and then I'd have fantasies of having sex with them. Then I'd purge, getting rid of the fat I got from eating and getting rid of the guilt. But all of this didn't just stay in my head. I'd go out of my way to run into my latest imagined partner in school. And then I'd be rude to him to "get back at him" for what he did to me. That was my way of getting rid of him.

She was not really delusional, but gradually her external behavior with a kind of twisted logic began to reinforce her internal fantasies. However, once her story became "public"—that is, once she began talking about it openly with her helper—she began to take back control of her life.

Help clients discuss the context of their concerns. Sometimes helping clients explore the background or context of the concern they bring up helps clarify things. This is a further application of the people-in-systems approach mentioned earlier (Conyne & Cook, 2004; Hutchison, 2003). Consider the following case.

> In a management development seminar, Tarik tells his counselor that he is a manager in a consulting firm. The firm is global, and he works in one of its offices in Southeast Asia. He says that he is already overworked, but now his new boss wants him to serve on a number of committees that will take away even more of his precious time. He is also having trouble with one of his subordinates, himself a manager, who Tarik says is undermining his authority in the wider team.

So far we have a garden-variety story, one that could be repeated thousands of time throughout the world.

The counselor, however, suspects there is more to this. Because the counselor is Canadian, he wants to learn about Tarik's Middle Eastern culture to get the full picture. He knows that there is an overlay of Western culture in these consulting firms, but he wants to deal with his client as a full person. And so in sharing highlights and by using a few probes, he learns enough about the background of the manager's story to cast a new light on the problem situation. Here is the fuller story that emerges.

> Tarik is not only a manager in the firm, but also a partner. However, he is a newly minted partner. The structure in these firms is relatively flat, but the culture is quite hierarchical. And so the clients he has been given to work with are, in large part, the "dogs" of the region. His boss is not only an American, but he has been there for only 4 months and, Tarik has heard, is going to stay for only 1 more year. Because the boss is near retirement, this is his "fling" in Asia. Though a decent man, he is quite distant and offers Tarik little help. This leads Tarik to believe that his real boss is his boss's boss, whom he can't approach because of company and cultural protocol. The subordinate who is giving Tarik trouble is also a partner. In fact, he has been a partner for several years, but has not been very successful. This man thought that he should have been made the manager of the unit Tarik is now running. He has been engaging in a bit of sabotage behind Tarik's back.

A search for some background quickly takes the client's story out of the "routine" category. Of course, you should not be looking for background just for the sake of looking. The right kind and amount of background provides both richness and context.

Assess the severity of the client's problems. Clients come to helpers with problems of every degree of severity. James Hicks (2005) has written an awarding-winning book in clear, straightforward language on the most common signs of mental illness. He provides a service that most books on abnormal psychology do not.

Objectively, problems run from the inconsequential to the life-threatening. Subjectively, however, a client can experience even a relatively inconsequential problem as severe. If a client thinks that a problem is critical, even though by objective standards the problem does not seem to be that bad, then for him or her it is critical. In such a case, the client's tendency to "catastrophize"—to judge a problem situation to be more severe than it actually is— itself becomes an important part of the problem situation. One of the counselor's tasks in this case will be to help the client put the problem in perspective or to teach him or her how to distinguish between degrees of problem severity. Howard (1991, p. 194) put it well:

> In the course of telling the story of his or her problem, the client provides the therapist with a rough idea of his or her orientation toward life, his or her plans, goals, ambitions, and some idea of the events and pressures surrounding the particular presenting problem. Over time, the therapist must decide whether this problem represents a minor deviation from an otherwise healthy life story. Is this a normal, developmentally appropriate adjustment issue? Or does the therapist detect signs of more thorough-going problems in the client's life story? Will therapy play a minor, supportive role to an individual experiencing a low point in his or her life course? If so, the orientation and major themes of the life will be largely unchanged in the therapy experience. But if the trajectory of the life story is problematic in some fundamental way, then more serious, long-term story repair might be indicated. So, from this perspective, part of the work between client and therapist can be seen as life-story elaboration, adjustment, or repair.

Savvy therapists not only gain an understanding of the severity of a client's problem or the extent of the client's unused resources, but also understand the limits of helping. What has been this client's highest lifetime level of functioning? What, then, are appropriate expectations? What Howard calls life-story adjustment or repair is not the same as attempting to redo the client's personality.

Years ago Mehrabian and Reed (1969) suggested the following formula as a way of determining the severity of any given problem situation. It is still useful today.

$$\text{Severity} = \text{Distress} \times \text{Uncontrollability} \times \text{Frequency}$$

The multiplication signs in the formula indicate that these factors are not just additive. Even low-level anxiety, if it is uncontrollable or persistent, can constitute a severe problem; that is, it can severely interfere with the quality of a client's life. In some cases, assessing for possible self-harm or harm to others is called for. The literature on suicide details possible self-harm signs that you need to look out for. You can also review the literature on violence in social relationships for signs of possible impending aggression. Even a casual reading of accounts of outbreaks of social violence tells as how easily signs can be missed.

Because clients' stories often unfold over time, assessment is ongoing. But this is true even in medical practice.

Help clients talk productively about the past. Some schools of psychology suggest that problem situations are not clear and fully comprehended until they are understood in the context of their historic roots. Therefore, helpers in these schools spend a great deal of time helping clients uncover the past. Others disagree with that point of view. Glasser (2000, p. 23) puts it this way: "Although many of us have been traumatized in the past, we are not the victims of our past unless we presently choose to be. The solution to our problem is rarely found in explorations of the past unless the focus is on past successes."

Fish (1995) suggested that attempts to discover the hidden root causes of current problem behavior may be unnecessary, misguided, or even counterproductive. Constructive change does not depend on causal connections in the past. There is evidence to support Fish's contention. Long ago Deutsch (1954) noted that it is often almost impossible, even in carefully controlled laboratory situations, to determine whether event B, which follows event A in time, is actually caused by event A. Trying to connect present complicated patterns of current behavior with complicated events that took place in the past is an exercise in frustration.

Therefore, asking clients to come up with causal connections between current unproductive behavior and past events could be an exercise in futility for a number of reasons. First, causal connections cannot be proved; they remain hypothetical. Second, there is little evidence suggesting that understanding past behavior causes changes in present behavior. Third, talking about the past often focuses mostly on what happened to clients (their experiences) rather than on what they did about what happened to them (their thoughts, intentions, decisions, and behaviors) and therefore interferes with the "bias toward action" clients need to manage current problems.

This is not to say that a person's past does not influence current behavior. Nor does it imply that a client's past should not be discussed. Quite often being stuck in the past is a function of rejection and resentment. But the fact that past experiences may well *influence* current behavior does not mean that they necessarily *determine* present behavior. Kagan (1996) has challenged what may be called the "scarred for life" assumption: "If orphans who spent their first years in a Nazi concentration camp can become productive adults and if young children made homeless by war can learn adaptive strategies after being adopted by nurturing families" (p. 901), that means that there is hope for us all. As you can imagine, this is one of those issues that members of the helping professions argue about endlessly. Therefore, this is not a debate that is to be settled with a few words here. Here are some suggestions for helping clients talk meaningfully and productively about the past.

Help clients talk about the past to make sense of the present. Many clients come expecting to talk about the past or wanting to talk about the past. There are ways of talking about the past that help clients make sense of the present. But making sense of the present needs to remain center stage. Thus, *how* the past is discussed is more important than whether it is discussed. The following man has been discussing how his interpersonal style gets him into trouble. His father, now dead, played a key role in the development of his son's style.

HELPER: So your father's unproductive interpersonal style is, in some ways, alive and well in you.

CLIENT: Until we began talking I had no idea about how alive and well it is. For instance, even though I hated his cruelty, it lives on in me in much smaller ways. He beat my brother. But now I just cut him down to size verbally. He told my mother what she could do and couldn't do. I try to get my mother to adopt my "reasonable" proposals "for her own good"—without, of course listening very carefully to her point of view. There's a whole pattern that I haven't noticed. I've inherited more than his genes.

HELPER: That's quite an inheritance. . . . But now what?

CLIENT: Well, now that I see what's happened, I'd like to change things. A lot of this is ingrained in me, but I don't think it's genetic in any scientific sense. I've developed a lot of bad habits.

It really does not make any difference whether the client's behavior has been "caused" by his father or not. In fact, by hooking the present into the past, if he feels in some way that his current nasty style is not his fault, then he has a new problem. Helping is about the future. Now that the problem has been named and is out in the open, it is possible to do something about it. It is about "bad habits," not sociobiological determinism.

Help clients talk about the past to be reconciled to or liberated from it. A potentially dangerous logic can underlie discussions of the past. It goes something like this: "I am what I am today because of my past. But I cannot change my past. So how can I be expected to change today?"

CLIENT: I was all right until I was about 13. I began to dislike myself as a teenager. I hated all the changes—the awkwardness, the different emotions, having to be as "cool" as my friends. I was so impressionable. I began to think that life actually must get worse and worse instead of better and better. I just got locked into that way of thinking. That's the same mess I'm in today.

That is not liberation talk. The past is still casting its spell. Helpers need to understand that clients may see themselves as prisoners of their past, but then, in the spirit of Kagan's earlier comments, help them move beyond such self-defeating beliefs.

The following case provides a different perspective. It is about the father of a boy who has been sexually abused by a minister of their church. He finds that he can't deal with his son's ordeal without revealing his own abuse by his father. In a tearful session he tells the whole story. In a second interview he has this to say:

CLIENT: Someone said that good things can come from evil things. What happened to my son was evil. But we'll give him all the support he needs to get through this. Though I had the same thing happen to me, I kept it all in until now. It was all locked up inside. I was so ashamed, and my shame became part of me. When I let it all out last week, it was like throwing off a dirty cloak that I'd been wearing for years. Getting it out was so painful, but now I feel so different, so good. I wonder why I had to hold it in for so long.

This is liberation talk. When counselors help or encourage clients to talk about the past, they should have a clear idea of what their objective is. Is it to learn from the past? Is it to be liberated from it? To assume that there is some "silver bullet" in the past that will solve today's problem is probably asking too much of the past.

Help clients talk about the past in order to prepare for action in the future. The well-known historian A. J. Toynbee had this to say about history: "History not used is nothing, for all intellectual life is action, like practical life, and if you don't use the stuff—well, it might as well be dead." As we will soon see, any discussion of problems or opportunities should lead to constructive action, starting with Stage I and going all the way through to implementation. The insights you help clients get from the past should in some way stir them to action. When one client, Christopher, realized how much his father and one of his high school teachers had done to make him feel inadequate, he made this resolve: "I'm not going to do anything to demean anyone around me. You know, up to now I think I have, but I called it something else—wit. I thought I was being funny when I was actually being mean." Help clients invest the past proactively in the future.

This does not mean that you should never let clients talk about the past unless they do it in a way that fits the guidelines reviewed here. Short (2006) puts it well: "I don't like to focus on the past or on a person's symptoms, but people shouldn't suffer alone, so I listen to these stories with respect and acceptance" (p. 72). To do otherwise would contribute to what Hansen (2005) calls the "devaluation of [clients'] inner subjective experiences by the counseling profession" (p. 83).

As clients tell their stories, search for resources, especially unused resources. Incompetent helpers concentrate on clients' deficits. Skilled helpers, as they listen to and observe clients, do not blind themselves to deficits, but they are quick to spot clients' resources, whether used, unused, or even abused. These resources can become the building blocks for the future.

Consider this example: Terry, a young woman in her late teens who has been arrested several times for street prostitution, is an involuntary, or "mandated," client. The charge this time is possession of drugs. She ran away from home when she was 16 and is now living in a large city on the East Coast. Like many other runaways, she was seduced into prostitution "by a halfway decent pimp." Now she is very street-smart. She is also cynical about herself, her occupation, and the world. She is forced to see a counselor as part of the probation process. As might be expected, Terry is quite hostile during the interview. She has seen other counselors and sees the interview as a game. The counselor already knows a great deal of the story because of the court record. The dialogue is not easy. Some of it goes like this:

TERRY: If you think I'm going to talk with you, you've got another think coming. What I do is my business.

COUNSELOR: You don't have to talk about what you do outside. We could talk about what we're doing here in this meeting.

TERRY: I'm here because I have to be. You're here either because you're dumb and like to do stupid things or because you couldn't get a better job. You people are no better than the people on the street. You're just more "respectable."

COUNSELOR: So nobody could really be interested in you.

TERRY: I'm not even interested in me!

COUNSELOR: So, if you're not interested in yourself, then no one else could be, including me.

Terry has obvious deficits. She is engaged in a dangerous and self-defeating lifestyle. But as the counselor listens to Terry, he spots many different resources. Terry is a tough, street-smart woman. The very virulence of her cynicism and self-hate, the very strength of her resistance to help, and her almost unchallengeable determination to go it alone are all signs of resources. Many of her resources are currently being used in self-defeating ways. They are misused resources, but they are resources nevertheless.

Helpers need a resource-oriented mind-set in all their interactions with clients. Contrast two different approaches.

CLIENT: I practically never stand up for my rights. If I disagree with what anyone is saying—especially in a group—I keep my mouth shut.

COUNSELOR A: So clamming up is the best policy. . . . What happens when you do speak up?

CLIENT (pausing): I suppose that on the rare occasions when I do speak up, the world doesn't fall in on me. Sometimes others do actually listen to me. But I still don't seem to have much impact on anyone.

COUNSELOR A: So speaking up, even when it's safe, doesn't get you much.

CLIENT: No, it doesn't.

Counselor A, sticking to sharing highlights, misses the resource mentioned by the client. Although it is true that the client habitually fails to speak up, he has some impact when he does speak. Others do listen, at least sometimes, and this is a resource. Counselor A emphasizes the deficit. Let's try another counselor.

COUNSELOR B: So when you do speak up, you don't get blasted, you even get a hearing. Tell me what makes you think you don't exercise much influence when you speak?

CLIENT (pauses): Well, maybe influence isn't the issue. Usually I don't want to get involved. Speaking up gets you involved.

Note that both counselors share a highlight, but they focus on different parts of the client's message. Counselor A emphasizes the deficit; Counselor B notes an asset and follows up with a probe. This produces a significant clarification of the client's problem situation. Now not wanting to get involved is an issue to be explored.

The search for resources is especially important when the story being told is bleak. I once listened to a man's story that included a number of bone-jarring life defeats—a bitter divorce, false accusations that led to his dismissal from a job, months of unemployment, serious health concerns, months of homelessness, and more. The only emotion the man exhibited during his story was depression. Toward the end of the session, we had this interchange:

HELPER: Just one blow after another, grinding you down.

CLIENT: Grinding me down, and almost doing me in.

HELPER: Tell me a little more about the "almost" part of it.

CLIENT: Well, I'm still alive, still sitting here talking to you.

HELPER: Despite all these blows, you haven't fallen apart. That seems to say something about the fiber in you.

At the word *fiber,* the client looked up, and there seemed to be a glimmer of something besides depression in his face. I put a line down the center of a newsprint pad. On the left side I listed the life blows the man had experienced. On the right I put "Fiber." Then I said, "Let's see if we can add to the list on the right side." We came up with a list of the man's resources. His "fiber" included his musical talent, his honesty, his concern for and ability to talk to others, and so forth. After about a half hour, he smiled weakly, but he did smile. He said that it was the first time he could remember smiling in months.

Help clients see every problem as an opportunity. Clients don't come with just problems *or* opportunities. They come with a mixture of both. Although there is no justification for romanticizing pain, the flip side of a problem is an opportunity. Here are a few examples.

- Kevin used his diagnosis of AIDS as a starting point for reintegrating himself into his extended family and challenging the members of his family to come to grips with some of their own problems, problems they had been denying for a long time.

- Beatrice used her divorce as an opportunity to develop a new approach to men based on mutuality. Because she was on her own and had to make her own way, she discovered that she had entrepreneurial skills. She started an arts and crafts company.

- Jerome, after an accident, used a long convalescence period to review and reset some of his values and life goals. He began to visit other patients in the rehabilitation center. This gave him deep satisfaction. He began to explore opportunities in the helping professions.

- Sheila used her incarceration for shoplifting—she called it "time out"—as an opportunity to finish her high school degree and get a head start on college.

- An actor suffering from a traumatic disability found new life by becoming a public advocate for those suffering a similar fate.

- A couple mourning the death of their only child started a day-care center in conjunction with other members of their church.

William C. Miller (1986) talked about one of the worst days of his life. Everything was going wrong at work. Projects were not working out, people were not responding, the work overload was bad and getting worse—nothing but failure all around. Later that day, over a cup of coffee, he took some paper, put the title "Lessons Learned and Relearned" at the top, and wrote down as many entries as he could. Some hours and seven pages later, he had listed 27 lessons. The day turned out to be one of the best of his life. So he began to keep a daily "Lessons Learned" journal. It helped him avoid getting caught up in self-blame and defeatism. Subsequently, on days when things were not working out, he would say to himself, "Ah, this will be a day filled with learnings!" Sometimes helping a client spot a small opportunity, be it the flip side of a problem or a standalone, provides enough positive-psychology leverage to put him or her on a more constructive tack.

In telling their stories, clients can move forward by "reauthoring" them, reinterpreting them, learning from them, and moving beyond them (Anderson & Goolishian, 1992; Angus & McLeod, 2004; Epston, White, & Murray, 1992).

Action Right From the Beginning

Shakespeare, in the person of Hamlet, talks about important enterprises—what he calls "enterprises of great pith and moment"—losing "the name of action." Helping, which is an enterprise "of great pith and moment," can lose the name of action. One of the principal reasons clients do not manage the problem situations of their lives effectively is their failure to act intelligently, forcefully, and prudently in their own best interests. A strength of positive psychology is its focus on personal agency and self-regulation (Bandura 2001; Caprara & Cervone, 2003), called a "bias toward action" in Chapter 3. Some of the language used—for instance, describing personality as "agentic"—is a bit daunting, but the underlying concept hits the target.

The importance of being proactive. Inactivity can be bad for body, mind, and spirit. Consider the following workplace example.

> A counselor at a large manufacturing concern realized that inactivity did not benefit injured workers. If they stayed at home, they tended to sit around, gain weight, lose muscle tone, and suffer from a range of psychological symptoms such as psychosomatic complaints unrelated to their injuries. Taking a people-in-systems approach (Egan & Cowan, 1979), she worked with management, the unions, and doctors to design temporary, physically light jobs for injured workers. In some cases, nurses or physical therapists visited these workers on the job. Counseling sessions helped to get the right worker into the right job. The workers, active again, felt better about themselves. Both they and the company benefitted from the program.

Counselors add value by helping their clients become proactive. Helping too often entails too much talking and too little action.

Here is a brief overview of a case that illustrates almost the magic of action. Marcus's "helpers" are his sister-in-law and a friend.

> Life was ganging up on Marcus. Only recently "retired" from his job as office manager of a brokerage firm because the managing partner did not think that he would fit into the new e-commerce strategy of the business, he discovered that he had a form of cancer, the course of which was unpredictable. He had started treatment and was relatively pain free, but he was always tired. He was also beginning to have trouble walking, but the medical specialists did not know why. It may or may not have been related to the cancer. Healthwise, his future was uncertain.
>
> After losing his job, he took a marketing position with a software company. But soon health problems forced him to give it up. Around the same time his daughter was expelled from high school for using drugs. She was sullen and uncommunicative and seemed indifferent to her father's plight. Marcus's wife and daughter did not get along well at all, and the home atmosphere was tense whenever they were together. Discussions with his wife about their daughter went nowhere. To top things off, one of his two married sons announced that he was getting a divorce form his wife and was going to fight for custody of their 3-year-old son.
>
> Given all these concerns, Marcus, though very independent and self-reliant, more or less said to himself, "I've got to take charge of my life or it could fall apart. And I could use some help." He found help in two people—Sarah, an intelligent, savvy, no-nonsense sister-in-law who was a lay minister of their church and, Sam, a friend of many years who was a doctor in family practice.
>
> Sam helped Marcus maintain his self-reliance by helping him find out as much as he could about his illness through the Internet. Without becoming preoccupied with his search, Marcus learned enough about his illness and possible treatments to partner with his doctors in choosing therapeutic interventions. He also drew up a living will and gave Sam power of attorney in order to avoid needlessly prolonging his life at the end. Sam also helped him pursue some intellectual interests—art, literature, theater—that Marcus hadn't had time for because of business.

Here's a man who, with some help from his friends, uses adversity as an opportunity to get more fully involved with life. He redefines life, in part, as good conversations with family and friends. Refusing to become a victim, he stays active. Marcus did die 2 years later, but he managed to live until he died.

Using the time between sessions productively. If helping proves to be short-term for whatever reason, then counselors must help clients use the time between sessions as productively as possible. "How can I leverage what I do within the session to have an impact on what the client does the rest of the week?" Or month, as the case may be. This does not deny that good things are happening within the session. On the contrary, it capitalizes on whatever learning or change takes place there. Wosket (2006; see Chapter 2), at the end of each session, asks the client to describe what has happened in the session and where they are. Then she asks the client if there is anything they could do between sessions to move things along before the next session. At the beginning of the next session, she asks the client to review what he or she did between sessions and what impact it had. If the client has done nothing, she asks, nonjudgmentally, "If you had thought of acting, what might you have done?" This reinforces the necessity-for-action theme.

Sixty-eight percent of practicing psychologists frequently use between-session assignments with their clients (Kazantzis & Deane, 2005) as a way of helping clients act on what they're learning in the sessions (Detweiler-Bedell & Whisman, 2005; Kazantzis, 2000; Scheel, Hanson, & Razzhavaikina, 2004; Tompkins, 2005; also, much of the May 2002, Vol. 58, *Journal of Clinical Psychology* was devoted to articles dealing with homework). Mahrer and his associates (1994) reviewed the methods helpers use to do this. They came up with 16 methods. Among them are the following:

- Mention some homework task, and ask the client to carry it out and report back because he or she is now a "new person."
- Wait until the client comes up with a postsession task, and then help the client clarify and focus it.
- Highlight the client's readiness or seeming willingness to carry out some task, but leave the final decision to the client.
- Use some contractual agreement to move the client to some appropriate activity.

Counselors provide help in defining the activity and custom fitting it to the client's situation. Some helpers exhort clients to carry out some action. Some even mandate some kind of action as a condition for further sessions—"I'll see you after you've attended your first Alcoholics Anonymous meeting."

For further suggestions on how to incorporate homework into your helping sessions, see Broder (2000). He not only offers many techniques but discusses why some clients resist the idea of homework and suggest ways of addressing such resistance. For instance, in conjunction with Albert Ellis, the originator of Rational Emotive Behavior Therapy, Broder has developed a series of audiotapes dealing with many of the problems clients encounter. Clients use these tapes to further their understanding of what they learned in the therapy session and to engage in activities that will help them resolve their conflicts (see http://www.therapistassistant.com).

There is no need to call what the client does between sessions homework. The term *homework* puts some clients off. It is also too "teacherish" for some helpers.

It sounds like an add-on rather than something that flows organically from what takes place within the helping sessions. But *flows* is a wide term. The principle behind homework is more important than the name, and the principle is clear. Use every stage and every task of the helping process as a stimulus for problem-managing and opportunity-developing action. Use the term *homework* if it works for you. "Assign" it if this works for you and your client. But have a clear picture of why you are assigning any particular task. Don't routinely assign homework for its own sake.

Homework often has a predetermined cast to it. But encouraging clients to act in their own behalf can be a much more spontaneous exercise. Take the case of Mildred.

> Mildred, 70, single, was a retired teacher. In fact, she had managed to find ways of teaching even after the mandatory retirement age. But she was ultimately "eased out." Given her savings, her pension, and social security, she was ready for retirement financially. But she had not prepared socially or emotionally. She was soon depressed and described life as "aimless." She wandered around for a few months and finally, at the urging of a friend, saw a counselor, someone about her own age. The counselor knew that Mildred had many internal resources, but they had all seemed to go dormant. In the second session, he said to her, "Mildred, you're allowing yourself to become a wreck. You'd never stand for this kind of behavior from your students. You need to get off your butt and seize life again. Tell me what you're going to do. I mean right after you get out of this session with me."
> This was a bit of shock treatment for Mildred. A revived Mildred looked at him and asked, "Well, what do you do?" They went on to discuss what life can offer after 70. Her first task, she decided, was to get into community in some way. The school had been her community, but it was gone. She went to the next counseling session with some options.

You can use homework assignments or find other, perhaps more organic, ways of helping clients move to action. Your role demands that you be a catalyst for client action. Box 10-3 provides some guidelines for doing the work of Task 1.

TASK 2: HELP CLIENTS DEVELOP NEW PERSPECTIVES AND REFRAME THEIR STORIES—"WHAT AM I OVERLOOKING OR AVOIDING?"

Often enough counselors need to use the challenging skills outlined in Chapters 7 and 8 to help clients follow the principles outlined in Task 1. When clients present themselves to helpers, there is no instant or easy way of reading what is in their hearts. This is revealed over the course of helping sessions. And some helpers are much better than others in discovering "what is really going on." As we have seen, clients approach storytelling in quite different ways. Let's consider a couple of shadow-side versions.

Help clients paint an objective picture. Clients who tell stories that are general, partial, and ambiguous may or may not have ulterior motives. For instance, one subtext in the shadows is, "If I tell my story too clearly and reveal myself warts and all, then I will be expected to do something about it." The accountability issue lurks in the background. Another issue is the accuracy of the story. At one end are clients who tell their stories as honestly as possible. At the other end are clients who, for whatever reason, lie. Or fudge. Anyone who has done any marital counseling doesn't have to prove this. They just watch it.

Box 10-3   Guidelines for Task 1

Establish a Working Alliance

- Develop a collaborative working relationship with the client.
- Use the relationship as a vehicle for social-emotional reeducation.
- Don't do for clients what they can do for themselves.

Help Clients Tell Their Stories

- Use a mix of tuning in, listening, empathy, probing, and summarizing, to help clients tell their stories, share their points of view, discuss their decisions, and talk through their proposals as concretely as possible.
- Use probes when clients get stuck, wander about, or lack clarity.
- Understand blocks to client self-disclosure and providing support for clients having difficulty talking about themselves.
- Help clients talk productively about the past.

Build Ongoing Client Assessment Into the Helping Process

- Get an initial feel for the severity of a client's problems and his or her ability to handle them.
- Note client resources, especially unused resources, and help the client work with these resources.
- Understand clients' problems and opportunities in the larger context of their lives.

Help Clients Move to Action

- Help clients develop an action orientation.
- Help clients spot early opportunities for changing self-defeating behavior or engaging in opportunity-development behavior.
- Find innovative forms of homework to help clients engage in problem-managing action.

Integrate Evaluation Into the Helping Process

- Keep an evaluative eye on the entire process with the goal of adding value through each interaction and making each session better.
- Find ways of getting clients to participate in and own the evaluation process.

Fudging seems to have something to do with self-image. Some clients are not especially concerned about what their helpers think of them. They have no particular need to be seen in a favorable light. Other clients ask themselves, at least subconsciously, "What will the helper think of me?" Some are extremely concerned about what their helpers might think of them and therefore skew their stories to present themselves in the best light. A. E. Kelly (2000b, 2002; also see Kelly, Kahn, & Coulter, 1996) sees therapy, at least in part, as a self-presentational process. She suggests that clients benefit by perceiving that their therapists have favorable views of them. Therefore, if therapy contributes to clients' positive identity development, then we should expect some fudging. Hiding some of the less desirable aspects of themselves, intentionally or otherwise (see Kelly, 2000a), becomes a means to an end. Hill, Gelso, and Mohr (2000) object to this hypothesis, suggesting that research shows that clients don't hide much from their therapists. Arkin and Hermann (2000) suggest that all of this is really more complicated than the others realize. Well.

Take an example. Relatively few male victims of childhood sexual abuse discuss the residue of that in adulthood (Holmes, Offen, & Waller, 1997). This used to be explained by the now-discounted myths that few males are sexually abused and that abuse has little impact on males. Rather, it seems that it is just harder for males, at least in North American society, to admit both the abuse and its effects. We know that some clients fudge and some clients lie. We can only hypothesize why. But we do know that the vagaries of client self-presentation muddy the waters.

Invite clients to own their problems and unused opportunities. It is all too common for clients to refuse to take responsibility for their problems and unused opportunities. Instead, there is a whole list of outside forces and other people who are to blame. Therefore, clients need to challenge themselves or be challenged to own the problem situation. Here is the experience of one counselor who had responsibility for about 150 young men in a youth prison within the confines of a larger central prison.

> I believe I interviewed each of the inmates. And I did learn from them. What I learned had little resemblance to what I had found when I read their files containing personal histories, including the description of their crimes. What I learned when I talked with them was that they didn't belong there. With almost universal consistency they reported a "reason" for their incarceration that had little to do with their own behavior. An inmate explained with perfect sincerity that he was there because the judge who sentenced him had also previously sentenced his brother, who looked very similar; the moment this inmate walked into the courtroom he knew he would be sentenced. Another explained with equal sincerity that he was in prison because his court-appointed lawyer wasn't really interested in helping him. (L. M. Miller, 1984, pp. 67–68)

This is perhaps an extreme form of a common phenomenon. But we don't have to go behind prison walls to find this lack of ownership. Lack of ownership is common.

Take the case of a client who feels that her business partner has been pulling a fast one on her. He's made a deal on his own. She's alarmed, but she hasn't done anything about it so far. Let's say that three different helpers, A, B, and C share a highlight. Consider how they differ.

HIGHLIGHT A: You feel angry because he unilaterally made the decision to close the deal on his terms.

HIGHLIGHT B: You're angry because your legitimate interests were ignored.

HIGHLIGHT C: You're furious because you were ignored, your interests were not taken into consideration, maybe you were even financially victimized, and you let him get away with it.

Challenging clients in terms of ownership means helping clients understand that in some situations they may have some responsibility for creating or at least perpetuating their problem situations. Statement C does precisely that—"You let him get away with it."

Not only problems but also opportunities need to be seized and owned by clients. As Wheeler and Janis (1980) noted, "Opportunities usually do not knock very loudly, and missing a golden opportunity can be just as unfortunate as missing a red-alert warning" (p. 18). Consider Tess and her brother, Josh.

Tess and Josh's father died years ago. Their mother died about a year ago. She left a small country cottage to both of them. They have been fighting over its use. Some of the fighting has been quite bitter. They have never been that close, but up till now they have just had squabbles, not all-out war. Without admitting it, Tess has been shocked by her own angry and bellicose behavior. But not shocked enough to do anything about it.

Then Josh has a heart attack. Tess knows that this is an opportunity to do something about their relationship. But she keeps putting it off. She even realizes that the longer she puts it off, the harder it will be to do something about it. In a counseling session, she says:

TESS: I thought that this was going to be our chance to patch things up, but he hasn't said anything.

COUNSELOR: So, nothing from his camp. . . . What about yours?

TESS: I think it might already be too late. We're falling right back into our old patterns.

COUNSELOR: I didn't think that's what you wanted.

TESS (angrily): Of course that's not what I wanted!. . . That's the way it is.

COUNSELOR: Tess, if someone put a gun to your head and said, "Make this work or I'll shoot," what would you do?

TESS (after a long pause): You mean it's up to me. . . .

COUNSELOR: If you mean that I'm assigning this to you as a task, then no. If you mean that you can still seize the opportunity no matter what Josh does, well. . . .

TESS: I think I'm really angry with myself. . . . I know deep down it's up to me . . . no matter what Josh does. And I keep putting it off.

Josh, of course, is not in the room. So it's about Tess's ownership of the opportunity. And she senses, now more deeply, that "missing a golden opportunity can be just as unfortunate as missing a red-alert warning."

Invite clients to state their problems as solvable. Jay Haley (1976, p. 9) said that if "therapy is to end properly, it must begin properly—by negotiating a solvable problem." Or exploring a realistic opportunity, someone might add. It is not uncommon for clients to state problems as unsolvable. This justifies a "poor-me" attitude and a failure to act.

UNSOLVABLE PROBLEM: In sum, my life is miserable now because of my past. My parents were indifferent to me and at times even unjustly hostile. If only they had been more loving, I wouldn't be in this mess. I am the failed product of an unhappy environment.

Of course, clients will not use this rather stilted language, but the message is common enough. The point is that the past cannot be changed. As we have seen, clients can change their attitudes about the past and deal with the present consequences of the past. Therefore, when a client defines the problem exclusively as a result of the past, the problem cannot be solved. "You certainly had it rough in the past and are still suffering from the consequences now" might be the kind of response that such a statement is designed to elicit. The client needs to move beyond such a point of view.

A solvable or manageable problem is one that clients can do something about. Consider a different version of the foregoing unsolvable problem.

SOLVABLE PROBLEM: Over the years I've been blaming my parents for my misery. I still spend a great deal of time feeling sorry for myself. As a result, I sit around and do nothing. I don't make friends, I don't involve myself in the community, I don't take any constructive steps to get a decent job.

This message is quite different from that of the previous client. The problem is now open to being managed because it is stated almost entirely as something the client does or fails to do. The client can stop wasting her time blaming her parents, because she cannot change them; she can increase her self-esteem through constructive action and therefore stop feeling sorry for herself; and she can develop the interpersonal skills and courage she needs to enter more creatively into relationships with others.

This does not mean that all problems are solvable by the client's direct action. A teenager may be miserable because his self-centered parents are constantly squabbling and seem indifferent to him. He certainly can't solve the problem by making them less self-centered, stopping them from fighting, or getting them to care for him more. But he can be helped to find ways to cope with his home situation more effectively by developing fuller social opportunities outside the home. This could mean helping him develop new perspectives on himself and family life and challenging him to act both internally and externally in his own behalf.

Invite clients to explore their "problem-maintenance structure." Pinsoff (1995) points out how important it is to explore with clients the "actions, biology, cognitions, emotions, object relations, and self-structures" (p. 7) that keep them mired in their problems. What he calls the "problem-maintenance structure" (p. 7) refers to the set of factors—including personal, social, organizational, community, and political factors—that keeps clients from identifying, exploring, and ultimately doing something about their problem situations and unused opportunities. "You say that you don't like the way you relate to people. Well, what is it that you don't like? And what is it that keeps that style in place? What is it that prevents you from doing something about it?"

Augustine, an Egyptian immigrant, rejoiced the day he got his green card. But the next 2 years were a letdown. He lived aimlessly, wandering from place to place and from one part-time job to another. All the great plans he had for his life "in the land of opportunity" disappeared. The lifestyle that he had let himself fall into left him miserable. He finally saw a counselor who helped him explore not only what had happened, but also the factors that kept him locked into a lifestyle he pretended to like but really hated. This forced him to look at personal, social, and cultural

Box 10-4   From Blind Spots to New Perspectives

These are the kinds of things clients will be able to say to themselves when you help them develop new perspectives.

- Here's a new angle. . . .
- Here's something I've not thought of. . . .
- Here's something I've overlooked. . . .
- To be completely honest. . . .
- Here's one way I've been fooling myself. . . .
- Here's an important piece of the puzzle. . . .
- Here's the real story. . . .
- Here's the complete story. . . .

factors he had been avoiding. An intelligent and personable university graduate, Augustine faced all sorts of unused opportunities. The work of helping clients "restructure" self-defeating defenses, cognitions, patterns of emotional expression, self-focused and outward-focused behavior, human relationships, and approaches to the environment (see Magnavita, 2005, Chapters 7–10) begins in Stage I.

Invite clients to move on to the next needed stage of the helping process. We have touched on this already in discussing probing. There is no reason to keep going over the same issue with clients. You can help clients challenge themselves to do the following:

- clarify problem situations by describing specific experiences, behaviors, and feelings when they are being vague or evasive
- talk about issues—problems, opportunities, goals, commitment, strategies, plans, actions—when they are reluctant to do so
- develop new perspectives on themselves, others, and the world when they prefer to cling to distortions
- review possibilities, critique them, develop goals, and commit themselves to reasonable agendas when they would rather continue wallowing in their problems
- search for ways of getting what they want, instead of just talking about what they would prefer
- spell out specific plans instead of taking a scattered, hit-or-miss approach to change
- persevere in the implementation of plans when they are tempted to give up
- review what is and what is not working in their pursuit of change "out there"

In sum, counselors can help clients challenge themselves to engage more effectively in all the stages and tasks of problem management during the sessions themselves and in the changes they are pursuing in everyday life. Box 10-4 outlines the kinds of things clients can say when you challenge them to develop new perspectives.

TASK 3: HELP CLIENTS WORK ON ISSUES THAT MAKE A DIFFERENCE—"WHAT PROBLEMS, IF MANAGED, WILL HELP ME MOST?"

Helping is expensive both financially and psychologically. It should not be undertaken lightly. Therefore, a word is in order about the what might be called the "economics" of helping. The term *leverage* is used to introduce the economics of helping. How can we help our clients get the most out of the helping process? Helpers need to ask themselves, "Am I adding value through each of my interactions with this client?" Clients need to be helped to ask themselves, "Am I working on the right things? Am I spending my time well in these sessions and between sessions?" The question here is not "Does helping help?" but "Is helping working in this situation? Is it worth it?"

It's important that helpers determine whether the client is ready to invest in constructive change or not. And to what degree. Change requires work on the part of clients. If they do not have the incentives to do the work, they might begin and then trail off. If this happens, it's a waste of resources. For instance, Helmut and Gretchen are mildly dissatisfied with their marriage and seem to be looking for a "psychological pill" that will magically make things better. There seem to be few incentives for the work required to reinvent the marriage. On the other hand, Beth is an intelligent but bored empty-nester. Her husband travels a great deal and she has too much time on her hands. She's quite dissatisfied with her current lifestyle. She has plenty of incentives to deal with her malaise and to identify and develop opportunities for a fuller life. Leverage comes from having a good client-helper relationship, working on the right problems and opportunities, choosing the right goals, and pursuing these goals through the right strategies—all ending up in constructive change.

Guidelines in the Search for Leverage

Clients often need help to get a handle on complex problem situations. A 41-year-old depressed man with a failing marriage, a boring, run-of-the-mill job, deteriorating interpersonal relationships, health concerns, and a drinking problem cannot work on everything at once. Priorities need to be set. The blunt questions go something like this: Where is the biggest payoff? Where should the limited resources of both client and helper be invested? Where to start?

> Andrea, a woman in her mid-30s, is referred to a neighborhood mental health clinic by a social worker. During her first visit, she pours out a story of woe both historical and current—brutal parents, teenage drug abuse, a poor marriage, unemployment, poverty, and the like. Andrea is so taken up with getting it all out that the helper can do little more than sit and listen.

Where is Andrea to start? How can the time, effort, and money invested in helping provide a reasonable return to her? What are the economics of helping in Andrea's case?

The following principles of leverage serve as guidelines for choosing issues to be worked on. These seven principles overlap; more than one may apply at the same time.

1. Determine whether or not helping is called for.

2. If there is a crisis, first help the client manage the crisis.

3. Begin with the problem that seems to be causing the client the most pain.

4. Begin with issues the client sees as important and is willing to work on.

5. Begin with some manageable subproblem of a larger problem situation.

6. Move as quickly as possible to a problem that, if handled, will lead to some kind of general improvement.

7. Focus on a problem for which the benefits will outweigh the costs.

Underlying all these principles is an attempt to make clients' initial experience of the helping process rewarding so that they will have the incentives they need to continue to work. These principles are guidelines not a set of step-by-step directives. The outcome is important: clients working on issues that will make a difference in their lives. Examples of the use and abuse of these principles follow.

1. Determine whether or not helping is called for. Relatively little is said in the literature about screening—that is, about deciding whether any given problem situation or opportunity deserves attention. The reasons are obvious. Helpers-to-be are rightly urged to take their clients and their clients' concerns seriously. They are also urged to adopt an optimistic attitude, an attitude of hope, about their clients. Finally, they are schooled to take their profession seriously and are convinced that their services can make a difference in clients' lives. For those and other reasons, the first impulse of the average counselor is to try to help clients no matter what the problem situation might be.

There is something very laudable in this. It is rewarding to see helpers prize people and express interest in their concerns. It is rewarding to see helpers put aside the almost instinctive tendency to evaluate and judge others and to offer their services to clients just because they are human beings. However, like other professions, helping can suffer from the "law of the instrument." A child, given a hammer, soon discovers that almost everything needs hammering. Helpers, once equipped with the models, methods, and skills of the helping process, can see all human problems as needing their attention. In fact, in many cases, counseling may be a useful intervention and yet a luxury expense that cannot be justified.

Under the term *differential therapeutics*, Frances, Clarkin, and Perry (1984) discussed ways of fitting different kinds of treatment to different kinds of clients. They also discussed the conditions under which "no treatment" is the best option. In the no-treatment category, they included clients who have a history of treatment failure or who seem to get worse from treatment, such as the following:

- criminals trying to avoid or diminish punishment by claiming to be suffering from psychiatric conditions
- patients with malingering or fictitious illness
- chronic nonresponders to treatment
- clients likely to improve on their own
- healthy clients with minor chronic problems
- reluctant and resistant clients who try one dodge after another to avoid help

Although helpers must make a decision in each case, and some might dispute one or more of the categories proposed, the possibility of no treatment deserves serious attention.

The no-treatment or no-further-treatment option can do a number of useful things: interrupt helping sessions that are going nowhere or are actually destructive;

keep both client and helper from wasting time, effort, and money; delay help until the client is ready to do the work required for constructive change; provide a "breather" period that allows the client to consolidate gains from previous treatments; provide the client with an opportunity to discover that he or she can do without treatment; keep helper and client from playing games with themselves and one another; and provide motivation for the client to find help in his or her own daily life. However, helping professionals' decision not to treat or to discontinue treatment that is proving fruitless is countercultural and therefore difficult to make.

It goes without saying that screening can be done in a heavy-handed way. Statements such as the following are not useful:

"Your concerns are actually not that serious."

"You should be able to work that through without help."

"I don't have time for problems as simple as that."

Whether such sentiments are expressed or implied, they obviously indicate a lack of respect and constitute a caricature of the screening process.

Practitioners in the helping and human service professions are not alone in grappling with the economics of treatment. Doctors face clients day in and day out with problems that run from the life-threatening to the inconsequential. Statistics suggest that more than half of the people who come to doctors have nothing physically wrong with them. Doctors consequently have to find ways to screen patients' complaints. I am sure that the best doctors find ways to do so that preserve their patients' dignity.

Effective helpers, because they are empathic, pick up clues relating to a client's commitment, but they don't jump to conclusions. They test the waters in various ways. If clients' problems seem inconsequential, they probe for more substantive issues. If clients seem reluctant, resistant, and unwilling to work, they challenge clients' attitudes and help them work through their resistance. But in both cases they realize that there may come a time, and it may come fairly quickly, to judge that further effort is uncalled for because of lack of results. It is better, however, to help clients make such a decision themselves or challenge them to do so. In the end, the helper might have to call a halt, but his or her way of doing so should reflect basic counseling values.

2. If there is a crisis, first help the client manage the crisis.

Although crisis intervention is sometimes seen as a special form of counseling (France, 2005), it can also be seen as a rapid application of the three stages of the helping process to the most distressing aspects of a crisis situation.

PRINCIPLE VIOLATED: Zachary, a student near the end of the second year of a 4-year doctoral program in counseling, gets drunk one night and is accused of sexual harassment by a student whom he met at a party. Knowing that he has never sexually harassed anyone, he seeks the counsel of a faculty member whom he trusts. The faculty member asks him many questions about his past, his relationship with women, how he feels about the program, and so on. Zachary becomes more and more agitated and then explodes: "Why are you asking me all these silly questions?" He stalks out and goes to a fellow student's house.

PRINCIPLE USED: Seeing his agitation, his friend says, "Good grief, Zach, you look terrible! Come in. What's going on?" He listens to Zachary's account of what has happened, interrupting very little, merely putting in a word here and there to let his friend know he is with him. He sits with Zachary when he falls silent or cries a bit, and then slowly and reassuringly "talks Zach down," engaging in an easy dialogue that gives his friend an opportunity gradually to find his composure again. Zach's student friend has a friend in the university's student services department. They call him up, go over, and have a counseling and strategy session on the next steps in dealing with the harassment charges.

The friend's instincts are much better than those of the faculty member. He does what he can to defuse the immediate crisis and helps Zach take the next crisis-management step. France (2005), in an overview of crisis intervention, maintains that problem solving is central to the process. I use a compressed version of the problem-management process outlined in this book to help clients defuse crises. In helping clients defuse crises, it is important to focus on the context in which the crisis takes place (Myer & Moore, 2006). Because the client is often only one of the stakeholders, a client-only focus can be simplistic and ineffective.

3. Begin with the problem that seems to be causing the client the most pain. Clients often come for help because they are hurting even though they are not in crisis. Their hurt, then, becomes a point of leverage. Their pain also makes them vulnerable. If it is evident that they are open to influence because of their pain, seize the opportunity, but move cautiously. Their pain may also make them demanding. They can't understand why you cannot help them get rid of it immediately. This kind of impatience may put you off, but it, too, needs to be understood. Such clients are like patients in the emergency room, each seeing himself or herself as needing immediate attention. Their demands for immediate relief may well signal a self-centeredness that is part of their character and therefore part of the broader problem situation. It may be that their pain is, in your eyes, self-inflicted, and ultimately you may have to challenge them on it. But pain, whether self-inflicted or not, is still pain. Part of your respect for clients is your respect for them as vulnerable.

PRINCIPLE VIOLATED: Rob, a man in his mid-20s, comes to a counselor in great distress because his wife has just left him. The counselor's first impression is that Rob is an impulsive, self-centered person with whom it would be difficult to live. The counselor immediately challenges him to take a look at his interpersonal style and the ways in which he alienates others. Rob seems to listen, but he does not return.

PRINCIPLE USED: Rob goes to a second counselor, who also sees a number of clues indicating self-centeredness and a lack of maturity and discipline. However, she listens carefully to his story, even though it is one-sided and might well be very complicated. She explores with him the incident that precipitated his wife's leaving. Instead of adding to his pain by making him come to grips with his selfishness, she focuses on what he wants for the future, especially the immediate future. Of course, Rob thinks that his wife's return is the most important part of a better future. She says, "I assume she would have to be comfortable with returning." He says, "Of course." She asks, "What could you do on your part to help make her more inclined to want to return?" His pain provides the incentive for working with the counselor on how he might need to change even in the short term.

The second helper does not use pain as a club. However guilty he might be, Rob didn't need his nose rubbed in his pain. The helper uses his pain as a point of leverage. What is Rob willing to do to rid himself of his pain and create the future he says he wants?

4. Begin with issues the client sees as important and is willing to work on. The frame of reference of the client is a point of leverage. Given the client's story, you may think that he or she has not chosen the most important issues for initial consideration. However, helping clients work on issues that are important in their eyes sends an important message: "Your interests are important to me."

PRINCIPLE VIOLATED: A woman comes to a counselor complaining about her relationship with her boss. She believes that he is sexist. Male colleagues not as talented as she get the best assignments and a couple of them are promoted. After listening to her story, the counselor has her explore her family background for "context." After listening, he believes that she probably has some leftover developmental issues with her father and an older brother that affect her attitude toward older men. He pursues this line of thinking with her. She is confused and feels put down. When she does not return for a second interview, the counselor tells himself that his hypothesis has been confirmed.

PRINCIPLE USED: The woman seeks out a lawyer who deals with equal-opportunity cases. The lawyer, older and not only smart but wise, listens carefully to her story and probes for missing details. Then he gives her a snapshot of what such cases involve when they go to litigation. Against that background, he helps her explore what she really wants. Is it more respect? more pay? a promotion? revenge? a different kind of boss? a better use of her talents? a job in a company that does not discriminate? Once she names her preferences, he discusses with her options for getting what she wants. Litigation is not one of them.

The first helper substituted his own agenda for hers. He turned out to be somewhat sexist himself. The second helper accepted her agenda and helped her broaden it. He saw no leverage in litigation, but he did suggest that her problem situation could be an opportunity to reset her career. He suspected that there were plenty of firms eager to employ people of her caliber.

5. Begin with some manageable subproblem of a larger problem situation. Large, complicated problem situations often remain vague and unmanageable. Dividing a problem into manageable bits can provide leverage. Most larger problems can be broken down into smaller, more manageable subproblems.

PRINCIPLE VIOLATED: Aaron and Ruth, in their mid-50s, have a 25-year-old son living with them who has been diagnosed as schizophrenic. Aaron is a manager in a manufacturing concern that is in economic difficulty. His wife has a history of panic attacks and chronic anxiety. All these problems have placed a great deal of strain on the marriage. They both feel guilty about their son's illness. The son has become quite abusive at home and has been stigmatized by people in the neighborhood for his "odd" behavior. Aaron and Ruth have also been stigmatized for "bringing him up wrong." They have been seeing a counselor who specializes in a "systems" approach to such problems. They are confused by his "everything is related to everything else" approach. They are looking for relief but are exposed to more and more complexity. They finally come to the conclusion that they do not have the internal resources to deal with the enormity of the problem situation and drop out.

PRINCIPLE USED: A couple of weeks pass before they screw up their courage to contact a psychiatrist, Fiona, whom a family friend has recommended highly. Although Fiona understands the systemic complexity of the problem situation, she also understands their need for some respite. She first sees the son and prescribes some antipsychotic medication. His odd and abusive behavior is greatly reduced. Even though Aaron and his wife are not actively religious, she also arranges a meeting with a rabbi from the community known for his ecumenical activism. The rabbi puts them in touch with an ecumenical group of Jews and Christians who are committed to developing a "city that cares" by starting up neighborhood groups. This is a positive-psychology approach. Involvement with one of these groups helps diminish their sense of stigma. They still have many concerns, but they now have some relief and better access to both internal and community resources to help them with those concerns.

The psychiatrist helps then target two manageable subproblems—the son's behavior and the couple's sense of isolation from the community. Some immediate relief puts them in a much better position to tackle their problems longer term. There are, indeed, serious family systems issues here, but theories and methodologies should not take precedence over the client's immediate needs.

6. Move as quickly as possible to a problem that, if handled, will lead to some kind of general improvement. Some problems, when addressed, yield results beyond what might be expected. This is the spread effect.

PRINCIPLE VIOLATED: Jeff, a single carpenter in his late 20s, comes to a community mental-health center with a variety of complaints, including insomnia, light drug use, feelings of alienation from his family, a variety of psychosomatic complaints, and temptations toward exhibitionism. He also has an intense fear of dogs, something that occasionally affects his work. The counselor sees this last problem as one that can be managed through the application of a behavior modification methodology. He and Jeff spend a fair amount of time in the desensitization of the phobia. Jeff's fear of dogs abates quite a bit and many of his symptoms diminish. But gradually the old symptoms reemerge. His phobia, though significant, is not related closely enough to his primary concerns to involve any kind of significant spread effect.

PRINCIPLE USED: Predictably, Jeff's major problems reemerge. One day he becomes disoriented and bangs a number of cars near his job site with his hammer. He is overheard saying, "I'll get even with you." He is admitted briefly to a general hospital's psychiatric ward. The immediate crisis is managed quickly and effectively. During his brief stay, he talks with a psychiatric social worker. He feels good about the interaction, and they agree to have a few sessions after he is discharged. In their talks, the focus turns to his isolation. This has a great deal to do with his lack of self-esteem—"Who would want to be my friend?" The social worker believes that helping Jeff to get back into community may well help with other problems. He has managed this problem by staying away from close relationships with both men and women. They discuss ways in which he can begin to socialize. Instead of focusing on the origins of Jeff's feelings of isolation, the social worker, taking an opportunity-development approach, helps him involve himself in mini-experiments in socialization. As Jeff begins to get involved with others, his symptoms begin to abate.

The second helper finds leverage in Jeff's lack of human contact. The mini-experiments in socialization reveal some underlying problems. One of the reasons that others shy away from him is his self-centered and abrasive interpersonal style. When he repeats his question—"Who would want to be my friend?"—the helper responds, "I bet a lot of people would . . . if you were a bit more concerned about them and a bit less abrasive. You're not abrasive with me. We get along fine. What's going on?"

7. Focus on a problem for which the benefits will outweigh the costs. This is not an excuse for not tackling difficult problems, but it is a call for balance. If you demand a great deal of work from both yourself and the client, then the expectation is that there will be some kind of reasonable payoff for both of you.

PRINCIPLE VIOLATED: Margaret discovers to her horror that her husband Hector is HIV-positive. Tests reveal that she and her recently born son have not contracted the disease. This helps cushion the shock. But she has difficulty with her relationship with Hector. He claims that he picked up the virus from a "dirty needle." But she didn't even know that he had ever used drugs. The counselor focuses on the need for the "reconstruction of the marital relationship." He tells her that some of this would be painful because it would mean looking at areas of their lives that they had never reviewed or discussed. But Margaret is looking for some practical help in reorienting herself to her husband and to family life. After three sessions, she decides to stop coming.

PRINCIPLE USED: Margaret still searches for help. The doctor who is treating Hector suggests a self-help group for spouses, children, and partners of HIV-positive patients. In the sessions, Margaret learns a great deal about how to relate to some-one who is HIV-positive. The meetings are very practical. The fact that Hector is not in the group helps. She begins to understand herself and her needs better. In the security of the group, she explores mistakes she has made in relating to Hector. Although she does not "reconstruct" either her relationship with Hector or her own personality, she does learn how to live more creatively with both herself and him. She realizes that there will probably be further anguish, but she also sees that she is getting better prepared to face that future.

Reconstructing both relationships and personality, even if possible, is a very costly and chancy proposition. The cost-benefit ratio is out of balance. Once more it may be a question of a helper more committed to his or her theories than the needs of clients. Margaret gets the help she needs from the group and from sessions she and Hector have with a psychiatric aide. The leverage mind-set that pervades these principles is second nature in effective helpers.

Leverage and Action

Helping clients identify and deal with high-leverage issues, if done well, should also help them move toward the little actions that lead to a more formal plan for constructive change.

One client, a man in his early 30s, discussed all sorts of problems and unused opportunities with a counselor. He skipped from one topic to another, usually settling on the issue that had caused him the most trouble the previous week. The counselor always listened attentively. At the beginning of the third session she said, "I have a hypothesis I'd like to explore with you." She went on to name what she saw as a thread that wove its way through most of the issues he discussed—his reluctance to commit himself fully to anyone or anything. He was thunderstruck, because he instantly recognized the truth. In the next few weeks, he began to explore his problems and concerns from this perspective. Time after time, he saw himself withdrawing when he should have been committing himself, whether to a project or to a person. Gradually he began to commit himself in little ways. For instance, when an intimate woman friend of his began talking about where their relationship was going, he said, "There is something in me that wants to change the subject or run away. But no. Let's begin talking about the future and what future we might have together." At least he committed himself to opening a discussion about the relationship, if not to the relationship itself.

Discussing issues that make a difference can galvanize some clients into using resources that have lain dormant for years. Box 10-5 pulls together some questions you can help clients ask themselves about leverage.

Box 10-5   Questions on Leverage

Here are some questions related to leverage that counselors can help clients ask themselves.

- What problem or opportunity should I really be working on?
- Which issue, if faced, would make a substantial difference in my life?
- Which problem or opportunity has the greatest payoff value?
- Which issue do I have both the will and the courage to work on?
- Which problem, if managed, will take care of other problems?
- Which opportunity, if developed, will help me deal with critical problems?
- What is the best place for me to start?
- If I need to start slowly, where should I start?
- If I need a boost or a quick win, which problem or opportunity should I work on?

The Shadow Side of Leverage

For obvious reasons, no one knows how much time is wasted in counseling sessions discussing problems that won't make much of a difference in clients' lives. Some helpers perhaps unknowingly encourage clients to discuss problems that fit their theories and their approaches to therapy more than the needs of clients. They fit the client to the "technology" rather than joining clients in the search for issues that will make a difference in clients' lives. Some counselors misuse the principles of or fail to spot opportunities for leverage:

- They begin with the framework of the client, but never get beyond it.
- They deal effectively with the issues the client brings up, but fail to challenge clients to consider significant issues the clients are avoiding.
- They help clients explore and deal with problems, but never help clients translate the problems into opportunities or move on to other unused opportunities.
- They recognize the client's pain as a source of leverage, but then overfocus on the pain or allow it to mask other dimensions of the client's problems.
- They help clients achieve small victories, but fail to help clients build on their successes.
- They help clients start with small, manageable problems, but fail to help clients face more demanding problems or to see the larger opportunities embedded in these problems.
- They help clients deal with a problem or opportunity in one area of life (for instance, self-discipline in an exercise program), but fail to help them generalize what they learn to other, more difficult areas of living (for instance, self-control in interpersonal relationships).

> **Box 10-6 Guidelines for Assuring Leverage**
>
> - Help clients focus on issues that have payoff potential for them.
> - Maintain a sense of movement and direction in the helping process.
> - Avoid unnecessarily extending the problem identification and exploration stage.
> - Move to other stages of the helping process as clients' needs dictate.
> - Encourage clients to act on what they are learning.

Client behaviors also contribute to the shadow side of leverage. Some clients, for whatever reasons, don't bring up their key concerns. Effective helpers can spot clues that suggest that clients are hiding some concerns, skirting important issues, or even lying (Newman & Strauss, 2003). Then, with tact, they can help clients approach these issues. But in the end, counselors are not mind readers. Clients can avoid talking about their real concerns if they want to. One way of handling this kind of avoidance behavior is to point out its possibility in the initial helping contract shared with the client. The contract could say something like this:

> Sometimes clients don't bring up things that really bother them. Or they are uncertain whether they want to tackle a particular problem. They feel that they are not ready. They may even feel caught and lie about their problems. Because counselors are not mind readers, all this might escape the counselor's notice. All I can say is that I am willing to help you explore any issue that you believe will make a difference in your life. I am quite aware of human foibles, both my own and those of my clients. I will do my best to deal with sensitive issues in a caring and respectful way.

Choose your own words. One client told me, "I'm glad that you mentioned that no-fault part about lying. In the last couple of sessions, I haven't told you the entire truth. But now I'm ready." Then we got down to business. Box 10-6 contains some guidelines for assuring leverage.

It should be clear by now that the three tasks outlined here—helping clients tell their stories, helping them develop new perspectives and reframe their stories, and helping them work on issues that make a difference—are interrelated, not sequential steps. They are ways of being with clients to help them get a clear, objective, and realistic picture of their concerns.

IS STAGE I SOMETIMES ENOUGH?

Some clients seem to need only the opportunity to tell their stories. That is, in a relatively limited amount of time, they meet with a counselor and get help in exploring their problem situation. They tell their story in greater or lesser detail, perhaps clear up some blind spots, develop some new perspectives, and then go off and manage quite well on their own. That is, they set goals for themselves, draw up a plan of action, and get on with it. Here are two ways in which clients may need only the first stage of the helping process.

A declaration of intent and the mobilization of resources. For some clients, the very fact that they approach someone for help may be sufficient to help them begin to pull together the resources needed to manage their problem situations more effectively. For these clients, going to a helper is a declaration, not of helplessness, but of intent: "I'm going to do something about what is bothering me."

> Gerard was a young man from Ireland living illegally in New York. Even though he had a community of Irish immigrants to provide him support, he felt, as he put it, "hunted." He moved from one low-paying job to another, always living with the fear that something terrible was going to happen. One day, he talked to a priest about his concerns. This was really the first time he shared the burden that was weighing him down. The priest listened carefully but gave him no advice. But the talk provided the stimulus Gerard needed. "I know that I've got to do something about my life," he said at the end of their discussion. He quickly decided to return to Ireland. This unleashed pent-up inner resources. Once in Ireland, Gerard earned a degree in computer technology, then moved to Britain, and eventually got a job with the English arm of a German computer firm. Just as important, he felt "whole" again.

Merely seeking help can trigger a resource-mobilization process in some clients. Then they begin to manage their lives quite well on their own.

Coming out from under self-defeating emotions. Some clients come to helpers because they are incapacitated, to varying degrees, by negative feelings and emotions. Often when helpers show such clients respect, listen carefully to them, and understand them in a nonjudgmental way, those self-defeating feelings and emotions subside. That is, the clients benefit from the counseling relationship as a process of social-emotional reeducation and repair. Once that happens, they are able to call on their own inner and environmental resources and begin to manage the problem situation that precipitated the incapacitating feelings and emotions. In short, they move to action. These clients, too, seem to be "cured" merely by telling their story.

> One woman, let's call her Katrina, was very depressed after undergoing a hysterectomy. During two relatively brief counseling sessions, Katrina was helped to sort through the emotions she was feeling, and she discovered that the predominant emotion was shame. She felt wounded and exposed to herself and guilty about her incompleteness. Once she saw what was going on, she was able to pick up her life once more.

Of course, not all clients move on once they are given the opportunity to identify and explore their concerns. Many need help in sorting out what they really need and want and how they can get it. And so in the next chapter we will move on to Stage II.

INTRODUCTION TO STAGES II AND III: DECISIONS, GOALS, AND PLANS

CHAPTER 11

Stages II and III are so important, and yet too often overlooked in the helping liter-ature, that it makes sense to set the stage for them. This is the purpose of Chapter 11, which deals with decision making, the solution-focused nature of helping, and the value of goal setting. Sometimes problems go away by discussing them, but most of the time they don't. Helping relationships are essential, but they are not magic. If clients are to manage their lives more effectively, they need to change—change the ways they think, the ways they act, the ways they deal with emotions. Although the kind of change varies from client to client, change is the constant. Indeed, change is at the heart of all human endeavors. If "change is the constant," as the cliché goes, then getting good at managing change—even better, getting out in front of change, leading change—is good for our clients. That means that clients must review possi-bilities for a better future, set goals, determine how to achieve these goals, and find ways of moving from planning to action—even better, learning how to incorporate life-enhancing action into the planning process. All of this requires clients to make decisions. "Aye, there's the rub," as the bard would say, because decision making is a tricky business. Your job as a counselor is to help clients make multiple "good" deci-sions. So we begin with a brief review of decision—to be followed by a review of the solution-focused nature of helping and the importance of goal setting.

HELP CLIENTS BECOME MORE EFFECTIVE DECISION MAKERS

James Harkin (2005) begins a review of three books (Rosenthal, 2005; Schwartz, 2004; Sunstein, 2005) on choice (or having too many choices) and decision making with the simple statement, "Human beings are not adept at making decisions." The three authors demonstrate why having choices is not all that it's cracked up to be. It can be enslaving as well as liberating. This is unfortunate because all of us go about making any number of decisions of varying importance every day. Our clients have to make many important decisions both during helping sessions and in their day-to-day lives.

Because the second overall goal for helping outlined in Chapter 1 is to help clients, either directly or indirectly, become better problem solvers and opportunity developers in their everyday lives, helping them become better decision makers is part of the package. If the encouragement of client self-responsibility is a key helping value, helping clients not only make good decisions but also become better decision makers is not an amenity but a necessity. Consider this case.

Alice's third marriage has just ended. Her first husband proved to be brutish and physically abusive. Her second husband, a "safe bet," was bland beyond repair. Her third husband turned out to be gay. Not a good track record. So her counselor is helping Alice explore the decisions she has made in developing and dissolving intimate relationships. Alice discovers that she makes poor decisions about people in general, for instance, by being too trusting too soon. Although she is horrified by all the mistakes she has made, she realizes that without these sessions she might well go on making the same kind of mistakes.

Where did Alice learn how to make decisions? Where do any of us learn to make reasonable, if not perfect, decisions? Few people receive any kind of formal training in decision making, as important as it seems to be. Learning how to make "good" decisions is left to chance in society. Like other skills—interpersonal communication, problem solving, parenting, and managing—everyone thinks that decision-making skills are very important. But when people are asked in what forum these critical skills are to be learned, they shrug their shoulders. Once more life itself is to be the teacher. Because clients make many important decisions both within counseling sessions and in their daily lives, it is odd that the helping professions have not been calling for mainstream scientific research into decision making. It seems shortsighted to promote empirically supported approaches to helping without promoting empirically sound approaches to decision making. There is a great deal of research on decision making. Comparatively little of it makes its way into helping.

Rational Decision Making

One of the reasons clients get into trouble in the first place is that they, like Alice, make poor decisions. We need only review our own experiences to see how often our decisions—or our failure to make decisions—got us into trouble. There are many decision points in the helping process. We have already seen a number of them. Clients must decide to come to a counseling interview in the first place, to talk about themselves, to talk about specific issues, to return for subsequent sessions, to respond to the helper's empathy, probes, and challenges, and to choose issues to work on. We are about to see that clients must also decide what they want, to set goals, to develop strategies, to make plans, and to implement those plans. Deciding—rather than letting the world decide for you—is at the heart of helping, as it is at the heart of living.

Decision making in its broadest sense is the same as problem solving. Indeed, this book could be called a decision-making approach to helping. In this chapter, however, the focus is on decision making in a narrower sense—the internal (mental) action of identifying alternatives or options and choosing from among them. It is a commitment to do or to refrain from doing something:

- "I have decided to discuss my career problems but not my sexual concerns."
- "I have decided to start a new business."
- "I have decided to ask the courts to remove artificial life support from my comatose wife."
- "I have decided to strike a better balance between work and home life."
- "I have decided not to undergo chemotherapy."
- "I have decided to stop putting myself down."
- "I have decided to move into a retirement home."

The commitment can be to an internal action—"I have decided to get rid of my preoccupation with my ex-wife"—or to an external action—"I have decided to confront my son about his drinking." Decision making in the fullest sense includes the implementation of the decision: "I made a resolution to give up smoking, and I haven't smoked for 3 years." "I decided that I was being too hard on myself, so I took a week off work and just enjoyed myself."

Decision making is often enough presented as a rational, linear process involving information gathering, analysis, and choice. What follows are the bare essentials of what may be called the Newtonian, or totally rational, approach to decision making (Baron, 2001; Galotti, 2002; Hastie & Dawes, 2001). Then we will see that the world is seldom that simple.

Information gathering. The first rational task is to gather information related to the particular issue or concern. A patient who must decide whether to have a series of chemotherapy treatments needs some essential information. What are the treatments like? What will they accomplish? What are the side effects? What are the consequences of not having them? What would another doctor say? And so forth. And there is a whole range of ways in which she might gather this information—Internet searches, books and articles, talking to doctors, talking to patients who have undergone treatment or who have refused treatment. Many patients today routinely mount extensive Internet searches on their medical conditions in order to make better informed decisions. This does not mean that the information they gather provides ready-made answers. Take Doug. Testing indicated that he might (or might not) have an aggressive form of prostate cancer. This motivated him to gather a great deal of information about prostate cancer, its diagnosis, and its treatment. But he found no ready-made answers. If he did have cancer, he would have to choose from among many different kinds of treatment, each with its downside.

Analysis. The next rational step is processing the information. This includes analyzing, thinking about, working with, discussing, meditating on, and immersing oneself in the information. Just as there are many ways of gathering information, so there are many ways of processing it. Effective information processing leads to a clarification and an understanding of the range of possible choices. "Now, let's see, what are the advantages and disadvantages of each of these choices?" is one way of analyzing information. Effective analysis assumes that decision makers have criteria, whether objective or subjective, for comparing alternatives. For instance, a patient wants to determine whether the weeks or months of life she will gain through a series of chemotherapy treatments will be worth the effort and discomfort.

Making a choice. Finally, decision makers need to make a choice—that is, commit themselves to some internal or external action that is based on the analysis: "After thinking about it, I have decided to sue for custody of the children." As indicated earlier, the fullness of the choice includes an action: "I had my lawyer file the custody papers this morning." There are also rational "rules" that can be used to make a decision. For instance, one rule, stated as a question, deals with the consequences of the decision: "Will it get me everything I want or just part?" Values also enter the picture because, from one point of view, values are criteria for making decisions. "Should I do X or Y? Well, what are my values?" The woman suing for the custody of the children says to herself, "I value fairness. I'm not going to try to extort a lot of money for child care. I'll make reasonable demands."

Counselors help clients engage in rational decision making; that is, they help clients gather information, analyze what they find, and then base action decisions on the analysis. This is fine when it works, but it does not always work that way.

The Shadow Side of Decision Making: Choices in Everyday Life

Thinking and reasoning are not always what they are supposed to be or seem to be in everyday life. And, when people get into trouble, thinking and reasoning can go even "further south" (Basic Behavioral Science Task Force of the National Advisory Mental Health Council, 1996). This means that decision making in everyday life, and in counseling, is not the straightforward rational process just outlined. Rather, it is an ambiguous, highly complicated process with a deep shadow side (Cosier & Schwenk, 1990; Etzioni, 1989; Gigerenzer, 2002; Gilovich, 1991; Heppner, 1989; Kaye, 1992; March, 1994; Plous, 1993; Russo & Schoemaker, 1990; Stroh & Miller, 1993; Whyte, 1991). For instance, Gati, Krausz, and Osipow (1996) discuss the messiness associated with making career decisions and list 10 ways in which such decisions can be flawed. There is no such thing as the perfect career decision.

Headlee and Kalogjera (1988) found evidence that some of the roots of the shadow side of decision making begin in childhood. Some children are allowed too many choices, whereas others are given too few. Moreover, in the early years, distortions of choice evolve because of racial, ethnic, sexual, religious, and other prejudices. By the time the child becomes an adult, these distortions are ingrained in the decision-making process and nobody thinks about them. The sources of possible distortion are myriad. In everyday life, decision making is often confused, covert, difficult to describe, unsystematic, and, at times, quite irrational. A shadow-side analysis of decision making as it is actually practiced reveals a less-than-rational application of the three dimensions outlined above.

Information gathering. Information gathering should lead to a clear definition of the issues to be decided. A client trying to decide whether to pursue a divorce needs information about that entire process. However, information gathering is practically never straightforward. Decision makers, for whatever reason, are often complacent and engage in perfunctory searches; they get too much, too little, inaccurate, or misleading information. The search for information is often clouded by emotion. In counseling, the client trying to decide whether to proceed with therapy may have already made up his or her mind and therefore may not be open to confirming or disconfirming information.

Because full, unambiguous information is seldom available, all decisions are at risk. In fact, there is no such thing as completely objective information. Information, especially in decision making, is often filtered, clouded, and distorted by the decision maker. Objective information takes on a subjective cast. People providing others with information also have their filters. Surgeons tend to push surgery as the "golden option" for prostate cancer. No wonder, Ackoff (1974) called human problem solving "mess management" (p. 21).

> Eloise wanted to make a decision about whether or not to marry her partner. One of the obstacles was conflicting careers. She didn't know whether she'd be in the same career 5 years from now; neither did he. Another obstacle lay in the fact that she knew little about his past. She thought it didn't matter. She liked him now. He knew that she was a nonpracticing Catholic but knew little about how her

Catholicism affected her or how it would affect them in the future, especially if they had children. Because religion was not currently an issue, he did not explore it. There were many other things they did not know about themselves and each other. They eventually did marry, but the marriage lasted less than a year. The damage done lasted a lot longer.

Granted, clients' stories are never complete, and information will always be partial and open to distortion. Yet, though counselors cannot help clients make information gathering perfect, they can help them make it at least "good enough" for problem management and opportunity development. Eloise and her partner never saw a counselor before they got married, but in retrospect it would have been wise for both of them to "take counsel" with themselves, with one another, with family, with friends. These people could have helped the couple in a friendly way to ask the right questions of themselves and one another.

Processing the information. Because it is impossible to separate the decision from the decision maker, the processing of information is as complex as the person making the decision. Factors affecting information analysis include clients' feelings and emotions, their values-in-use, which often differ from their espoused values, their assumptions about "the way things work," and their level of motivation.

There is no such thing as full, objective processing of gathered information. Poorly gathered information is often subjected to further mistreatment. Because of their biases, clients focus on bits and pieces of the information they have gathered rather than seeing the full picture. Furthermore, few clients have the time or the patience to spell out all possible choices related to the issue at hand, together with the pros and cons of each. Therefore, some say, most decisions are based not on evidence but on taste: "I like it. It sounds good."

> Jamie was in a high-risk category for AIDS because of occasional drug use and sexual promiscuity. Once, when he was busted for drug use, he had to attend a couple of sessions on AIDS awareness. He listened to all the information, but he processed it poorly. These were problems for "other people." He engaged in risky sexual behavior "only occasionally." He was sure that his sexual partners were "clean." One or two "harmless mistakes" were not going to do him in. He had read an article that said that it was actually difficult to "catch" HIV. He knew others who engaged in much riskier behavior than he and "nothing had happened to them." He'd be "more careful," though it was not clear what that meant as far as his behavior was concerned. He was in good health, and "healthy people can take a lot."

Jamie distorted information and rationalized away most of the risk of his current lifestyle. He was living, not on the edge, but on a precipice.

Up to a point, counselors can help clients overcome inertia and biases and tackle the work of analysis. For instance, if a client says his values have "matured" but he still automatically makes decisions based on his former values, then he can be challenged to use these more mature values in his decision making. One client, trying to make a decision about a career change, kept moving toward options in the helping professions even though he had become quite interested in business. There was something in him that kept saying, "You have to choose a helping profession. Otherwise you will be a traitor." The counselor helped him see his bias. In the end, the client became a consultant, then a manager, and then a senior manager. But he still had to salve his conscience by noting both to himself and to others that "running a successful business is an important contribution to society."

Choice and execution.   A host of strange things can happen on the way to executing a decision. People sometimes do the following:

- They skip the analysis stage and move quickly to choice. "Let's put mother in a nursing home. That will make it easier for all of us to get involved." This kind of thinking led to a poor choice of a nursing home and literally years of family infighting.

- They ignore the analysis and base the decision on something else entirely; the analysis was nothing but a sham, because the decision criteria, however covert, were already in place. Reiner goes through an extended analysis of the reasons for becoming an entrepreneur and starting his own business, starts a business but folds it within 2 months at great financial and psychological cost, and accepts an offer from a large firm. He ignored the fact that security was his main driver.

- They engage in what Janis and Mann (1977) called "defensive avoidance." That is, they procrastinate, attempt to shift responsibility, or rationalize delaying a choice. Dad says, "I know that it makes sense to sell this big house and move into a retirement village, but what do the kids want? And what if we don't like it? We might run into people we don't like. We'd better take a closer look at this." The looking takes years.

- They confuse confidence in decision making with competence. Tom, 76, gets a physical and discovers that he has a high PSA score. "I know what I want. If it's prostate cancer, I'm going for surgery and get it over with." At his age, it is quite likely that he will die of something else. "Watch waiting" might be the best option.

- They panic and seize upon a hastily contrived solution that gives promise of immediate relief. The choice may work in the short term but have negative long-term consequences. Tess and Lars panic and arrange for an abortion when they find out that she is pregnant. "We've got to do this quickly." Tess is traumatized by speed of the decision and the operation. Lars is not very sympathetic. Their relationship collapses. The next 2 years are rocky for both of them.

- They are swayed by a course of action that is most salient at the time or by one that comes highly recommended, even though it is not right for them. Imogene, single, gives up her child for adoption. Later she bitterly regrets her decision.

- They let enthusiasm and other emotions govern their choices. Ben is so elated to be offered a promotion he says yes right away. Only later does he realize that he was not cut out to be a manager.

- They announce a choice to themselves or to others, but then do nothing about it. Bert and Linda tell their teenage children that they want to involve them more in household decision making, then do nothing about it. The kids become even more resentful.

- They translate the decision into action only halfheartedly. Sandra, grieving over the loss of her husband, decides to renew her social life. But she often fails to return phone calls, cancels engagements, and leaves get-togethers shortly after arriving, offering what seem to others rather lame excuses.

- They decide one thing but do another. Ted decides to turn down a job offer because it's "not for me," but ends up taking it anyway.

The fact that choices do not necessarily make life easier for self and others explains a great deal of the shadow side of decision making. It is clear that counselors cannot help clients avoid all the pitfalls involved in making decisions, but they can help clients minimize them.

In summary, then, pure-form rational, linear decision making has probably never been the norm in human affairs. Decision making goes on at more than one level. There is, as it were, the rational decision-making process in the foreground and an emotional or impulsive decision-making process in the background. Gelatt (1989) called for an approach to decision making that factors in these shadow-side realities: "What is appropriate now is a decision and counseling framework that helps clients deal with change and ambiguity, accept uncertainty and inconsistency, and utilize the nonrational and intuitive side of thinking and choosing" (p. 252). Positive uncertainty means, paradoxically, being positive (comfortable and confident) in the face of uncertainty (ambiguity and doubt)—feeling both uncertain about the future and positive about the uncertainty. Stages II and III together with an understanding of the shadow side of these two stages provide methodologies clients can use to make decisions, explore their consequences, and act on them. This is not a total decision-making cure. But then nothing is.

Making Smarter Decisions

Hammond, Keeney, and Raiffa (1998, 1999) offer a guide to making better decisions, taking into account, of course, the shadow side of decision making. In their article (1998), they focus on hidden traps in decision making and how to handle them. Here are some of them.

The status quo bias. Clients often have a bias toward perpetuating the status quo, which is seen as the "safe" option even when it is not. Jeff, 27, is a self-harming client, who had been sexually abused by a male relative during his teen years. He is a loner steeped in guilt and self-loathing, often depressed. Though he has been cutting himself for several years, he has never been suicidal. Like many self-harming clients, the physical pain he experiences either during or, more often, after cutting himself gives him some relief from his emotional pain. Although helpers differ in their understanding of self-injury and its treatment, most tend to see it as a dysfunctional way of dealing with emotional pain (Conterio & Lader, 1998; Ferentz, 2002; Kennedy, 2004; Levenkron, 1999; Turner, 2002). Some helpers see it is an addiction, whereas others see it as addiction-like behavior.

This means that self-harming clients tend to opt for the status quo. They resist the idea that they should give up this behavior and often enough leave counseling because they feel they are not being understood. One counselor focused almost exclusively on the reduction and ultimate elimination of Jeff's cutting behavior. After several verbal battles with his counselor, Jeff quit therapy 2 months after he started. A second counselor tried to help Jeff find more creative substitutes for his cutting behavior. Although Jeff found this approach more appealing, the focus was still on his self-harming behavior. In the end, his status-quo bias persisted and he quit. A third counselor, believing that Jeff's self-injury behavior, whether addictive or not, was a symptom rather than the central issue, focused on Jeff's negative feelings about himself and gradually helped him revise his view of himself and get back into some kind of community. It took 2 years, but Jeff's cutting behavior gradually decreased and finally disappeared.

The third counselor focused, as she put it, on the "self-healing person" inside of Jeff, which helped him revise the way he thought about himself. Better thinking coupled with small but meaningful external actions did the trick.

The confirming-evidence trap. If I have secretly—hidden more or less even from myself—decided to do or to avoid doing something, I can begin looking for evidence that will confirm my choice or avoid evidence that will challenge it. Sheila, a college junior, was seeing a counselor because she was both shy and perfectionistic. She grew tired of counseling quickly because the counselor seemed to be trying to determine whether she was shy *because* she was perfectionistic—staying away from people gave her time to "get things done right"—or whether she was perfectionistic *because* she was shy—her "high standards" meant that she had very few friends. Or maybe it was something else. Sheila threw in the towel.

A second counselor, after hearing her recapitulate her experience with the first counselor, said during their second session, "Well, what do you really want, Sheila?" She remained totally self-possessed, paused, and then said, "I want to leave school. My mother is dying. She needs me. She might have 6 months. She might have 2 years. No matter. My place is at her side." It was something else. When the counselor asked, "Is this what your mother wants?" Sheila replied, "That's not the issue." In a later session the counselor, after finding out that Sheila's decision was not seconded by her mother, father, or any of her three younger brothers, tried to come at it from various angles. Sheila was very bright. She amassed evidence supporting her decision from every source—psychology, sociology, the Bible, theology, and her commitment "to my family and myself." No one wanted to badger her into changing her mind, but everyone wanted her to do the "right thing" for the "right reason." In the end, her family and the counselor knew that she would make up her own mind.

Here are some things the counselor might do to help Sheila reach a sound decision.

- Help Sheila examine all the evidence with equal vigor. She was being very intellectual about it all. She did a great job with the evidence from the human and godly sciences, but attended very little to what the significant people in her life were saying.

- Get someone Sheila respects to act as devil's advocate. Sheila had friends—family friends, her doctor, her minister, and so forth. Or, better yet, get Sheila herself to play her own devil's advocate. Reverse roles. The counselor plays Sheila, and Sheila becomes the helper.

- Have Sheila take a closer look at her motives. What does she really want? Is there something behind her leaving school besides her mother's illness? Are shyness and perfectionism part of the picture? Is leaving school her way of being both "alone" in some sense and "perfect"? If her mother were not sick, would she still be leaving school? It might be helpful for Sheila to know what she is doing and what is moving her to do it.

- If Sheila seeks advice from others, help her frame her questions so that they don't merely invite confirmation of what she has already decided. Sheila could say, "Here's what I'm thinking of doing. Grill me on it, will you?"

It might well be that Sheila wants to leave school because she wants to be at her mother's side during these difficult times. It may have nothing to do with either isolation or perfectionism. And if Sheila truly thinks it's right for herself, so be it. After all, it is her life.

In their book, *Smart Choices: A Practical Guide to Making Better Decisions*, Hammond, Keeney, and Raiffa (1999) lay out a system for making smart choices. The eight elements of the system highlight the eight most common and most serious errors in decision making (p. 189):

1. Working on the wrong problem
2. Failing to identify key objectives
3. Failing to develop a range of good, creative alternatives
4. Overlooking crucial consequences of alternatives
5. Giving inadequate thought to tradeoffs
6. Disregarding uncertainty
7. Failing to account for risk tolerance
8. Failing to plan ahead when decisions are linked over time

As you can see, when put positively ("Working on the right problem" and so forth), these are also elements of the problem-management process, most of which are addressed in Stages II and III.

Finally, because helpers themselves are human, they do not escape the shadow side of decision making. As you can see, helpers make decisions throughout the helping process. Pfeiffer, Whelan, and Martin (2000, p. 429), after reviewing the decision-making research, comment:

> When examined as a whole, this research suggests that people tend to preferentially attend to information, gather information, and interpret information in a manner that supports, rather than tests, their decisions about another person. Therapists may not be exempt from this tendency, particularly given the often complex and ambiguous nature of clients' problems.

No matter how empathic you are, as a helper you will still make hypotheses about your clients throughout the helping process and base some of your decisions on these hypotheses. Your challenge is to continually test these hypotheses against the reality of your clients in the context of their lives. Theories are theories, not substitutes for clients.

SOLUTION-FOCUSED HELPING

Back in 1989, O'Hanlon and Weiner-Davis (1989, p. 6; see also W. H. O'Hanlon, Weiner-Davis, & B. O'Hanlon, 2003) claimed that a trend "away from explanations, problems, and pathology, and toward solutions, competence, and capabilities" was emerging in the helping professions. An earlier study showed that clients were interested in solutions to their problems and feeling better, whereas many helpers were often more concerned about the origin of problems and transforming them through insight (Llewelyn, 1988). Over the years solution-focused or outcome-focused therapies have aimed at tackling this disconnect (Bloom, 1997; de Jong & Berg, 2002; de Shazer, 1985, 1988; Duncan, Miller, & Sparks, 2004; Fish, 1995; Guterman, 2006; Manthei, 1997; Metcalf, 1998; Miller, Hubble, & Duncan, 1996; Murphy, 1997;

O'Connell, 2005; O'Connell & Palmer, 2003; Presbury, Echterling, & McKee, 2002; Walter & Peller, 1992; Zimmerman, Prest, & Wetzel, 1997).

Even today too many approaches to helping still focus on Stage I activities. Too many helper training programs still emphasize—or overemphasize—the exploration methods of Stage I. Communication skills are required in every approach to helping, but limiting their use to Stage I endeavors is a waste. Intensive discussion of problem situations is often based on a "working through" mentality, whereas action or "solution" approaches are based on the assumption that many problems need to be dealt with or even "transcended" rather than worked through. At any rate, the goal of helping, as stated in Chapter 1, is "problems managed," not just "problems explored and understood" and "opportunities developed," not just "opportunities identified and discussed."

Common Factors in Solution-Focused Approaches to Helping

Solution-focused therapies have a common philosophy and approach to helping that fits in very well with the approach taken in this book. They are, each in its own way, problem-management and, especially, opportunity-development approaches. Here is a quick overview of the principles these approaches have in common.

Positive philosophy. In relating with clients, focus on resources rather than deficits; on success and rather than failure; on credit rather than blame; on solutions rather than problems. Use common sense. Don't let theory get in the way of helping clients.

View of clients. Clients are people like the rest of us. See them as people with complaints about life, not symptoms. Don't assume that they will arrive ambivalent about change and resistant to therapy. Clients have a reservoir of wisdom, learned and forgotten, but still available. Clients have resources and strengths to resolve complaints. Clients will have their own view of life just as everyone else. Respect the reality they construct, even though you might have to help them move beyond it. In a way, clients are experts in their own lives. Help them feel competent to solve their own problems. When helpers see clients as problems to be solved, they impoverish them and take their power away.

Past successes rather than failures. Past trauma did happen. However, if you help clients dwell on it, they will become captives of it. In fact, many will arrive as captives. They need to liberate themselves from the past. Clients should get an organized or integrated view of past bad experiences, but it is usually not that helpful to explore origins and causation. Looking for deep, underlying causes for symptoms is a mistake. Focus on clients' ability to survive problem situations. Getting at causes does not usually resolve a complaint. Resolving the complaint resolves it. That said, clients have more confidence and comfort in their journey to the future when they carry forward helpful parts of the past. The best things that clients can bring forward are past successes, what they have done that works. If clients carry part of the past forward, it should be what is best about the past. As Bushe (1995) put it, this is "an attempt to generate a collective image of a new and better future by exploring the best of what is and has been."

Role of helper. Helpers are consultants, catalysts, guides, facilitators, assistants. It is very helpful to adopt, at least temporarily, the client's world view. This helps lessen reluctance and resistance. Sharing empathic understanding helps demonstrate your appreciation of the client's world. Your job is to notice and amplify the client's life-giving forces and any sign of constructive change. Become a detective for good things. Develop an "appreciative ear." Listen to problems, but listen even more to the opportunity buried within the problem. Use questions that inspire and encourage clients to give positive examples. Questions should stimulate dialogue. Remember that questions are not just questions; they are interventions.

The discovery phase: Help clients explore and exploit competencies, successes, and "normal times." Help them identify ways of thinking, behaving, and interacting that have worked in past. And, because clients are not continually manifesting problem behavior, help them explore the times when they are free of such behavior. The "free" times point toward solutions. Help them identify what has been working during these misery-free periods and capitalize on it. How can they amplify what has been working? Have them recall successes from the past, for instance, when they have handled disagreements more creatively. When your marriage was good, what was it like? What did you do when you successfully resisted the urge to drink? Catch clients being competent and resourceful and help them take a good look at themselves at such times. Notice competencies revealed in a client's story and behavior. There are things that work in every client. Recognize and discuss these competencies because they are strengths that clients can build on. What is the client like when performing well?

The nature of problems and how we talk about things. Like the rest of us, clients become what they talk about. If you always encourage them to talk about problems, they run the risk of becoming "problem people." Then helping turns into remedying pathology and deficits. What clients focus on becomes their chronic reality. Help them see their problems as "complaints." We all have complaints. In other words, help them "normalize" their problems; they are the ordinary difficulties of life. For instance, overeating is showing too much enthusiasm for the wonderful texture, taste, and comfort of food. Hyperactivity is energy that at times gets the better of us and interferes with rest, relaxation, and relationships. A perfectionist is person who loves quality but who goes too far.

Help clients see problems as external to themselves, not things that define and control their lives. Problems are intruders that get the best of us at times. Problems are complaints that bother us rather than define us. Ordinarily, it is not necessary to know a great deal about the complaint in order to resolve it. So be careful about the questions you ask. They should not keep clients mired in problem talk because problem talk can keep clients immersed in frustration, impotence, and even despair.

Insight into resources rather than problems. Insight is a two-edged sword. Insight into the origin of problems is not necessary for change. Too often insights are about problems, not about solutions. Therefore, they keep clients defining themselves in terms of the problem. On the other hand, encourage insights that focus on resources and preferred outcomes. Help clients define themselves in terms of strengths and successes. Engage them in generating "positive outcome" scenarios.

Constructive dreaming: Possibilities for a better future. The principle is this: The future we anticipate is the future we create. The helper and the client should partner in the systematic search for possibilities and potential. The client's imagination needs to be "provoked" to discover new ways of approaching life's critical issues.

Questions should stimulate clients to think as creatively as possible about this better future. What images capture your hopes for your future? What can you do to keep these hopeful images alive? If you no longer needed help, how would that show up in your actions? If you did the right thing for yourself this week and were filmed, what would the film's highlights show? On the assumption that you would like to move forward, what might you do to push the envelope a bit in the coming week?

Designing solutions. Clients define the goal. There is no one correct way to live one's life. Help clients actively design solutions that will turn their possibilities into life-enhancing realities. Solutions do not have to be complex. Rather, help clients look for simple solutions to complex problems. Often a small change is all that is necessary. Nor do solutions have to take care of everything. Troubles don't have to be totally solved. Help clients find systemic solutions. A change in one part of the system can produce good results in another part. Look for interventions that break up patterns of self-limiting behavior. Don't hesitate to design solutions that get rid of symptoms. Getting rid of symptoms is not shallow, useless, or dangerous. Part of the solution should be clients' ability to grapple with future problems on their own. Clients should leave therapy with identified tools to do so.

Delivering: Getting things done. Implementation is everything. The pace of change will be different for each client. Some need help to ease themselves into solutions gradually. The smallest action is a step forward. On the other hand, rapid change is possible. Don't shortchange clients. Solutions often require that clients develop new ways of relating to their social environment. Where will support come from? Who needs to be engaged to make things work?

Criticisms of the Solution-Based Approach

Of course, there have been some criticisms of a solution-focused approach. For instance, some say that it runs the risk of being a don't-worry-just-be-happy approach. It's too pie-in-the-sky. Others say, let's get real. Change comes from dealing with problems. People are used to dealing with problems, so this approach might be disorienting for some clients.

On the other hand, traditional approaches to helping, including problem-solving approaches, also come in for their knocks. Critics of traditional approaches to helping say that they are often painfully slow, ask clients to look back at yesterday's failures, and speculate about the causes of problems. They place blame, promote defensiveness, use deficit- rather than resource-focused language, and rarely result in a new focus on life.

Perhaps both sets of criticism, put as baldly as in the last two paragraphs, is unfair. Forget blaming. Wedding the upbeat approach of solution-focused therapies to a problem-management and opportunity-development approach to problem situations while borrowing the best from traditional approaches faces down both sets of criticism. Once more, there is no need for either/or. Any helping approach worth its salt must be solution-focused. Clients need better lives, not good helping sessions. The problem-management and opportunity-development approach provides the backbone of helping. But the use of its stages and tasks are dictated by client need. Any methodology or treatment that is ethical and works can be incorporated. Evidence, drawn from theory, research, practice, or common sense, is important. Flexibility is the key. The arbiter is client-focused social intelligence. Do what is best for the client.

Big S and Small s Solutions

This book offers a solution-focused approach to helping. There is, however, a semantic problem with the word *solution*. It means two distinct things. First and foremost it means an end state—results, accomplishments. Take Pinta. Her eating was out of control. She was killing herself with food. But now she is eating moderately and has lost a lot of weight. A new approach to eating is in place. This is a solution in the end-state sense. This is a Big S solution.

However, *solution* also means a strategy a client uses to get to the end state. For instance, Pinta joined a 12-step program for overeaters and faithfully attended the meetings. "My eating is out of control," she said at the first meeting. In the group, she learned a variety of ways to get back in control of her eating. Joining the group and using the strategies she learned in the group were Small s solutions. Small s solutions are the means client take to achieve Big S. They were activities or strategies that helped her get to the end state she now enjoys. Stage II of the helping process deals with Big S solutions—end states, accomplishments, goals, outcomes. Stage III focuses on Small s solutions—means, strategies, actions.

The distinction is not inconsequential. Many approaches to problem solving confuse the two. Or they pay little attention to solutions as end states—what the client really wants and needs—and talk mostly about solutions in terms of the strategies clients must use to "solve" the problem. They leap from problem or unused opportunity to action without linking action to outcome. The correct logic is this. Link solutions-as-goals to the problem situation or unused opportunity. Pinta's problem is overeating. Her solution-as-goal is "healthy exercise and nutritional habits consistently in place." Her solution-as-means is the 12-step group together with the self-regulation strategies it sponsors. Neither problems nor strategies should be the main driver of action. Life-enhancing goals should drive action.

The Solution-Focused Nature of Brief Therapy

Helping can be "lean and mean" and still be most human. As part of a research program, a colleague of mine experimented, quite successfully, with shortening the counseling "hour." At one point, he would begin an experimental session by saying, "We have 5 minutes together. Let's see what we can get done." He was totally with each "client." It was amazing how much he and his clients could get done in a short time. The helping professions, driven by political and the financial dynamics of managed care, are focusing more and more on results-oriented, evidence-based "brief psychotherapy." But even if that were not the case, helpers would still owe their clients value for money. Helping does not always deliver everything a clients wants. Helping that achieves only partial results may, at times, be the best that client and helper can do together. *Brief* does not mean unprofessional.

Brief therapies are of their very essence solution-focused. If there is little time, most of it had better be spent focusing on a better future. In fact, many books and articles on brief therapy have "solution-focused" in their titles and vice versa. Research has demonstrated that brief interventions can produce both substantive and lasting changes. Solution-focused therapists help clients craft fairly well-defined goals that can be accomplished in a reasonable time frame. Brief therapy can be brief but still comprehensive (Lazarus, 1997). Asay and Lambert (1999, p. 42), after reviewing the research on brief therapy, drew the following conclusions:

> The beneficial effects of therapy can be achieved in short periods (5 to 10 sessions) with at least 50% of clients seen in routine clinical practice. For most clients, therapy will be brief. . . . In consequence, therapists need to organize their work to optimize outcomes within a few sessions. Therapists also need to develop and practice intervention methods that assume clients will be in therapy for fewer than 10 sessions.

They also found that there are three categories of clients who do poorly in brief therapy: clients who are poorly motivated and hostile, clients who come with a history of poor relationships, and clients who expect to be passive recipients of help as if it were a medical procedure. Their research showed that about 20% to 30% of clients require treatment lasting more than 25 sessions. So how long should therapy last? Length should serve clients and outcomes, not theories. Efficiency—that is, the wise use of resources—and effectiveness—that is, successful outcomes—are guidelines.

Nick and Janet Cummings (2000) reviewed the evidence for what they call "time-sensitive" psychotherapy.

> There is a growing body of evidence that time-sensitive psychotherapies are effective with a large number of patients, perhaps as many as 85% of all those seen in the usual practice (Austad, 1996; Budman & Steenbarger, 1997; Cummings, Pallak, & Cummings, 1996; Hoyt, 1993), but it is not advocated that all psychotherapy be brief psychotherapy. Even Miller (1996b), an outspoken critic of short-term therapy, begrudgingly concedes that there is a place for time-sensitive psychotherapies and that psychotherapists of the future must be trained in both short-term and long-term interventions to know when to use one or the other. (p. 44)

The Cummings advocate "clinically determined" therapy rather than therapy determined by the economic needs of either clinicians or managed-care enterprises. Effective and efficient helping (Cummings, Budman, & Thomas, 1998), rather than the length of helping, should remain center stage.

Asking "How long should therapy take?" is like asking "How long is a piece of string?" I have known people who were in therapy most of their lives without changing much. I know others who reset the course of their lives in a single session. Although the whole helping process can take a long time, substantial helping can also take place in a short time. Consider the following scenario.

> Lara, a social worker in a tough urban neighborhood, gets a call from a local minister. He tells her that a teenager fears for his life because of neighborhood gang activity and needs to see someone. She sees the boy, listens to his concerns, feels his anxiety. Although the minister thought that the boy was under some immediate threat, Lara realizes that his anxiety stems from a recent shooting that had nothing to do with him directly. She talks with him to get some idea what his daily comings and goings are like. It is soon clear that he is not the kind of person who courts trouble. She then talks him through the kinds of prudent things he needs to do to avoid trouble "in a neighborhood like ours" and in their dialogue points out that he is already doing most of those things. She gently challenges his false perception that the recent shooting somehow involved him directly. All of this helps allay his fears. But he discovers there are a couple of things he needs to do differently, like having "good" friends and not letting himself be a lone target on the streets. She also points out that he is more likely to be hit by lightning than a stray bullet. The boy calms down as he begins to realize that in many ways his life is in his own hands.

Lara listens to the boy's immediate concerns and elicits from him a picture of what his daily life in the community looks like (Stage I). Realizing that he needs some relief from his anxiety (Stage II), she "talks him down." Through their dialogue the boy learns some things he can do to assure his safety (Stage III). She also helps him dispel a couple of blind spots—that is, that he is an immediate target of gang activity and that he is in imminent danger of being hit by a stray bullet. So even if you have only have only one meeting with a client, it would be a mistake to assume that only Stage I things could happen.

Helpers would do well to develop a "whole-process" mentality about helping— any part of any stage or task can be invoked at any time in any session if it proves beneficial for the client. Think of the helping process as a hologram—that laser-generated three-dimensional image that seems to float in space. In a hologram, the whole is found in each of its parts. The helping model is more like a hologram than a tool kit— it works best when the whole is found in each of its parts. The hologram is centered on the client, not the helping process.

HELP CLIENTS DISCOVER AND USE THEIR POWER THROUGH GOAL SETTING

Goal setting has a great deal of power. First, it is allied with hope. Second, it has power in its own right. Let's start with hope and move on to the power of goals.

The Psychology of Hope

Hope as part of human experience is as old as humanity. Who of us has not started sentences with "I hope . . . "? Who of us has not experienced hope or lost hope? Hope also has a long history as a religious concept. St. Paul said, "Hope that centers around things you can see is not really hope," thus highlighting the element of uncertainty. If you know that tomorrow you will receive an Oscar, you can no longer hope for it. You know it's a sure thing. Hope plays a key role in both developing and implementing possibilities for a better future. An Internet search reveals that scientific psychology has not always been interested in hope (R. S. Lazarus, 1999; Stotland, 1969), but things are beginning to change. Rick Snyder, who, as we have seen earlier, has written extensively about the positive and negative uses of excuses in everyday life (Snyder & Higgins, 1988; Snyder, Higgins, & Stucky, 1983), has become a kind of champion for hope (1994, 1995, 1997, 1998; McDermott & Snyder, 1999; Snyder, McDermott, Cook, & Rapoff, 1997; Snyder, Michael, & Cheavens, 1999). Indeed, he linked excuses and hope in an article entitled "Reality negotiation: From excuses to hope and beyond" (1989). He has also developed scales for measuring both dispositional hope (Snyder et al., 1991) and state hope (Snyder et al., 1996).

The nature of hope. Jerome Groopman (2004), who holds a chair of medicine at Harvard Medical School, in a very moving book on the anatomy of hope, defines it "as the elevating feeling we experience when we see—in the mind's eye—a path to a better future. Hope acknowledges the significant obstacles and deep pitfalls along that path. True hope has no room for delusion" (p. xiv). His search for a scientific basis for understanding the key role hope plays in dealing with illness takes him to the "biology" of hope. His book also shows how counseling is at the heart of medical practice.

Snyder, on the other hand, starts with the premise that human beings are goal directed. According to Snyder, hope is the process of thinking about one's goals—*Serena is determined that she will give up smoking, drinking, and soft drugs now that she is pregnant*—of having the will, desire, or motivation to move toward these goals—*Serena is serious about her goal because she has seen the damaged children of mothers on drugs, and she is also, at heart, a decent, caring person*—and of thinking about the strategies for accomplishing one's goals—*Serena knows that two or three of her friends will give her the support she needs and she is willing to join an arduous 12-step program to achieve her goal.* Serena is hopeful. If we say that Serena has "high hopes," we mean that her goal is clear, her sense of agency (or urgency) is high, and that she is realistic in planning the pathways to her goal. Both a sense of agency and some clarity around pathways are required.

Hope, of course, has emotional connotations. But it is not a free-floating emotion. Rather, it is the by-product or outcome of the work of setting goals, developing a sense of agency, and devising pathways to the goal. Serena feels a mixture of positive emotions—elation, determination, satisfaction—knowing that "the will" (agency) and "the way" (pathways) have come together. Success is in sight even though she knows that there will be barriers—for instance, the ongoing lure of tobacco, wine, and soft drugs.

The benefits of hope. Snyder (1995, pp. 357–358) has combed the research literature in order to discover the benefits of hope as he defines it. Here is what he has found.

> The advantages of elevated hope are many. Higher as compared with lower hope people have a greater number of goals, have more difficult goals, have success at achieving their goals, perceive their goals as challenges, have greater happiness and less distress, have superior coping skills, recover better from physical injury, and report less burnout at work, to name but a few advantages.

Counselors who do not spend a significant part of their time helping clients develop possibilities, clarify goals, devise strategies or pathways, and develop the sense of agency needed to bring all of this to fruition are certainly shortchanging their clients. Because Stages II and III deal with possibilities, goals, commitment, pathways, and overcoming barriers, they could be named "ways of nurturing hope."

The Power of Goal Setting

Goal setting, whether it is called that or not, is part of everyday life. We all do it all the time.

> Why do we formulate goals? Well, if we didn't have goals, we wouldn't do anything. No one cooks a meal, reads a book, or writes a letter without having a reason, or several reasons, for doing so. We want to get something we want through our actions or we want to prevent or avoid something we don't want. *These desires are beacons for our actions;* [italics added] they tell us which way to go. When formalized into goals, they play an important role in problem solving. (Dorner, 1996, p. 49)

Even not setting goals is a form of goal setting. If we don't name our goals, that does not mean that we don't have any. Instead of overt goals, then, we have a set of covert goals. These are our default goals. They may be enhancing or limiting. We don't like the sagging muscles and flab we see in the mirror. But not deciding to get into better shape is a decision to continue to allow the fitness program to drift.

Because life is filled with goals—chosen goals or goals by default—it makes sense to make them work for us rather than against us. Goals at their best mobilize our resources; they get us moving. They are a critical part of the self-regulation system. If they are the right goals for us, they get us headed in the right direction. There is a massive amount of sophisticated theory and research on goals and goal setting (Karoly, 1999; Locke & Latham, 1984, 1990, 2002). In their 2002 *American Psychologist* article, Locke and Latham summarize 35 years of empirical research on goal setting. According to this research, helping clients set goals empowers them in the following four ways.

Goals help clients focus their attention and action. A counselor at a refugee center in London described Simon, a victim of torture in a Middle Eastern country, to her supervisor as aimless and minimally cooperative in exploring the meaning of his brutal experience. Her supervisor suggested that she help Simon explore possibilities for a better future instead of focusing on the hell he had gone through. The counselor started one session by asking, "Simon, if you could have one thing you don't have, what would it be?" Simon response was immediate. "A friend," he said. During the rest of the session, he was totally focused. What was uppermost in his mind was not the torture but the fact that he was so lonely in a foreign country. When he did talk about the torture, it was to express his fear that torture had "disfigured" him, if not physically, then psychologically, thus making him unattractive to others.

Goals help clients mobilize their energy and effort. Clients who seem lethargic during the problem-exploration phase often come to life when asked to discuss possibilities for a better future. A patient in a long-term rehabilitation program who had been listless and uncooperative said to her counselor after a visit from her minister, "I've decided that God and God's creation and not pain will be the center of my life. This is what I want." That was the beginning of a new commitment to the arduous program. She collaborated more fully in exercises that helped her manage her pain. Clients with goals are less likely to engage in aimless behavior. Goal setting is not just a "head" exercise. Many clients begin engaging in constructive change after setting even broad or rudimentary goals.

Goals provide incentives for clients to search for strategies to accomplish them. Setting goals, a Stage II task, leads naturally into a search for means to accomplish them, a Stage III task. Lonnie, a woman in her 70s who had been described by her friends as "going downhill fast," decided, after a heart-problem scare that proved to be a false alarm, that she wanted to live as fully as possible until she died. She searched out ingenious ways of redeveloping her social life, including remodeling her house and taking in two young women from a local college as boarders.

Clear and specific goals help clients increase persistence. Not only are clients with clear and specific goals energized to do something, but they also tend to work harder and longer. An AIDS patient who said that he wanted to be reintegrated into his extended family managed, against all odds, to recover from five hospitalizations to achieve what he wanted. He did everything he could to buy the time he needed.

Clients with clear and realistic goals don't give up as easily as clients with vague goals or with no goals at all.

One study (Payne, Robbins, & Dougherty, 1991) showed that high-goal-directed retirees were more outgoing, involved, resourceful, and persistent in their social settings than low-goal-directed retirees. The latter were more self-critical, dissatisfied, sulky, and self-centered. People with a sense of direction don't waste time in wishful thinking. Rather, they translate wishes into specific outcomes toward which they can work. Picture a continuum. At one end is the aimless person; at the other, a person with a keen sense of direction. Your clients may come from any point on the continuum. Taz knows that he wants to become a better supervisor but needs help in developing a program to do just that. On the other hand, Lola, one of Taz's colleagues, doesn't even know whether this is the right job for her and does little to explore other possibilities. Any given client may be at different points with respect to different issues—for instance, mature in seizing opportunities for education but aimless in developing sexual maturity. Most of us have had directionless periods in one area of life or another at one time or another.

Setting goals, whether formally or informally, provides clients with a sense of direction. People with a sense of direction tend to experience the following:

- have a sense of purpose
- live lives that are going somewhere
- have self-enhancing patterns of behavior in place
- focus on results, outcomes, and accomplishments
- avoid mistaking random actions for accomplishments
- have a defined rather than an aimless lifestyle

Although the motivational value of goal setting is incontrovertible, the number of people who disregard problem-managing and opportunity-developing goal setting and its advantages are legion. The challenge for counselors is to help clients do it well.

As you can well understand, not all theory and research is easily translated into practical advice for helpers. There is also an extensive self-help literature dealing with solution-focused problem solving and goal setting and implementation in everyday life (for instance, Blair, 2000a, 2000b; D. Ellis, 1999; K. Ellis, 1998; Jackson & McKergow, 2002; Secunda, 1999; Tracy, 2003). There is a great deal of practical wisdom—admittedly too often intermingled with hype—to be mined from the self-help literature, and helpers-to-be would be doing themselves a disservice if they were to turn their noses up at it. Once you are fortified with the best in theory and research, you should be able to spot the best practical advice for both helpers and clients in the popular literature.

CHAPTER

12

Stage II: Help Clients Set Viable Goals

The Three Tasks of Stage II

Task 1: Help Clients Discover Possibilities for a Better Future— "How Can My Life Be Better?"

Skills for Identifying Possibilities for a Better Future

Possible Selves

Creativity and Helping

Divergent Thinking

Brainstorming: A Tool for Divergent Thinking

Suspend your own judgment, and help clients suspend theirs

Encourage clients to come up with as many possibilities as possible

Help clients use one idea to stimulate others

Help clients let themselves go and develop some "wild" possibilities

Future-Oriented Probes

Exemplars and Models as a Source of Possibilities

Cases Featuring Possibilities for a Better Future

The case of Brendan: Dying better

The Washington family case

Task 2: Help Clients Move From Possibilities to Choices—"What Solutions Are Best for Me?"

Help Clients Design and Shape Their Goals

Help clients state what they need and want as outcomes or accomplishments

Help clients move from broad aims to clear and specific goals

Help clients establish goals with substance, goals that make a difference

Help clients set goals that are prudent

Help clients formulate realistic goals

Help clients set goals that can be sustained

Help clients choose goals that have some flexibility

Help clients choose goals consistent with their values

Help clients establish realistic time frames for accomplishing goals

Help Clients Distinguish Needs From Wants

Help Clients Spot Emerging Goals

Help Clients Set Adaptive Goals

Adaptive goals

Strategic self-limitation

Coping as an Important Goal

Positive reappraisal

Problem-focused coping

Infusing ordinary events with positive meaning

A Bias for Action as a Metagoal

Task 3: Help Clients Commit Themselves—"What Am I Willing to Pay for What I Want?"

Help Clients Commit Themselves to a Better Future

Help clients set goals that are worth more than they cost

Help clients set appealing goals

Help clients embrace and own the goals they set

Help clients deal with competing agendas

Great Expectations: Encourage Client Self-Efficacy—"I Can, I Will"

The nature of self-efficacy

Helping clients develop self-efficacy

Stage II and Action

The Shadow Side of Goal Setting

The Three Tasks of Stage II

In many ways Stages II and III together with the Action Arrow are the most important parts of the helping model because they are about "solutions." It is here that counselors help clients develop and implement programs for constructive change. The payoff for identifying and clarifying both problem situations and unused opportunities lies in doing something about them. The skills helpers need to help clients to do precisely that—engage in constructive change—are reviewed and illustrated in Stages II and III. In these stages, counselors help clients ask and answer the following two commonsense but critical questions. *What do you want?* and *What do you have to do to get what you want?*

Problems can make clients feel hemmed in and closed off. To a greater or lesser extent they have no future, or the future they have looks troubled. But, as Gelatt (1989) noted, "The future does not exist and cannot be predicted. It must be imagined and invented" (p. 255). The interrelated tasks of Stage II outline three ways in which helpers can partner with their clients with a view to exploring and developing this better future.

- **Task 1—Possibilities.** "What possibilities do I have for a better future?" "What are some of the things I think I want?" "What about my needs?" In helping clients move from problems to solutions, counselors help them develop a sense of hope.

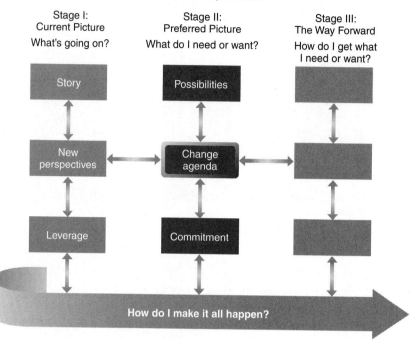

FIGURE 12-1
The Helping Model—Stage II

- **Task 2—Choices.** "What do I really want and need? What solutions are best for me?" Here counselors help clients craft a viable *change agenda* from among the possibilities. Helping them shape this agenda is the central task of helping.
- **Task 3—Commitment.** "What am I willing to pay for what I want?" Help clients discover incentives for commitment to their change agenda. This is a further look at the economics of personal change discussed in Task 3 of Stage I.

Figure 12-1 highlights these three tasks of the helping process. Without minimizing in any way what counselors can help their clients accomplish through Stage-I skills and interventions—that is, problem and opportunity clarification; the development of new, more constructive perspectives of self, others, and the world; the choice of high-leverage issues to work on—the real power of helping lies in helping clients set goals and move to accomplish them.

TASK 1: HELP CLIENTS DISCOVER POSSIBILITIES FOR A BETTER FUTURE—"HOW CAN MY LIFE BE BETTER?"

The goal of Task 1 is to help clients develop a sense of direction by exploring possibilities for a better future. I once was sitting at the counter of a late-night diner when a young man sat down next to me. The conversation drifted to the problems he was having with a friend of his. I listened for a while and then asked, "Well, if your relationship was just what you wanted it to be, what would it look like?" It took him a bit to get started, but eventually he drew a picture of the kind of relationship he could live with. Then he stopped, looked at me, and said, "You must be a professional." I believe he thought that because this was the first time in his life that anyone had ever asked him to describe some possibilities for a better future.

Too often the exploration and clarification of problem situations are followed, almost immediately, by the search for solutions in the secondary sense—actions that will help deal with the problem or develop the opportunity. But in many ways, outcomes are more important than actions. *What will be in place* once those actions are completed? As we saw in the last chapter, failure to specify outcomes is one of the major decision-making traps. The outcome is a solution with a big S, whereas the actions leading to this outcome constitute a solution with a small s. There is great power in visualizing outcomes, just as there is a danger in formulating action strategies before getting a clear idea of desired outcomes. Stage II is about identifying or visualizing desired results, outcomes, or accomplishments. Task 1 in this stage is about envisioning possibilities. Stage III is about developing strategies, actions, and plans for delivering those outcomes. From another point of view, Stages II and III are about hope.

SKILLS FOR IDENTIFYING POSSIBILITIES FOR A BETTER FUTURE

At its best, counseling helps clients move from problem-centered mode to "discovery" mode. Discovery mode involves creativity and divergent thinking. However, according to Sternberg and Lubart (1996), creativity is one of those topics in which psychology has underinvested. Dean Simonton (2000) reviews advances in our understanding and use of creativity as part of positive psychology. According to Taylor, Pham, Rivkin, and Armor, however, not just any kind of mental stimulation will do. Mental stimulation is helpful to the degree that it "provides a window on the future by enabling people to envision possibilities and develop plans for bringing those possibilities about. In moving oneself from a current situation toward an envisioned future one, the anticipation and management of emotions and the initiation and maintenance of problem-solving activities are fundamental tasks" (1998, p. 429). This kind of thinking moves in the same direction as Snyder's. Not just fantasy. Not just rumination. The full problem-management and opportunity development framework helps clients, to use Simonton's phrase, "harness the imagination."

Possible Selves

One of the characters in Gail Godwin's (1984) novel *The Finishing School* warns against getting involved with people who have "congealed into their final selves." Clients come to helpers, not necessarily because they have congealed into their final selves—if this is the case, why come at all?—but because they are stuck in their current selves. Counseling is a process of helping clients get "unstuck" and develop a sense of direction. Markus and Nurius used the term *possible selves* to represent "individuals' ideas of what they might become, what they would like to become, and what they are afraid of becoming" (1986, p. 954; see also Cross & Markus, 1991, 1994; Frazier, Hooker, Johnson, & Kaus, 2000; Knox, Funk, Elliott, & Bush, 1998; Robinson, Davis, & Meara, 2003). Consider the case of Ernesto. He was very young but very stuck for a variety of sociocultural and emotional reasons.

A counselor first met Ernesto in the emergency room of a large urban hospital. He was throwing up blood into a pan. He was a member of a street gang, and this was the third time he had been beaten up in the last year. He had been so severely beaten this time that it was likely that he would suffer permanent physical damage. Ernesto's lifestyle was doing him in, but it was the only one he knew. He was in need of a new way of living, a new scenario, a new way of participating in city life. This time he was hurting enough to consider the possibility of some kind of change.

The counselor worked with Ernesto, not by helping him explore the complex socio-cultural and emotional reasons he was in this fix, but principally by helping him explore his "possible selves" in order to discover a different purpose in life, a different direction, a different lifestyle.

Task 1 is about "possible selves." The notion of possible selves has captured the imagination of many helpers and of those interested in human development such as teachers (Cameron, 1999; Cross & Marcus, 1994; Hooker, Fiese, Jenkins, Morfei, & Schwagler, 1996; Strauss & Goldberg, 1999). Enter the term *possible selves* into an Internet search engine, and you will find all sorts of examples of how helpers and teachers have been using this concept. In Task 1, your job is to help clients discover their possible selves.

Remember that identifying and developing an opportunity can help clients manage a problem by transcending rather than "solving" it. More than 40 million people with physical challenges live in the United States. Brenda, a victim of a hit-and-run driver, is one of them. The accident left her with a back condition that cut short her career as a fitness instructor and left her understandably depressed. She was now one of the physically challenged who, as noted by Cartwright, Arredondo, and D'Andrea (2004), "possess strengths that are often overlooked by persons in the dominant cultural group in general and by many counselors who work with these person in particular" (p. 24).

Some of Brenda's friends now saw her as "broken" and shied away from spending time with her. Unfortunately, her counselor also fell into this category. He was sympathetic, rather than empathic, and expected little from her. Once she realized that this relationship was one of the main sources of her depression, she switched counselors. Her new counselor helped Brenda use inner resources to focus on opportunities rather than loss. She found ways of coping with her chronic pain. She had always been interested in design, but never had the time to pursue this interest. Now she wanted to experiment with becoming an interior designer. She took online courses and purchased design software. She became a successful interior designer, now mentally and imaginatively active, and she did whatever she could physically. She became so engaged in her new career that there was no need to mourn the loss of her previous one. Her new career also brought some new friends. She still lived with her bad back, but transcending was her way of coping.

Creativity and Helping

One of the myths of creativity is that some people are creative and others are not. Like the rest of us, clients can be more creative than they are. It is a question of finding ways to help them be so. Of course, counselors cannot help clients be more creative unless they themselves are creative about the helping process itself. Carson and Becker (2004; see also Carson & Becker, 2003), in reviewing a group of articles in a special 2002 issue of the *Journal of Clinical Activities, Assignments, & Handouts in Psychotherapy Practice*, edited by L. Hecker (see Hecker & Kottler, 2002), suggest that "being able to access our own creativity at peak levels in an effort to help clients tap their own creative problem-solving abilities (internal and relational) and creative resources is a prerequisite to effective therapy" (p. 111). Clients, they say, should be a source of creativity for helpers and vice versa. Stages II and III help clients tap into their dormant creativity. A review of the requirements for creativity (see Cole & Sarnoff, 1980; Robertshaw, Mecca, & Rerick, 1978, pp. 118–120) shows, by implication, that people in trouble often fail to use whatever creative resources they might have. The creative person is characterized by the following:

- optimism and confidence (whereas clients are often depressed and feel powerless)
- acceptance of ambiguity and uncertainty (whereas clients may feel tortured by ambiguity and uncertainty and want to escape from them as quickly as possible)
- a wide range of interests (whereas clients may be people with a narrow range of interests or whose normal interests have been severely narrowed by anxiety and pain)
- flexibility (whereas clients may have become rigid in their approach to themselves, others, and the social settings of life)
- tolerance of complexity (whereas clients are often confused and looking for simplicity and simple solutions)
- verbal fluency (whereas clients are often unable to articulate their problems, much less their goals and ways of accomplishing them)
- curiosity (whereas clients may not have developed a searching approach to life or may have been hurt by being too venturesome)
- drive and persistence (whereas clients may be all too ready to give up)
- independence (whereas clients may be quite dependent or counterdependent)
- nonconformity or reasonable risk taking (whereas clients may have a history of being very conservative and conformist or may get into trouble with others and with society precisely because of their particular brand of nonconformity)

A review of some of the principal obstacles or barriers to creativity (see Azar, 1995) brings further problems to the surface. Innovation is hindered by the following:

- Fear—clients are often quite fearful and anxious.
- Fixed habits—clients may have self-defeating habits or patterns of behavior that may be deeply ingrained.
- Dependence on authority—clients may come to helpers looking for the "right answers" or be quite counterdependent (the other side of the dependence coin) and fight efforts to be helped with a variety of games.
- Perfectionism—clients may come to helpers precisely because they are hounded by this problem and can accept only ideal or perfect solutions.
- Problems with social networks—being "different" sets clients apart when they want to belong.

These are not definitive sets. Their purpose is to help you think about creativity in helping. It is easy to say that imagination and creativity are most useful in Stages II and III, but it is another thing to help clients stimulate their own, perhaps dormant, creative potential. Although there is a very rich body of literature on creativity, including creativity in the helping professions (Carson & Becker, 2003; Carson, Becker, Vance, & Forth, 2003; Sternberg, Grigerenko, & Singer, 2004), only a small percentage of this creativity seems to find its way into the helping process. Perhaps one sign of creativity's increasing importance is the formation of the Association for Creativity in Counseling, the newest division in the American Counseling Association (Kennedy, 2004). But it has always been important.

Divergent Thinking

Many people habitually take a convergent-thinking approach to problem solving—that is, they look for the "one right answer." Such thinking has its uses, of course. However, many of life's problem situations are too complex to be handled by convergent thinking. Such thinking limits the ways in which people use their own and environmental resources.

Divergent thinking, on the other hand, assumes that there is always more than one answer. De Bono (1992) calls it "lateral thinking." It is related to curiosity, "a positive emotional-motivational system associated with the recognition, pursuit, and self-regulation of novelty and challenge" (Kashdan, Rose, & Fincham, 2004, p. 291). In helping, divergent thinking means "more than one way to manage a problem or develop an opportunity." Unfortunately, as helpful as it can be, divergent thinking is not always rewarded in our culture and sometimes is even punished. For instance, students who think divergently can be thorns in the sides of teachers. Some teachers feel comfortable only when they ask questions in such a way as to elicit the "one right answer." When students who think divergently give answers that are different from the ones expected—even though their responses might be quite useful (perhaps more useful than the expected responses)—the students may be ignored, corrected, or punished. Students then may generalize their experience and end up thinking that it is simply not a useful form of behavior. Consider the following case.

> Quentin wanted to be a doctor, so he enrolled in the premed program at school. He did well but not well enough to get into medical school. When he received the last notice of refusal, he said to himself, "Well, that's it for me and the world of medicine. Now what will I do?" When he graduated, he took a job in his brother-in-law's business. He became a manager and did fairly well financially, but he never experienced much career satisfaction. He was glad that his marriage was good and his home life rewarding, because he derived little satisfaction from his work.

Not much divergent thinking went into handling this problem situation. No one asked Quentin what he really wanted. For Quentin, becoming a doctor was the "one right career." He didn't give serious thought to any other career related to the field of medicine, even though there are dozens and dozens of interesting and challenging jobs in the field of health care.

The case of Caroline, who also wanted to become a doctor but failed to get into medical school, is quite different from that of Quentin.

> Caroline thought to herself, "Medicine still interests me; I'd like to do something in the health field." With the help of a medical career counselor, she reviewed the possibilities. Even though she was in premed, she had never realized that there were so many medical careers. She decided to take whatever courses and practicum experiences she needed to become a nurse. Then, while working in a clinic in the hills of Appalachia—an invaluable experience for her—she managed to get an M.A. in

family-practice nursing by attending a nearby state university part time. She chose this specialty because she thought that it would enable her to be closely associated with delivery of a broad range of services to patients and would also enable her to have more responsibility for the delivery of these services.

When Caroline graduated, she entered private practice as a nurse practitioner with a doctor in a small Midwestern town. Because the doctor divided his time among three small clinics, Caroline had a great deal of responsibility in the clinic where she practiced. She also taught a course in family-practice nursing at a nearby state school and conducted workshops in holistic approaches to preventive medical self-care. Still not satisfied, she began and finished a doctoral program in practical nursing. She taught at a state university and continued her practice. Needless to say, her persistence paid off with an extremely high degree of career satisfaction.

A successful professional career in health care remained Caroline's aim throughout. A great deal of divergent thinking and creativity went into the elaboration of that aim into specific goals and coming up with the courses of action to accomplish them. But for every success story, there are many more failures. Quentin's case is probably the norm, not Caroline's. For many, divergent thinking is either uncomfortable or too much work.

Brainstorming: A Tool for Divergent Thinking

One excellent way of helping clients think divergently and more creatively is brainstorming. Brainstorming is a simple idea-stimulation technique for exploring the elements of complex situations. Brainstorming in Stages II and III is a tool for helping clients develop both possibilities for a better future and ways of accomplishing goals.

There are certain rules that help make this technique work: suspend judgment, produce as many ideas as possible, use one idea as a takeoff point for others, get rid of normal constraints to thinking, and produce even more ideas by clarifying items on the list. Here, then, are the rules.

Suspend your own judgment, and help clients suspend theirs. When brainstorming, do not let clients criticize the ideas they are generating and, of course, do not criticize them yourself. There is some evidence that this rule is especially effective when the problem situation has been clarified and defined and goals have not yet been set. In the following example, a woman whose children are grown and married is looking for ways of putting meaning into her life.

CLIENT: One possibility is that I could become a volunteer, but the very word makes me sound a bit pathetic.

HELPER: Add it to the list. Remember, we'll discuss and critique them later.

Having clients suspend judgment is one way of handling the tendency on the part of some to play a "Yes, but" game with themselves. That is, they come up with a good idea and then immediately show why it isn't really a good idea, as in the preceding example. By the same token, avoid saying such things as "I like that idea," "This one is useful," "I'm not sure about that idea," or "How would that work?" Premature approval and criticism cut down on creativity. A marriage counselor was helping a couple brainstorm possibilities for a better future. When Nina said, "We will stop bringing up past hurts," Tip, her husband, replied, "That's your major weapon when we fight. You'll never be able to give that up." The helper said, "Add it to the list. We'll look at the realism of these possibilities later on."

Encourage clients to come up with as many possibilities as possible. The principle is that quantity ultimately breeds quality. Some of the best ideas come along later in the brainstorming process. Cutting the process short can be self-defeating. In the following example, a man in a sex-addiction program has been brainstorming activities that might replace his preoccupation with sex.

CLIENT: Maybe that's enough. We can start putting it all together.

HELPER: It doesn't sound like you were running out of ideas.

CLIENT: I'm not. It's actually fun. It's almost liberating.

HELPER: Well, let's keep on having fun for a while.

CLIENT (pausing): Ha! I could become a monk.

Later on, the counselor, focusing on this "possibility," asked, "What would a modern-day monk who's not even a Catholic look like?" This helped the client explore the concept of sexual responsibility from a completely different perspective and to rethink the place of religion and service to others in his life. And so, within reason, the more ideas the better. Helping clients identify many possibilities for a better future increases the quality of the possibilities that are eventually chosen and turned into goals. In the end, however, do not invoke this rule for its own sake. Possibility generation is not an end in itself. Use your clinical judgment, your social intelligence, to determine when enough is enough. If a client wants to stop, often it's best to stop.

Help clients use one idea to stimulate others. This is called piggybacking. Without criticizing the client's productivity, encourage him or her both to develop strategies already generated and to combine different ideas to form new possibilities. In the following example, a client suffering from chronic pain is trying to come up with possibilities for a better future.

CLIENT: Well, if there is no way to get rid of all the pain, then I picture myself living a full life without pain at its center.

HELPER: Expand that a bit for me.

CLIENT: The papers are filled with stories of people who have been living with pain for years. When they're interviewed, they always look miserable. They're like me. But every once in a while there is a story about someone who has learned how to live creatively with pain. Very often they are involved in some sort of cause which takes up their energies. They don't have time to be preoccupied with pain.

When one client with multiple sclerosis brought of this possibility: "I'll have a friend or two with whom I can share my frustrations as they build up," the helper asked, "What would that look like?" The client replied, "Not just a complaining session or just a poor-me thing. It would be a normal part of a give-and-take relationship. We'd be sharing both joys and pain of our lives like other people do."

Help clients let themselves go and develop some "wild" possibilities. When clients seem to be "drying up" or when the possibilities being generated are quite pedestrian, you might say, "Okay, now draw a line under the items on your list and write the word *wild* under the line. Now let's see if you can come up with some really wild possibilities." Later it is easier to cut suggested possibilities down than to expand them. The wildest possibilities often have within them at least a kernel of an idea that will work. In the following example, an older single man who is lonely is exploring possibilities for a better future.

CLIENT: I can't think of anything else. And what I've come up with isn't very exciting.

HELPER: How about getting a bit wild? You know, some crazy possibilities.

CLIENT: Well, let me think. . . . I'd start a commune and would be living in it. . . . And. . . .

Clients often need permission to let themselves go even in harmless ways. They repress good ideas because they might sound foolish. Helpers need to create an atmosphere in which such apparently foolish ideas will be not only accepted but also encouraged. Help clients come up with conservative possibilities, liberal possibilities, radical possibilities, and even outrageous possibilities.

It's not always necessary to use brainstorming explicitly. As helper, you can keep these rules in mind and then by sharing highlights and using probes, you can get clients to brainstorm even though they don't know that's what they're doing. A brainstorming mentality, not its ritualistic practice, is useful throughout the helping process. Sir Peter Hall, who directed 30 Shakespeare plays on major stages over 50 years, said that directors should not present the cast with a finished concept on the first day, but rather lead them on a voyage of discovery during the weeks of rehearsal and then edit the findings (Hall, 2005). There is no one right way in the theater of life either. Helpers lead clients on their own "voyage of discovery."

Future-Oriented Probes

One way of helping clients invent the future is to ask them, or get them to ask themselves, future-oriented questions related to their current unmanaged problems or undeveloped opportunities. The following questions are different ways of helping clients find answers to the questions "What do you want?" and "What do you need?" These questions focus on outcomes—that is, on what will be in place after the clients act.

- **What would this problem situation look like if you were managing it better?** Ken, a college student who has been a "loner," has been talking about his general dissatisfaction with his life. In response to this question, he said, "I'd be having fewer anxiety attacks. And I'd be spending more time with people rather than by myself."

- **What changes in your present lifestyle would make sense?** Cindy, who described herself as a "bored homemaker," replied, "I would not be drinking as much. I'd be getting more exercise. I would not sit around and watch the soaps all day. I'd have something meaningful to do."

- **What would you be doing differently with the people in your life?** Lon, a graduate student at a university near his parents' home, realized that he had not yet developed the kind of autonomy suited to his age. He mentioned these possibilities: "I would not be letting my mother make my decisions for me. I'd be sharing an apartment with one or two friends."

- **What patterns of behavior would be in place that are not currently in place?** Bridget, a depressed resident in a nursing home, had this suggestion: "I'd be engaging in more of the activities offered here in the nursing home." Rick, who is suffering from lymphoma, said, "Instead of seeing myself as a victim, I'd be on the web finding out every last thing I can about this disease and how to deal with it. I know there are new treatment options. And I'd also be getting a second or a third opinion. You know, I'd be managing my lymphoma instead of just suffering from it."

- **What current patterns of behavior would be eliminated?** Bridget, a resident in a nursing home, added these to her list, "I would not be putting myself down for incontinence I cannot control. I would not be complaining all the time. It gets me and everyone else down!"

- **What would you have that you don't have now?** Sissy, a single woman who has lived in a housing project for 11 years, said, "I'd have a place to live that's not rat-infested. I'd have some friends. I wouldn't be so miserable all the time." Drew, a man tortured by perfectionism, mused, "I'd be wearing sloppy clothes, at least at times, and like it. More than that, I'd have a more realistic sense of the world and my place in it. The world is messy, it's chaotic much of the time. I'd find the beauty in the chaos."

- **What accomplishments would be in place that are not in place now?** Ryan, a divorced man in his mid-30s, said, "I'd have my degree in practical nursing. I'd be doing some part-time teaching. I'd be close to someone that I'd like to marry."

- **What would this opportunity look like if you developed it?** Enid, a woman with a great deal of talent who has been given one modest promotion in her company but who feels like a second-class citizen, had this to say: "In 2 years I'll be an officer of this company or have a very good job in another firm."

It is a mistake to suppose that clients will automatically gush with answers. Ask the kinds of questions just listed, or encourage them to ask themselves the questions, but then help them answer them. Create the therapeutic dialogue around possibilities for a better future. Many clients don't know how to use their innate creativity. Thinking divergently is not part of their mental lifestyle. You have to work with clients to help them produce some creative output. Some clients are reluctant to name possibilities for a better future because they sense that this will bring more responsibility. They will have to move into action mode.

Exemplars and Models as a Source of Possibilities

Some clients can see future possibilities better when they see them embodied in others. You can help clients brainstorm possibilities for a better future by helping them identify exemplars or models. By models, I don't mean superstars or people who do things perfectly. That would be self-defeating. In the next example, a marriage counselor is talking with a middle-aged, childless couple. They are bored with their marriage. When he asked them, "What would your marriage look like if it looked a little better?" he could see that they were stuck.

COUNSELOR: Maybe the question would be easier to answer if you reviewed some of your married relatives, friends, or acquaintances.

WIFE: None of them have super marriages. (Husband nods in agreement.)

COUNSELOR: No, I don't mean super marriages. I'm looking for things you could put in your marriage that would make it a little better.

WIFE: Well, Fred and Lisa are not like us. They don't always have to be doing everything together.

HUSBAND: Who says we have to be doing everything together? I thought that was your idea.

WIFE: Well, we always are together. If we weren't always together, we wouldn't be in each other's hair all the time.

COUNSELOR: All right, who else do you know who are doing things in their marriage that appeal to you? Anyone.

HUSBAND: You know Ron and Carol do some volunteer work together. Ron was saying that it gets them out of themselves. I bet they have better conversations because of it.

COUNSELOR: Now we're cooking. . . . What else? What couple do you find the most interesting?

Even though it was a somewhat torturous process, these two people were able to come up with a range of possibilities for a better marriage. The counselor had them write them down so they wouldn't lose them. At this point, the purpose was not to get the clients to commit themselves to these possibilities but to identify them.

In the following case, the client finds herself making discoveries by observing people she had not identified as models at all.

> Fran, a somewhat withdrawn college junior, realizes that when it comes to interpersonal competence, she is not ready for the business world she intends to enter when she graduates. She wants to do something about her interpersonal style and a few nagging personal problems. She sees a counselor in the Office of Student Services. After a couple of discussions with him, she joins a "lifestyle" group on campus that includes some training in interpersonal skills. Even though she expands her horizons a bit from what the members of the group say about their experiences, behaviors, and feelings, she tells her counselor that she learns even more by watching her fellow group members in action. She sees behaviors that she would like to incorporate in her own style. A number of times she says to herself in the group, "Ah, there's something I never thought of." Without becoming a slavish imitator, she begins to incorporate some of the patterns she sees in others into her own style.

Models or exemplars can help clients name what they want more specifically. Models can be found anywhere: among the client's relatives, friends, and associates, in books, on television, in history, in movies. Counselors can help clients identify models, choose those dimensions of others that are relevant, and translate what they see into realistic possibilities for themselves.

Lockwood and Kunda (1999) have shown that under normal circumstances, individuals can be inspired by role models so that their motivation and self-evaluations are enhanced. But not always. Bringing up role models with people who have been reviewing "best past selves" has a way of deflating people. Their best can pale in comparison with the model. This is important because in solution-focused therapies, reviewing past successes is an important part of the process. In addition, if people are asked to come up with ideas about their "best possible selves" and then are asked to review what they like about a role model, their ability to draw inspiration from the role model is impaired. In sum, using role models as sources of inspiration certainly works, but it can be tricky.

Cases Featuring Possibilities for a Better Future

Here are a couple of cases that illustrate how helping clients develop possibilities for a better future had a substantial impact.

The case of Brendan: Dying better. Brendan, a heavy drinker, had extensive and irreversible liver damage, and it was clear that he was getting sicker. But he wanted to "get some things done" before he died. Brendan's action orientation helped a great deal. Over the course of a few months, a counselor helped him to name some of the things he wanted before he died or on his journey toward death. Brendan came up with the following possibilities:

- "I'd like to have some talks with someone who has a religious orientation, like a minister. I want to discuss some of the 'bigger' issues of life and death."
- "I don't want to die hopeless. I want to die with a sense of meaning."
- "I want to belong. You know, to some kind of community, people who know what I'm going through, but who are not sentimental about it. People not disgusted with me because of the way I've done myself in."
- "I'd like to get rid of some of my financial worries."
- "I'd like a couple of close friends with whom I could share the ups and downs of daily life. With no apologies."
- "As long as possible, I'd like to be doing some kind of productive work, whether paid or not. I've been a flake. I want to contribute even if just in an ordinary way."
- "I need a decent place to live, maybe with others."
- "I need decent medical attention. I'd like a doctor who has some compassion. One who could challenge me to live until I die."
- "I need to manage these bouts of anxiety and depression better."
- "I want to be get back with my family again. I want to hug my dad. I want him to hug me."
- "I'd like to make peace with one or two of my closest friends. They more or less dropped me when I got sick. But at heart, they're good guys."
- "I want to die in my home town."

Of course, Brendan didn't name all these possibilities at once. Through understanding and probes, the counselor helped Brendan name what he needed and wanted and then helped him stitch together a set of goals from these possibilities (Stage II) and ways of accomplishing them (Stage III). Box 12-1 outlines the kinds of questions you can help clients ask themselves to discover possibilities for a better future.

The Washington family case. This case is more complex because it involves a family. Not only does the family as a unit have its wants and needs, but also each of the individual members has his or her own. Therefore, it is even more imperative to review possibilities for a better future so that competing needs can be reconciled.

Lane, the 15-year-old son of Troy and Rhonda Washington, was hospitalized with what was diagnosed as an "acute schizophrenic attack." He had two older brothers, both teenagers, and two younger sisters, one 10 and one 12, all living at home. The Washingtons lived in a large city. Although both parents worked, their combined income still left them pinching pennies. They also ran into a host of problems associated with their son's hospitalization—the need to arrange ongoing help and care for Lane, financial burdens, behavioral problems among the other siblings, marital conflict, and

Box 12-1   Questions for Exploring Possibilities

Help clients ask themselves these kinds of questions:

- What are my most critical needs and wants?
- What are some possibilities for a better future?
- What outcomes or accomplishments would take care of my most pressing problems?
- What would my life look like if I were to develop a couple of key opportunities?
- What should my life look like a year from now?
- What should I put in place that is currently not in place?
- What are some wild possibilities for making my life better?

stigma in the community ("They're a funny family with a crazy son"; "What kind of parents are they?"). To make things worse, they did not think the psychiatrist and the psychologist they met at the hospital took the time to understand their concerns. They felt that the helpers were trying to push Lane back out into the community; in their eyes, the hospital was "trying to get rid of him." "They give him some pills and then give him back to you" was their complaint. No one explained to them that short-term hospitalization was meant to guard the civil rights of patients and avoid the negative effects of longer-term institutionalization.

When Lane was discharged, his parents were told that he might have a relapse, but they were not told what to do about it. They faced the prospect of caring for Lane in a climate of stigma without adequate information, services, or relief. Feeling abandoned, they were very angry with the mental health establishment. They had no idea what they should do to respond to Lane's illness or to the range of family problems that had been precipitated by the episode. By chance, the Washingtons met someone who had worked for the National Alliance for the Mentally Ill (NAMI), an advocacy and education organization. This person referred them to an agency that provided support and help.

What does the future hold for such a family? With help, what kind of future can be fashioned? Social workers at the agency helped the Washingtons identify both needs and wants in seven areas (see Bernheim, 1989).

- **The home environment.** The Washingtons needed an environment in which the needs of all the family members are balanced. They didn't want their home be an extension of the hospital. They wanted Lane taken care of, but they wanted to attend to the needs of the other children and to their own needs as well.

- **Care outside the home.** They wanted a comprehensive therapeutic program for Lane. They needed to review possible services, identify relevant services, and arrange access to those services. They needed to find a way of paying for all this.

- **Care inside the home.** They wanted all family members to know how to cope with Lane's residual symptoms. He might be withdrawn or aggressive, but they needed to know how to relate to him and help him handle behavioral problems.

- **Prevention.** Family members needed to be able to spot early warning symptoms of impending relapse. They also needed to know what to do when they saw those signs, including such things as contacting the clinic or, in the case of more severe problems, arranging for an ambulance or getting help from the police.

- **Family stress.** They needed to know how to cope with the increased stress all of this would entail. They needed forums for working out their problems. They wanted to avoid family blowups, and when blowups occurred, they wanted to manage them without damaging the social fabric of the family.

- **Stigma.** They wanted to understand and be able to cope with whatever stigma might be attached to Lane's illness. For instance, when taunted for having a "crazy brother," the children needed to know what to do and what not to do. Family members needed to know whom to tell, what to say, how to respond to inquiries, and how to deal with blame and insults.

- **Limitation of grief.** They needed to know how to manage the normal guilt, anger, frustration, fear, and grief that go with problem situations like this.

Bernheim's schema constituted a useful checklist for stimulating thinking about possibilities for a better future. The Washingtons first needed help in developing these possibilities. Then they needed help in setting priorities and establish goals to be accomplished. This is the work of Task 2.

When it comes to serious mental illness in a family, Marsh and Johnson (1997) focus not just on family burden but also on family resilience and the internal and external resources that support such resilience. This is, of course, a positive psychology approach. They list the ways in which helpers can assist families in (p. 233):

1. Understanding and normalizing the family experience of mental illness
2. Focusing on the strengths and competencies of their family and relatives
3. Learning about mental illness, the mental health system, and community resources
4. Developing skills in stress management, problem solving, and communication
5. Resolving their feelings of grief and loss
6. Coping with the symptoms of mental illness and its repercussions for their family
7. Identifying and responding to the signs of impending relapse
8. Creating a supportive family environment
9. Developing realistic expectations for all members of the family
10. Playing a meaningful role in their relative's treatment, rehabilitation, and recovery
11. Maintaining a balance that meets the needs of all members of the family

They outline a number of interventions strategies that can help families meet these objectives:

- **Family interventions** that stress the role of the family as a support system rather than the cause of mental illness.

- **Family support and advocacy groups** such as the National Alliance for the Mentally Ill. These groups provide support, education, and encourage advocacy for improved services.

- **Family consultation,** which can aid in helping families determine their own goals and make informed choices regarding their use of available services.
- **Family education** with respect to information about mental illness, caregiving, the mental health system, community resources, and the like.
- **Family psychoeducation,** which focuses on such things as coping strategies and stress management.

In all of this, you can see the spirit of the solution-focused philosophy discussed in Chapter 11.

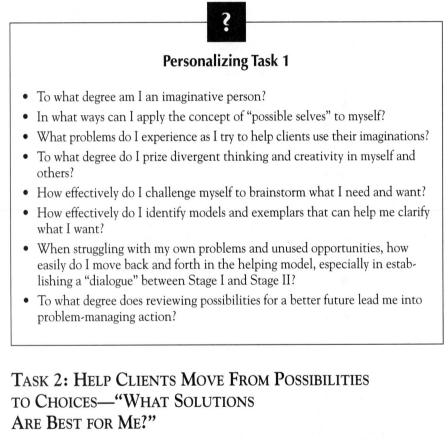

?

Personalizing Task 1

- To what degree am I an imaginative person?
- In what ways can I apply the concept of "possible selves" to myself?
- What problems do I experience as I try to help clients use their imaginations?
- To what degree do I prize divergent thinking and creativity in myself and others?
- How effectively do I challenge myself to brainstorm what I need and want?
- How effectively do I identify models and exemplars that can help me clarify what I want?
- When struggling with my own problems and unused opportunities, how easily do I move back and forth in the helping model, especially in establishing a "dialogue" between Stage I and Stage II?
- To what degree does reviewing possibilities for a better future lead me into problem-managing action?

TASK 2: HELP CLIENTS MOVE FROM POSSIBILITIES TO CHOICES—"WHAT SOLUTIONS ARE BEST FOR ME?"

In Chapter 3 of her book, Wosket (2006) recalls something I said during a lecture she attended. Stage I of the helping model is about *failed solutions*. Stage II is about solutions as *problem-managing accomplishments*. Stage III is about solutions as *strategies*. Once possibilities for a better future have been developed, clients need to make some choices—that is, they need to choose one or more of those possibilities and turn them into a program for constructive change. Task 1 is, in many ways, about *creativity*, getting rid of boundaries, thinking beyond one's limited horizon, moving outside the box. Task 2 is about *innovation*—that is, turning possibilities into a practical program for change. If implemented, a goal constitutes the "solution" with a Big S for the client's problem or opportunity. Consider the following case.

Bea, an African American woman, was arrested when she went on a rampage in a bank and broke several windows. She had exploded with anger because she felt that she had been denied a loan mainly because she was black and a single mother. In discussing the incident with her minister, she comes to see that she has become very prone to anger. Almost anything can get her going. She also realizes that venting her anger as she had done in the bank led to a range of negative consequences. But she is constantly "steamed up" about "the system." To complicate the picture, she tends to take her anger out on those around her, including her friends and her two children. The minister helps her look at four possible ways of dealing with her anger—venting it, repressing it, channeling it, or simply giving up and ignoring the things she gets angry at, including the injustices around her. Giving up is not in her makeup. Merely venting her anger seems to do little but make her more angry. Repressing her anger, she reasons, is just another way of giving up, and that is demeaning. And she's not very good at repressing anyway. The "channeling" option needs to be explored.

In the end, Bea takes a positive psychology approach to dealing with her frustrations. She joins a political action group involved in community organizing. She learns that she can channel her anger without giving up her values or her intensity. She also discovers that she is good at influencing others and getting things done. She begins to feel better about herself. The "system" doesn't seem to be such a fortress any more.

Because goals can be highly motivational, helping clients set realistic goals is one of the most important tasks of the helping process.

Help Clients Design and Shape Their Goals

Practical goals do not usually leap out fully formed. They need to be shaped, or "designed." Effective counselors add value by engaging clients in the kind of dialogue that will help them design, choose, craft, shape, and develop their goals. Goals are specific statements about what clients want and need.

The goals that emerge through this client-helper dialogue are more likely to be workable if they have the following characteristics. They need to be:

- stated as **outcomes** rather than activities
- **specific** enough to be verifiable and to drive action
- **substantive** and challenging
- both **venturesome** and **prudent**
- **realistic** in regard to resources needed to accomplish them
- **sustainable** over a reasonable time period
- **flexible** without being wishy-washy
- **congruent** with the client's **values**
- set in a **reasonable time frame**

Just how this package of goal characteristics will look in practice will differ from client to client. There is no one formula. From a practical point of view, these characteristics can be seen as "tools" that counselors can use to help clients design and shape or reshape their goals. Ineffective helpers will get lost in the details of these characteristics. Effective helpers will keep them in the back of their mind and, in a second-nature manner, turn them into helpful "sculpting" probes at the right time. These characteristics, then, take on life through the following flexible principles.

Help clients state what they need and want as outcomes or accomplishments.
The goal of counseling, as emphasized again and again, is neither discussing nor planning nor engaging in activities. Helping is about solutions with a Big S. "I want to start doing some exercise" is an activity rather than an outcome. "Within 6 months I will be running three miles in less than 30 minutes at least four times a week" is an outcome. It is a pattern of behavior that will be in place by a certain time. If a client says, "My goal is to get some training in interpersonal communication skills," then she is stating her goal as a set of activities—solution with a small s—rather than as an accomplishment. But if she says that she wants to become a better listener as a wife and mother, then she is stating her goal as an accomplishment, even though "better listener" needs further clarification. Goals stated as outcomes provide direction for clients.

You can help clients describe what they need and want by using this "past-participle" approach—drinking stopped, number of marital fights decreased, anger habitually controlled. Stating goals as outcomes or accomplishments is not just a question of language. Helping clients state goals as accomplishments rather than activities helps them avoid directionless and imprudent action. If a woman with breast cancer says that she thinks she should join a self-help group, she should be helped to see what she wants to get out of such a group. Joining a group and participating in it are activities. She wants support. She wants to feel supported. Goals, at their best, are expressions of what clients need and want. Clients who know what they want are more likely to work not just harder but also smarter.

Consider the case of Chester, a former Marine suffering from posttraumatic stress disorder.

> Chester was involved in the Kosovo peacekeeping effort. During a patrol, he and three of his buddies shot and killed four civilians who were out to kill their neighbors. Afterward he began acting in strange ways, wandering around at times in a daze. He was sent home and given a medical discharge. Although he seemed to recover, he lived an aimless life. He went to college but dropped out during the first semester. He became rather reclusive but never really engaged in odd behavior. Rather, he was sinking into the landscape. He moved in and out of a number of low-paying jobs. He also became less careful about his appearance. He said to a counselor, "You know, I used to be very careful about the way I dressed. Kind of proud of myself in the Marine tradition. Don't get me wrong; I'm not a bum and don't smell or anything, but I'm not myself." The whole direction of Chester's life was wrong; he was headed for serious trouble. He was bothered by thoughts about the war and had taken to sleeping whenever he felt like it, day or night, "just to make it all stop."

Ed, Chester's counselor, had a good relationship with Chester. He helped Chester tell his story and challenged some of his self-defeating thinking. He went on to help Chester focus on what he wanted from life. They moved back and forth between Stage I and Stage II, between problems and possibilities for a better future. Eventually, Chester began talking about his real needs and wants—that is, what he needed to accomplish to "get back to his old self." Here is an excerpt from their dialogue.

CHESTER: I've got to stop hiding in my hole. I'm going to get out and see people more. I'm going to stop feeling so damn sorry for myself. Who wants to be with a nothing!

COUNSELOR: What will Chester's life look like a year or two from now?

CHESTER: One thing for sure. He will be seeing women again. He might not be married, but he will probably have a special girlfriend. And she will see him as an ordinary guy.

Here Chester talks about changes as patterns of behavior that will be in place. He is painting a picture of what he wants to be. He is designing some goals. The counselor's probe reinforces this outcome approach.

Help clients move from broad aims to clear and specific goals. Specific rather than general goals tend to drive behavior. Therefore, broad goals need to be translated into more specific goals and tailored to the needs and abilities of each client. Skilled helpers use probes to help clients move from the general to the specific.

Chester said that he wanted to become "more disciplined." His counselor helped him make that more specific.

COUNSELOR: What areas do you want to focus on?

CHESTER: Well, if I'm going to put more order in my life, I need to look at the times I sleep. I've been going to bed whenever I feel like it and getting up whenever I feel like it. It was the only way I could get rid of those thoughts and the anxiety. But I'm not nearly as anxious as I used to be. Things are calming down.

COUNSELOR: So more disciplined means a more regular sleep schedule because there's no particular reason now for not having one.

CHESTER: Yeah, sleeping whenever I want is just a bad habit. And I can't get things done if I'm asleep.

Chester goes on to translate "more disciplined" into other problem-managing needs and wants related to school, work, and his appearance. Greater discipline, once translated into specific patterns of behavior, will have a decidedly positive impact on his life.

Counselors often add value by helping clients move from good intentions and vague desires to broad aims and then on to quite specific goals.

Good intentions. "I need to do something about this" is a statement of intent. However, even though good intentions are a good start, they need to be translated into aims and goals. In the following example, the client, Jon, has been discussing his relationship with his wife and children. The counselor has been helping him see that his "commitment to work" is perceived negatively by his family. Jon is open to challenge and is a fast learner.

JON: Boy, this session has been an eye-opener for me. I've really been blind. My wife and kids don't see my investment—rather, my overinvestment—in work as something I'm doing for them. I've been fooling myself, telling myself that I'm working hard to get them the good things in life. In fact, I'm spending most of my time at work because I like it. My work is mainly for me. It's time for me to realign some of my priorities.

The last statement is a good intention, an indication on Jon's part that he wants to do something about a problem now that he sees it more clearly. It may be that Jon will now go out and put a different pattern of behavior in place without further help from the counselor. Or he may benefit from some help in realigning his priorities.

Broad aims. A broad aim is more than a good intention. It has content—that is, it identifies the area in which the client wants to work and makes some general statement about that area. Let's return to the example of Jon and his overinvestment in work.

JON: I don't think I'm spending so much time at work in order to run away from family life. But family life is deteriorating because I'm just not around enough. I must spend more time with my wife and kids. Actually, it's not just a case of must. I want to.

Jon moves from a declaration of intent to an aim or a broad goal, spending more time at home. But he still has not created a picture of what that would look like.

Specific goals. To help Jon move toward greater specificity, the counselor uses such probes as "Tell me what 'spending more time at home' will look like."

JON: I'm going to consistently spend 3 out of 4 weekends a month at home. During the week I'll work no more than 2 evenings.

COUNSELOR: So you'll be at home a lot more. Tell me what you'll be doing with all this time.

Notice how much more specific Jon's statement is than "I'm going to spend more time with my family." He sets a goal as a specific pattern of behavior he wants to put in place. But his goal as stated deals with quantity, not quality. The counselor's probe is really a challenge. It's not just the amount of time Jon is going to spend with his family but also the kinds of things he will be doing. Quality time, some call it. Though a client trying to come to grips with work-life balance once said to me, "My family, especially my kids, don't make the distinction between quantity and quality. For them quantity is quality. Or there's no quality without a chunk of quantity." This warrants further discussion because maybe the family wants a relaxed rather than an intense Jon at home.

This example brings up the difference between *instrumental* goals and *higher-order* or *ultimate* goals. Jon's ultimate goal is "a good family life." Such a goal, once spelled out, will differ from family to family and from culture to culture. Think of your own definition. Therefore, when Jon says that one of his goals is spending more time at home, he is talking about an instrumental goal. Unless he's there, he can't do things with his wife and kids. But although just "being there" is a goal because it is a pattern of behavior *in place*, it is certainly not Jon's ultimate goal. But Jon is not worried about the ultimate goal. When he is there, they have a rich family life together. That's not the problem. However, because instrumental goals are *strategies* for achieving higher-order goals, it's important to make sure that the client has clarity about the higher-order goal. If Jon was spending a lot of time at the office because he didn't like being with his wife and kids or because there was a great deal of conflict at home, then his higher-order goal would be something like "experiencing the stimulation of an exciting workplace" (if home life was dull) or "peace of mind" (if home life was full of conflict). When you are helping clients design and shape instrumental goals, make sure they can answer the "instrumental-for-what?" question.

Helping clients move from good intentions to more and more specific goals is a shaping process. Consider the example of a couple whose marriage has degenerated into constant bickering, especially about finances.

- **Good intention.** "We want to straighten out our marriage."
- **Broad aim.** "We want to handle our decisions about finances in a much more constructive way."
- **Specific goal.** "We try to solve our problems about family finances by fighting and arguing. We'd like to reduce the number of fights we have and begin making mutual decisions about money. We yell instead of talking things out. We need to set up a month-by-month budget. Otherwise, we'll be arguing about money we don't even have. We'll have a trial budget ready the next time we meet with you."

Having sound household finances is a fine goal. In fact, it's a goal in itself. Reducing unproductive conflict is also a fine goal. In this case, however, installing a sound, fair, and flexible household budget system is also instrumental to establishing peace at home. Declarations of intent, broad goals, and specific goals can all drive constructive behavior, but specific goals have the best chance. Is it possible to get clients to be too specific about their goals? Yes, if they get lost in the planning details and crafting the goal becomes more important than the goal itself.

If the goal is clear enough, the client will be able to determine progress toward the goal. For many clients, being able to measure progress is an important incentive. If goals are stated too broadly, it is difficult to determine both progress and accomplishment. "I want to have a better relationship with my wife" is a very broad goal, difficult to verify. "I want to socialize more, you know, with couples we both enjoy" comes closer, but "socialize more" needs more clarity.

It is not always necessary to count things to determine whether a goal has been reached, though sometimes counting is helpful. Helping is about living more fully, not about accounting activities. At a minimum, however, desired outcomes need to be capable of being verified in some way. For instance, a couple might say something like "Our relationship is better, not because we've stop squabbling. In fact, we've discovered that we like to squabble. But life is better because the meanness has gone out of our squabbling. We accept each other more. We listen more carefully, we talk about more personal concerns, we are more relaxed, and we make more mutual decisions about issues that affect us both." This couple does not need a scientific experiment to verify that they have improved their relationship.

Help clients establish goals with substance, goals that make a difference. Outcomes and accomplishments are meaningless if they do not have the required impact on the client's life. The goals clients choose should have substance to them— that is, some significant contribution toward managing the original problem situation or developing some opportunity.

> Vitorio ran the family business. His son, Anthony, worked in sales. After spending a few years learning the business and getting an MBA part time at a local university, Anthony wanted more responsibility and authority. His father never thought that he was "ready." They began arguing quite a bit, and their relationship suffered from it. Finally, a friend of the family persuaded them to spend time with a consultant-counselor who worked with small family businesses. He spent relatively little time listening to their problems. After all, he had seen this same problem over and over again—the reluctance and conservatism of the father, the pushiness of the son.
>
> Vitorio wanted the business to stay on a tried-and-true course. Anthony wanted to be the company's marketer, to move it into new territory. After a number of discussions with the consultant-counselor, they settled on this scenario: A "marketing department" headed by Anthony would be created. He could divide his time between sales and marketing as he saw fit, provided that he maintained the current level of sales. Vitorio agreed not to interfere. They would meet once a month with the consultant-counselor to discuss problems and progress. Vitorio insisted that the consultant's fee come from increased sales. After some initial turmoil, the bickering decreased dramatically. Anthony easily found new customers, although they demanded modifications in the product line, which Vitorio reluctantly approved. Both sales and margins increased to the point that they needed another person in sales.

Not all issues in family businesses are handled as easily. In fact, a few years later, Anthony left the business and founded his own. But the goal package they worked out—the deal they cut—made quite a difference both in the father-son relationship and in the business.

Second, goals have substance to the degree that they help clients "stretch" themselves. As Locke and Latham (1984, pp. 21, 26) noted, "Extensive research . . . has established that, within reasonable limits, the . . . more challenging the goal, the better the resulting performance. . . . People try harder to attain the hard goal. They exert more effort. . . . In short, people become motivated in proportion to the level of challenge with which they are faced. . . . Even goals that cannot be fully reached will lead to high effort levels, provided that partial success can be achieved and is rewarded. Consider the following case.

> A young woman became a quadriplegic because of an auto accident. In the begin-
> ning, she was full of self-loathing—"The accident was all my fault; I was just stupid."
> She was close to despair. Over time, however, with the help of a counselor, she came
> to see herself, not as a victim of her own "stupidity," but as someone who could bring
> hope to young people with life-changing afflictions. In her spare time, she visited
> young patients in hospitals and rehabilitation centers, got some to join self-help
> groups, and generally helped people like herself to manage an impossible situation in
> a more humane way. One day she said to her counselor, "The best thing I ever did was
> to stop being a victim and become a fellow traveler with people like myself. The last
> 2 years, though bitter at times, have been the best years of my life." She had set her
> goals quite high—becoming an outgoing helper instead of remaining a self-centered
> victim, but they proved to be quite realistic.

Of course, when it comes to goals, "challenging" should not mean "impossible." There seems to be a curvilinear relationship between goal difficulty and goal performance. If the goal is too easy, people see it as trivial and ignore it. If the goal is too difficult, it is not accepted. However, this difficulty-performance ratio differs from person to person. What is small for some is big for others. To deal with the variations, see the section below on "adaptive" goals.

Help clients set goals that are prudent. Although the helping model described in this book encourages a bias toward client action, action needs to be both directional *and* wise. Discussing and setting goals should contribute to both direction and wisdom. The following case begins poorly but ends well.

> Harry was a sophomore in college who was admitted to a state mental hospital
> because of some bizarre behavior at the university. He was one of the disc jockeys for
> the university radio station. College officials noticed him one day when he put on an
> attention-getting performance that included rather lengthy dramatizations of
> grandiose religious themes. In the hospital, the counselors soon discovered that this
> quite pleasant, likable young man was actually a loner. Everyone who knew him at
> the university thought that he had many friends, but in fact he did not. The campus
> was large, and his lack of friends went unnoticed.
> Harry was soon released from the hospital but returned weekly for therapy. At one
> point he talked about his relationships with women. Once it became clear to him that
> his meetings with women were perfunctory and almost always took place in groups—
> he had imagined that he had a rather full social life with women—Harry launched a
> full program of getting involved with the opposite sex. His efforts ended in disaster,
> however, because Harry had some basic sexual and communication problems. He
> also had serious doubts about his own worth and therefore found it difficult to make
> a gift of himself to others. He ended up in the hospital again.
> The counselor helped Harry get over his sense of failure by emphasizing what
> Harry could learn from the "disaster." With the therapist's help, Harry returned to the

problem-clarification and new-perspectives part of the helping process and then established more realistic short-term goals regarding getting back "into community." The direction was the same—establishing a realistic social life—but the goals were now more prudent because they were "bite-size." Harry attended socials at a local church where a church volunteer provided support and guidance.

Harry's leaping from problem clarification to action without taking time to discuss possibilities and set reasonable goals was part of the problem rather than part of the solution. His lack of success in establishing solid relationships with women actually helped him see his problem with women more clearly. There are two kinds of prudence—playing it safe is one; doing the wise thing is the other. Problem management and opportunity development should be venturesome. They are about making wise choices rather than playing it safe.

Help clients formulate realistic goals. Setting stretch goals can help clients energize themselves. They rise to the challenge. On the other hand, goals set too high can do more harm than good. Locke and Latham (1984, p. 39) put it succinctly:

> Nothing breeds success like success. Conversely, nothing causes feelings of despair like perpetual failure. A primary purpose of goal setting is to increase the motivation level of the individual. But goal setting can have precisely the opposite effect if it produces a yardstick that constantly makes the individual feel inadequate.

A goal is realistic if the client has access to the resources needed to accomplish it, the goal is under the client's control, and external circumstances do not prevent its accomplishment.

Resources: Help clients choose goals for which the resources are available. It does little good to help clients develop specific, substantive, and verifiable goals if the resources needed for their accomplishment are not available. Consider the case of Rory, who has had to take a demotion because of merger and extensive restructuring. He now wants to leave the company and become a consultant.

INSUFFICIENT RESOURCES: Rory does not have the assertiveness, marketing savvy, industry expertise, or interpersonal style needed to become an effective consultant. Even if he did, he does not have the financial resources needed to tide him over while he develops a business.

SUFFICIENT RESOURCES: Challenged by the outplacement counselor, Rory changes his focus. Graphic design is an avocation of his. He is not good enough to take a technical position in the company's design department, but he does apply for a supervisory role in that department. He is good with people, very good at scheduling and planning, and knows enough about graphic design to discuss issues meaningfully with the members of the department.

Rory combines his managerial skills with his interest in graphic design to move in a more realistic direction. The move is challenging, but it can have a substantial impact on his work life. For instance, the opportunity to hone his graphic design skills will open up further career possibilities.

Control: Help clients choose goals that are under their control. Sometimes clients defeat their own purposes by setting goals that are not under their control. For instance, it is common for people to believe that their problems would be solved if only other people would not act the way they do. In most cases, however, we do not have any direct control over the ways others act. Consider the following example.

> Tony, a 16-year-old boy, felt that he was the victim of his parents' inability to relate to each other. Each tried to use him in the struggle, and at times he felt like a Ping-Pong ball. A counselor helped him see that he could probably do little to control his parents' behavior but that he might be able to do quite a bit to control his *reactions* to his parents' attempts to use him. For instance, when his parents started to fight, he could simply leave instead of trying to "help." If either tried to enlist him as an ally, he could say that he had no way of knowing who was right. Tony also worked at creating a good social life outside the home. That helped him weather the tensions he experienced when at home.

Tony needed a new way of managing his interactions with his parents to minimize their attempts to use him as a pawn in their own interpersonal game. Goals are not under clients' control if they are blocked by external forces that they cannot influence. "To live in a free country" may be an unrealistic goal for a person living in a totalitarian state because he cannot change internal politics, nor can he change emigration laws in his own country or immigration laws in other countries. "To live as freely as possible in a totalitarian state" might well be an aim that could be translated into realistic goals.

Help clients set goals that can be sustained. Clients need to commit themselves to goals that have staying power. One separated couple said that they wanted to get back together again. They did so only to get divorced again within 6 months. Their goal of getting back together again was achievable but not sustainable. Perhaps they should have asked themselves, "What do we need to do not only to get back together but also to stay together? What would our marriage have to look like to become and remain workable?" In discretionary-change situations, the issue of sustainability needs to be visited early on.

Many Alcoholics Anonymous–like programs work because of their one-day-at-a-time approach. The goal of being, say, drug-free has to be sustained only over a single day. The next day is a new era. In a previous example, Vitorio and Anthony's arrangement had enough staying power to produce good results in the short term. It also allowed them to reset their relationship and to improve the business. The goal was not designed to produce a lasting business arrangement because, in the end, Anthony's aspirations were bigger than the family business.

Help clients choose goals that have some flexibility. In many cases, goals have to be adapted to changing realities. Therefore, there might be some trade-offs between goal specificity and goal flexibility in uncertain situations. Napoleon noted this when he said, "He will not go far who knows from the first where he is going." Sometimes making goals too specific or too rigid does not allow clients to take advantage of emerging opportunities.

> Even though he liked the work and even the company he worked for, Jessie felt like a second-class citizen. He thought that his supervisor gave him most of the dirty work and that there was an undercurrent of prejudice against Hispanics in his department. Jessie wanted to quit and get another job, one that would pay the same relatively good wages he was now earning. A counselor helped Jessie challenge his choice. Even though the economy was booming, the industry in which Jessie was working was in recession. There were few jobs available for workers with Jessie's set of skills.

The counselor helped Jessie choose an interim goal that was more flexible and more directly related to coping with his present situation. The interim goal was to use his time preparing himself for a better job outside this industry. In 6 months to a year he could be better prepared for a career in a still healthy economy. Jessie began volunteering for special assignments that helped him learn some new skills and took some crash courses dealing with computers and the Internet. He felt good about what he was learning and more easily ignored the prejudice.

Counseling is a living, organic process. Just as organisms adapt to their changing environments, clients' choices need to be adapted to their changing circumstances.

Help clients choose goals consistent with their values. Although helping is a process of social influence, it remains ethical only if it respects, within reason, the values of the client. Values are criteria we use to make decisions. Helpers may challenge clients to reexamine their values, but they should not encourage clients to perform actions that are not in keeping with their values.

The son of Vincente and Consuela Garza is in a coma in the hospital after an automobile accident. He needs a life-support system to remain alive. His parents are experiencing a great deal of uncertainty, pain, and anxiety. They have been told that there is practically no chance that their son will ever come out of the coma. One possibility is to terminate the life-support system. The counselor should not urge them to terminate the life-support system if that is counter to their values. She can help them explore and clarify their values. In this case, the counselor suggests that they discuss their decision with their clergyman. In doing so, they find out that the termination of the life-support system would not be against the tenets of their religion. Now they are free to explore other values that relate to their decision.

Some problems involve a client's trying to pursue contradictory goals or values. Chester, the ex-Marine, wanted to get an education, but he also wanted to make a decent living as soon as possible. The former goal would put him in debt, but failing to get a college education would lessen his chances of securing the kind of job he wanted. The counselor helps him identify and use his values to consider some trade-offs. Chester chooses to work part time and go to school part time. He chooses an office job instead of one in construction. Even though the latter pays better, it would be much more exhausting and would leave him with little energy for school.

Help clients establish realistic time frames for accomplishing goals. Goals that are to be accomplished "sometime or other" probably won't be accomplished at all. Therefore, helping clients put some time frames in their goals can add value. Greenberg (1986) talked about immediate, intermediate, and final outcomes. Here's what they look like when applied to Janette's problem situation. She suffers in a variety of ways because she lets others take advantage of her. She needs to become more assertive and to stand up for her own rights.

- **Immediate outcomes** are changes in attitudes and behaviors evident in the helping sessions themselves. For Janette, the helping sessions constitute a safe forum for her to become more assertive. In her dialogues with her counselor, she learns and practices the skills of being more assertive.

- **Intermediate outcomes** are changes in attitudes and behaviors that lead to further change. It takes Janette a while to transfer her assertiveness skills both to the workplace and to her social life. She chooses relatively safe situations to practice being more assertive. For instance, she stands up to her mother more.

- **Final outcomes** refer to the completion of the overall program for constructive change through which problems are managed and opportunities developed. It takes more than 2 years for Janette to become assertive in a consistent, day-to-day way.

The next example deals with a young man who has been caught shoplifting. Here, too, there are immediate, intermediate, and final outcomes.

> Jensen, a 22-year-old on probation for shoplifting, was seeing a counselor as part of a court-mandated program. An immediate need in his case was overcoming his resistance to his court-appointed counselor and developing a working alliance with her. Because of the counselor's skills and her unapologetic caring attitude that had some toughness in it, he quickly came to see her as "on his side." Their relationship became a platform for establishing further goals. An intermediate outcome was a change in attitude. Brainwashed by what he saw on television, Jensen thought that America owed him some of its affluence and that personal effort had little to do with it. The counselor helped him see that his entitlement attitude was unrealistic and that hard work played a key role in most payoffs. There were two significant final outcomes in Jensen's case. First, he made it through the probation period free of any further shoplifting attempts. Second, he acquired and kept a job that helped him pay his debt to the retailer.

Taussig (1987) talked about the usefulness of setting and executing minigoals early in the helping process. Consider the case of Gaston.

> Gaston, a 16-year-old school dropout and loner, was arrested for arson. Though he lived in the inner city and came from a single-parent household, it was difficult to discover just why he had turned to arson. He had torched a few structures that seemed relatively safe to burn. No one was injured. Was his behavior a cry for help? Social rage expressed in vandalism? Just a way to get some kicks? His assigned social worker found these questions too speculative to be of much help. Instead of looking for the root causes of Gaston's malaise, she tried to help him set some simple goals that appealed to him and that could be accomplished relatively quickly.
>
> One goal was social support. The counselor helped Gaston join a social club at a local youth center. A second goal was having a role model. Gaston struck up a friendship with one of the more active members of the center, a dropout who had gotten a high school equivalency degree. He also received some special attention from one of the adult monitors of the center. This was the first time he had experienced the presence of a strong adult male in his life. A third goal was broadening his view of the world. A group of college students who did volunteer work in both the black and the white communities invited Gaston and a couple of the other boys to help them in a housing facility for the elderly located in a white neighborhood. This was the first time he had been engaged in any kind of work outside the black community. The experience helped him push back the walls a bit. He saw white people with real needs. The accomplishment of these minigoals helped Gaston become a bit more realistic about the world around him. He enjoyed the camaraderie of the volunteer group and began experiencing himself in a new, more constructive way.

It is not suggested here that goal setting is a facile answer to intractable social problems. But the achievement of sequenced minigoals can go a long way toward making a dent in these problems.

There is no such thing as a set time frame for every client. Some goals need to be accomplished now, some soon; others are short-term goals; still others are long-term. Consider the case of a priest who had been unjustly accused of child molestation.

- A *"now" goal:* some immediate relief from debilitating anxiety attacks and keeping his equilibrium during the investigation and court procedures

Box 12-2   Questions for Choosing and Shaping Goals

Help clients ask themselves these kinds of questions in order to shape their goals.

- Is the goal stated in outcome or results language?
- Is the goal specific enough to drive behavior? How will I know when I have accomplished it?
- If I accomplish this goal, will it make a difference? Will it really help manage the problems and opportunities I have identified?
- Does this goal have "bite" while remaining prudent?
- Is it doable?
- Can I sustain this goal over the long haul?
- Does this goal have some flexibility?
- Is this goal in keeping with my values?
- Have I set a realistic time frame for the accomplishment of the goal?

- A *"soon" goal:* obtaining the right kind of legal aid
- A *short-term goal:* winning the court case
- A *long-term goal:* reestablishing his credibility in the community and learning how to live with those who would continue to suspect him

There is no particular formula for helping all clients choose the right mix of goals at the right time and in the right sequence. Although helping is based on problem-management principles, it remains an art.

It is not always necessary, then, to make sure that each goal in a client's program for constructive change has all the characteristics outlined in this chapter. For some clients, identifying broad goals is enough to kick-start the entire problem-management and opportunity-development process. They shape the goals themselves. For others, some help in formulating more specific goals is called for. The principle is clear: Help clients develop goals that have some sort of agency—if not urgency—built in. In one case, this may mean helping a client deal with clarity; in another, with substance; in still another, with realism, values, or time frame. Box 12-2 outlines some questions that you can help clients ask themselves to choose and shape the most useful goals.

Help Clients Distinguish Needs From Wants

In some cases, what clients want and what they need coincide. The lonely person wants a better social life and needs some kind of community to live a more engaging human life. In other cases, clients might not want what they need. The alcoholic may need a life of total abstention but wants to drink moderately. Brainstorming possibilities for a better future should focus on the package of needs and wants that makes sense for this particular client. Consider the case of Irv.

Irv, a 41-year-old entrepreneur, collapsed one day at work. He had not had a physical in years. He was shocked to learn that he had both a mild heart condition and multiple sclerosis. His future was uncertain. The father of one of his wife's friends had multiple sclerosis but had lived and worked well into his 70s. But no one knew what the course of the disease would be. Because he had made his living by developing and then selling small businesses, he wanted to continue to do this, but it was too physically demanding. What he needed was a less physically demanding work schedule. Working 60–70 hours per week, even though he loved it, was no longer in the cards. Furthermore, he had always plowed the money he received from selling one business into starting up another. But now he needed to think of the future financial well-being of his wife and three children. Up to this point, his philosophy had been that the future would take care of itself. It was very wrenching for him to move from a lifestyle he wanted to one he needed.

Irv was a voluntary client who had to look at needs instead of wants. Involuntary clients often need to be challenged to look beyond their wants to their needs. One woman who voluntarily led a homeless life was attacked and severely beaten on the street. But she still wanted the freedom that came with her lifestyle. When challenged to consider the kinds of freedom she wanted, she admitted that freedom from responsibility was at the core. "I want to do what I want to do when I want to do it." It was her choice to live the way she wanted. The counselor helped her explore the consequences of her choices and tried to help her look at other options. How could she be "free" and not at risk? Was there some kind of trade-off between what she wanted and what she needed? In the end, of course, the decision was hers.

In the following case, the client, dogged by depression, was ultimately able to integrate what he wanted with what he needed.

Milos had come to the United States as a political refugee. The last few months in his native land had been terrifying. He had been jailed and beaten. He got out just before another crackdown. Once the initial euphoria of having escaped had subsided, he spent months feeling confused and disorganized. He tried to live as he had in his own country, but the North American culture was too invasive. He thought he should feel grateful, and yet he felt hostile. After 2 years of misery, he began seeing a counselor. He had resisted getting help because "back home" he had been "his own man."

In discussing these issues with a counselor, it gradually dawned on him that he *wanted* to reestablish links with his native land but that he *needed* to integrate himself into the life of his host country. He saw that the accomplishment of both these broad aims would be very freeing. He began finding out how other immigrants who had been here longer than he had accomplished this goal. He spent time in the immigrant community, which differed from the refugee community. In the immigrant community, there was a long history of keeping links to the homeland culture alive. But the immigrants had also adapted to their adopted country in practical ways that made sense to them. The friends he made became role models for him. The more active he became in the immigrant community, the more his depression lifted.

In this case, goals responded to a mixture of needs and wants. If Milos had focused only on one or the other, he would have remained unhappy.

Help Clients Spot Emerging Goals

It is not always a question of *designing* and *setting* goals in an explicit way. Rather, goals can naturally emerge through the client-helper dialogue. Often when clients talk about problems and unused opportunities, possible goals and action strategies bubble up. Once clients are helped to clarify a problem situation through a combination of probing, empathic highlights, and challenge, they begin to see more clearly what they want and what they have to do to manage the problem. Indeed, some clients must first act in some way before they find out just what they want to do. After goals begin to emerge, counselors can help clients clarify them and find ways to implement them. However, "emerge" should not mean that clients wait around until "something comes up." Nor should it mean that clients try many different solutions in the hope that one of them will work. These kinds of "emergence" tend to be self-defeating.

Although goals do often emerge, explicit goal setting is not to be underrated. Taussig (1987) showed that clients respond positively to goal setting even when goals are set very early in the counseling process. A client-centered, "no one right formula" approach seems to be best. Although all clients need focus and direction in managing problems and developing opportunities, what focus and direction will look like will differ from client to client.

Help Clients Set Adaptive Goals

Collins and Porras (1994) coined the term "big, hairy, audacious" goals (BHAGs) for "super-stretch" goals. However, the term fits better into the hype of business than the practicalities of helping. It is true that some clients are looking for big goals. They believe, and perhaps rightly so, that without big goals their lives will not be substantially different. But even clients who choose goals that can be called "big" in one way or another need a bit-by-bit approach to achieving these goals. It is usually better to take big goals and divide them up into smaller pieces lest the big goal on its own seems too daunting. The term "within reasonable limits" will differ from client to client.

Adaptive goals. Although difficult or "stretch" goals are often the most motivational, this is not true in every case. Some clients choose to make very substantive changes in their lives, but others take a more modest approach. Wheeler and Janis (1980, p. 98) cautioned against the search for the "absolute best" goal all the time: "Sometimes it is more reasonable to choose a satisfactory alternative than to continue searching for the absolute best. The time, energy, and expense of finding the best possible choice may outweigh the improvement in the choice." Consider the following case.

> Joyce, a near-middle-aged buyer for a large retail chain, centered most of her nonworking life on her aging mother. Joyce had even turned down promotions because the new positions would have demanded more travel and longer hours. Her mother had been pampered by her now-deceased husband and her three children and allowed to have her way all her life. She now played the role of the tyrannical old woman who constantly feels neglected and who can never be satisfied. Though Joyce knew that she could live much more independently without abandoning her mother, she found it very difficult to move in that direction. Guilt stood in the way of any change in her relationship with her mother. She even said that being a virtual slave to her mother's whims was not as bad as the guilt she experienced when she stood up to her mother or "neglected" her.
>
> The counselor helped Joyce experiment with a few new ways of dealing with her mother. For instance, Joyce went on a 2-week trip with friends even though her mother objected, saying that it was ill-timed. Although the experiments were successful in that no harm was done to Joyce's mother and Joyce did not experience excessive guilt, counseling did not help her restructure her relationship with her

mother in any substantial way. The experiments, however, did give her a sense of greater freedom. For instance, she felt freer to say no to her mother's demand. This provided enough slack, it seems, to make Joyce's life more livable.

In this case, counseling helped the client fashion a life that was "a little bit better," though not as good as the counselor thought it could be. When asked, "What do you want?" Joyce had in effect replied, "I want a bit more slack and freedom, but I do not want to abandon my mother." Joyce's "new" lifestyle did not differ dramatically from the old. But perhaps it was enough for her. It was a case of choosing a satisfactory alternative rather than the best.

Leahey and Wallace (1988, p. 216) offered the following example of another client in adaptive mode.

> "For the last five years, I've thought of myself as a person with low self-esteem and have read self-help books, gone to therapists, and put things off until I felt I had good self-esteem. I just need to get on with my life, and I can do that with excellent self-esteem or poor self-esteem. Treatment isn't really necessary. Being a person with enough self-esteem to handle situations is good enough for me."

The following client, putting a more positive spin on the problem situation itself, takes a more adaptive route (Weisz, Rothbaum, & Blackburn, 1984, p. 964).

> "I would say that I am completely cured. . . . I can still pinpoint these conditions which I had thought to be symptoms. . . . These worries and anxieties make me prepare thoroughly for the daily work I have to do. They prevent me from being careless. They are expressions of the desire to grow and to develop."

In some cases, clients will be satisfied with "surface" solutions such as the elimination of symptoms. For instance, a couple is satisfied with reducing and managing the petty annoyances both of them experience in their relationship. Yet the very structure of the relationship may be problematic because some fundamental inequalities or inequities are built into the relationship. But they don't want to do much about restructuring the relationship in order to avoid the annoyances they experience.

Some helpers, reviewing these examples, would be disappointed. Others would see them as legitimate examples of adapting to, rather than changing, reality. However, all these clients did act to achieve some kind of goal, however minimal. They did *something* about the way they thought and behaved. And they felt that their lives were better because of it.

Strategic self-limitation. Robert Leahy (1999) relates the kinds of reluctance and resistance reviewed in Chapter 9 to goal setting under the rubric of "strategic self-limitation." Reluctant and resistant behaviors serve the purpose of setting limits on change. All change carries some risk and uncertainty, and these can be distressing in themselves. Putting up barriers to change limits both risk and uncertainty. It is the client's way of saying, "Enough is enough. I don't want to engage in a change program that will lead to further effort, stress, failure, and regret." The strategies such clients use are the ordinary ones—attacking the therapist, failing to do homework assignments, emotional volatility, getting mired in a "this won't work" mentality, and so forth. Even though helpers may point out to clients the ways they are engaging in what Leahy calls "self-handicapping," they don't choose goals for clients. There is a huge difference between best possible goals and goals that are possible for this client in this set of circumstances.

The main point, however, is that helping clients cope with the adversities of life does not mean that you are shortchanging them. When you help them adapt rather than conquer, you are not failing. Neither are they. When it comes to outcomes, there is no one universal rule of success.

Coping as an Important Goal

An article suggested that Condoleeza Rice's challenge in the Middle East was to help parties in conflict "to adjust expectations without squashing hope" (*Economist*, February 5, 2005, p. 12). Not a bad definition of coping, I said to myself as I read the article. Choosing an adaptive, rather than a stretch, goal has been associated with *coping* (Coyne & Racioppo, 2000; Folkman & Moskowitz, 2000; Lazarus, 2000; Snyder, 1999). All human beings cope rather than conquer at times. In fact, in human affairs as a whole, coping probably outstrips conquering. And sometimes people have no other choice. It's cope or succumb. For some, coping has a bad reputation because it seems to be associated with mediocrity. But in many difficult situations helping clients cope is one of the best things helpers can do.

Coping often has an enormous upside. A young mother with three children has just lost her husband. Someone asks, "How's she doing?" The response, "She's coping quite well." She's not letting her grief get the better of her. She is taking care of the children and helping them deal with their sense of loss. She's moving along on all the tasks that a death in a family entails. At this stage, what could be more positive than that?

From a positive psychology point of view, Folkman and Moskowitz (2000) see positive affect as playing an important role in coping. And so they ask how counselors can help clients generate positive affect and sustain it in the face of chronic stress. They suggest three ways.

Positive reappraisal. Help clients reframe a situation to see it in a positive light. For instance, Victor, recovering from multiple injuries received in a bicycle accident, sees the entire rehabilitation process as "one big daunting glob." Taken as a whole, it looks undoable. However, the rehabilitation counselor first helps Victor picture the overall goal of the rehabilitation process. She encourages him to see himself engaging fully in the ordinary tasks of everyday life, even riding a bicycle. That is, she helps him separate the very desirable end state from the arduous set of activities that will get him there. Victor does not have to cope with the "big glob" ever. He needs to cope with each day. Victor is rebuilding his body. Every day he is doing something to forge a link in the recovery chain. Each week he is helped to see that there is something he can now do that he was not able to do the previous week. Victor has low moments, of course. But he also has moments of positive affect that keep him going.

Problem-focused coping. Help clients deal with problems one at a time as they arise. For instance, Agnes is caring for her husband who has multiple sclerosis (MS). There is a certain unpredictability and uncontrollability associated with her husband's disease. However, she does not have to cope with his MS. Rather, each day or each week or each stage brings its own set of problems. Her counselor can help her "pursue realistic, attainable goals by focusing on specific proximal tasks or problems related to caregiving" (Folkman & Moskowitz, 2000, p. 650). Agnes is heartened by the very fact that she faces and deals with each problem as it arises. The sense of mastery and control she experiences is accompanied by positive affect. Even in the face of great stress, she is buoyed enough to move on to the next task or stage with grace.

Infusing ordinary events with positive meaning. In one study, Folkman and Moskowitz (2000) asked the participants, all caregivers for people with AIDS, to describe something they did or something they experienced that made them feel good and helped them get through the day. More than 99% of the caregivers interviewed talked about some such event. The point is that even during times of great stress, people note and remember positive events. The events were not "big deals." Rather, they were "ordinary events," such things as having dinner with a friend, seeing some flowers in a hospital room, or receiving a compliment from someone. But these events together with the positive affect they produced helped them get through the day.

Lazarus (2000) adds a note of caution to all of this. He notes that so-called positively valenced emotions such as love and hope are often mixed with negative feelings and are therefore experienced as distressing. It is painful for caregivers to see those they love in pain. And so-called negatively valenced emotions such as anger are not unequivocally negative. Anger can be experienced as positive or is often mixed with positive feelings. Although counselors can help clients under great stress do things that will increase the kind of positive affect that makes their lives more livable, there are limits. In other words, Lazarus is cautioning us to use but be careful with positive-psychology approaches.

A Bias for Action as a Metagoal

Although clients set goals that are directly related to their problem situations, there are also metagoals, or superordinate goals that would make them more effective in pursuing the goals they set and in leading fuller lives. The overall goal of helping clients become more effective in problem management and opportunity development was mentioned in Chapter 1. Another metagoal is to help clients become more effective "agents" in life—doers rather than mere reactors, preventers rather than fixers, initiators rather than followers—in keeping with the "bias toward action" value outlined in Chapter 3.

> Lawrence was liked by his superiors for two reasons. First, he was competent—he got things done. Second, he did whatever they wanted him to do. They moved him from job to job when it suited them. He never complained. However, as he matured and began to think more of his future, he realized that there was a great deal of truth in the adage "If you're not in charge of your own career, no one is." After a session with a career counselor, he outlined the kind of career he wanted and presented it to his superiors. He pointed out to them how this would serve both the company's interests and his own. At first they were taken aback by Lawrence's assertiveness, but then they agreed. Later, when they seemed to be sidetracking him, he stood up for his rights. Assertiveness was his bias for action.

The doer is more likely to pursue stretch, rather than adaptive, goals in managing problems. The doer is also more likely to move beyond problem management to opportunity development.

Personalizing Task 2

Review goals you have set for yourself in your efforts to manage problem situations or develop unused opportunities. Then answer the following questions.

- To what degree do I choose specific goals from among a number of possibilities?
- How well do I challenge myself to translate good intentions into broad goals and broad goals into specific, actionable goals?
- To what extent do I shape my goals so that they have the characteristics outlined in Box 12-2?
- How effectively do I establish goals for myself that take into consideration both my needs and my wants?
- To what degree do I become aware of goals that are naturally emerging from the process of trying to manage a problem situation or identify and develop an opportunity?
- How well do I identify a range of options when the future is both risky and uncertain?
- How effectively do I choose for myself the right mix of adaptive and stretch goals?
- How well do I explore the consequences of the goals I am setting?
- What do I do to make a bias toward action one of my metagoals?

TASK 3: HELP CLIENTS COMMIT THEMSELVES—"WHAT AM I WILLING TO PAY FOR WHAT I WANT?"

After reviewing a number of books on human evolution and where the human race seems to be headed, Tickell (2005, p. W5) says with some sorrow, "If there is optimism about human ability to cope, there is pessimism about the human will to do so." Thus the necessity of addressing commitment. As mentioned earlier, Task 3 is not really a sequential step but rather a dimension of the goal-setting process. Clients may formulate goals, but that does not mean that they are willing to pay for them. Once clients state what they want and set goals, the battle is joined, as it were. It is as if the client's "old self" or old lifestyle begins vying for resources with the client's potential "new self" or new lifestyle. On a more positive note, history is full of examples of people whose strength of will to accomplish some goal has enabled them to do seemingly impossible things.

A woman with two sons in their 20s was dying of cancer. The doctors thought she could go at any time. However, one day she told the doctor that she wanted to live to see her older son get married 6 months hence. The doctor talked vaguely about "trusting in God" and "playing the cards she had been dealt." Against all odds, the woman lived to see her son get married. Her doctor was at the wedding. During the reception, he said to her and, "Well, you got what you wanted. Despite the way things are going, you must be deeply satisfied." She looked at him wryly and said, "But, Doctor, my second son will get married someday."

Although the job of counselors is not to encourage clients to heroic efforts, counselors should not undersell clients, either.

In this task, which is usually interrelated with the other two tasks of Stage II, counselors help their clients pose and answer such questions as:

- Why should I pursue this goal?
- Is it worth it?
- Is this where I want to invest my limited resources of time, money, and energy?
- What competes for my attention?
- What are the incentives for pursuing this agenda?
- How strong are competing agendas?

Again, there is no formula. Some clients, once they establish goals, race to accomplish them. At the other end of the spectrum are clients who, once they decide on goals, stop dead in the water. Furthermore, the same client might speed toward the accomplishment of one goal and drag her feet on another. Or start out fast and then slow to a crawl. The job of the counselor is to help clients face up to their commitments.

Help Clients Commit Themselves to a Better Future

There is a difference between initial commitment to a goal and an ongoing commitment to a strategy or plan to accomplish the goal. The proof of initial commitment lies in goal-accomplishing action. For instance, one client who chose as a goal a less abrasive interpersonal style began to engage in an "examination of conscience" each evening to review what his interactions with people had been like that day. In doing so, he discovered, somewhat painfully, that in some of his interactions he actually moved beyond abrasiveness to contempt. That forced him back to a deeper analysis of the problem situation and the blind spots associated with it. Being dismissive of people he did not like or who were "not important" had become ingrained in his interpersonal lifestyle.

There is a range of things you can do to help clients in their initial commitment to goals and the kind of action that is a sign of that commitment. Counselors can help clients by helping them make goals appealing, by helping them enhance their sense of ownership, and by helping them deal with competing agendas.

Help clients set goals that are worth more than they cost. Here we revisit the "economics" of helping. Cost-effectiveness could have been included in the characteristics of workable goals outlined earlier in this chapter, but it is considered here instead because of its close relationship to commitment. Some goals that can be accomplished carry too high a cost in relation to their payoff. It may sound overly technical to ask whether any given goal is "cost-effective," but the principle remains important. Skilled counselors help clients budget rather than squander their resources—work, time, emotional energy.

Eunice discovered that she had a terminal illness. In talking with several doctors, she found out that she would be able to prolong her life a bit through a combination of surgery, radiation treatment, and chemotherapy. However, no one suggested that these would lead to a cure. She also found out what each form of treatment and each combination would cost, not so much in monetary terms, but in added anxiety and pain. Ultimately she decided against all three because no combination of them promised much for the quality of the life that was being prolonged. Instead, with the help of a doctor who was an expert in hospice care, she developed a scenario that would ease both her anxiety and her physical pain as much as possible.

It goes without saying that another patient might have made a different decision. Costs and payoffs are relative. Some clients might value an extra month of life no matter what the cost.

Because it is often impossible to determine the cost-benefit ratio of any particular goal, counselors can add value by helping clients understand the consequences of choosing a particular goal. For instance, if a client sets her sights on a routine job with minimally adequate pay, this outcome might well take care of some of her immediate needs but prove to be a poor choice in the long run. Helping clients foresee the consequences of their choices may not be easy. Another woman with cancer felt she was no longer able to cope with the sickness and depression that came with her chemotherapy treatments. She decided abruptly one day to end the treatment, saying that she didn't care what happened. No one helped her explore the consequences of her decision. Eventually, when her health deteriorated, she had second thoughts about the treatments, saying, "There are still a number of things I must do before I die." But it was too late. Some reasonable challenge on the part of a helper might have helped her make a better decision.

The balance-sheet method outlined in Chapter 13 is also a tool you can use selectively as a way of helping clients choose best-fit strategies for accomplishing their goals. It can help clients weigh costs against benefits both in choosing goals and in choosing programs to implement goals.

Help clients set appealing goals. Just because goals will help in managing a problem situation or develop an opportunity and are cost-effective does not mean that they will automatically appeal to the client. Setting appealing goals is common sense, but it is not always easy to do. For instance, for many if not most addicts, a drug-free life is not immediately appealing, to say the least.

> A counselor tries to help Chester work through his resistance to giving up prescription drugs. He listens and is empathic. He also challenges the way Chester has come to think about drugs and his dependency on them. One day the counselor says something about "giving up the crutch and walking straight." In a flash Chester sees himself not as a drug addict but as a "cripple." A friend of his had lost a leg in a land-mine explosion in Iraq. He remembered how his friend had longed for the day when he could be fitted with a prosthesis and throw his crutches away. The image of "throwing away the crutch" and "walking straight" proved to be very appealing to Chester.

An incentive is a promise of a reward. As such, incentives can contribute to developing a climate of hope around problem management and opportunity development. A goal is appealing if there are incentives for pursuing it. Counselors need to help clients in their search for incentives throughout the helping process. Ordinarily, negative goals—giving up something that is harmful—need to be translated into positive goals—getting something that is helpful. It was much easier for Chester to commit himself to returning to school than to giving up prescription drugs, because school represented something he was getting. Images of himself with a degree and of holding some kind of professional job were solid incentives. The picture of him "throwing away the crutch" proved to be an important incentive in cutting down on drug use.

Help clients embrace and own the goals they set. Earlier, we discussed how important it is for clients to "own" the problems and unused opportunities they talk about. It is also important for them to own the goals they set. It is essential that the goals chosen be the client's rather than the helper's or someone else's. Various kinds of probes can be used to help clients discover what they want to do to manage some dimension of a problem situation more effectively.

For instance, Carl Rogers, in a film of a counseling session (Rogers, Perls, & Ellis, 1965), is asked by a woman what she should do about her relationship with her daughter. He says to her, "I think you've been telling me all along what you want to do." She knew what she wanted the relationship to look like, but she was asking for his approval. If he had given it, the goal would, to some degree, have become his goal instead of hers. At another time he asks, "What is it that you want me to tell you to do?" This question puts the responsibility for goal setting where it belongs—on the shoulders of the client. In the following case, the helper challenges a client to take responsibility for setting goals.

> Cynthia was dealing with a lawyer because of an impending divorce. Discussions about what would happen to the children had taken place, but no decision had been reached. One day she came in and said that she had decided on mutual custody. She wanted to work out such details as which residence, hers or her husband's, would be the children's principal one and so forth. The lawyer asked her how she had reached her decision. She said that she had been talking to her husband's parents—she was still on good terms with them—and that they had suggested this arrangement. The lawyer challenged Cynthia to take a closer look at her decision. "Let's start from zero," he said, "and you tell me what kind of living arrangements *you* want and why." He did not think that it was wise to help her carry out a decision that was not her own.

Choosing goals suggested by others enables clients to blame others if they fail to reach the goals. Also if they simply follow other people's advice, they often fail to explore the down-the-road consequences.

From compliance to ownership. Commitment to goals can take different forms—compliance, buy-in, and ownership. The least useful is mere compliance. "Well, I guess I'll have to change some of my habits if I want to keep my marriage afloat" does not augur well for sustaining changes in behavior. But it may be better than nothing. Buy-in is a level up from compliance. "Yes, these changes are essential if we are to have a marriage that makes sense for both of us. We say we want to preserve our marriage, but now we have to prove it to ourselves." This client has moved beyond mere compliance. But sometimes, like mere compliance, buy-in alone does not provide enough staying power because it depends too much on reason. "This is logical" is far different from "This is what I really want!" Ownership is a higher form of commitment. It means that the client can say, "This goal is not someone else's, it's not just a good idea; it is mine, it is what I want to do." Consider the following case.

> A counselor worked with a manager whose superiors had intimated that he would not be moving much further in his career unless he changed his style in dealing with the members of his team and other key people with whom he worked within the organization. At first the manager resisted setting any goals. "What they want me to do is a lot of hogwash. It won't do anything to make the business better," was his initial response. One day, when asked whether accomplishing what "they" wanted him to do would cost him that much, he pondered a few moments and then said, "No, not really." That got him started. He moved beyond resistance.
>   With a bit of help from the counselor, he identified a few areas of his managerial style that could well be "polished up." Within a few months he got much more into

the swing of things. Given the favorable response to his changed behavior he had gotten from the people who reported to him, he was able to say, "Well, I now see that this makes sense. But I'm doing it because it has a positive effect on the people in the department. It's the right thing to do." Buy-in had arrived. A year later, he moved up another notch. He became much more proactive in finding ways to improve his style. He delegated more, gave people feedback, asked for feedback, held a couple of managerial retreats, joined a human-resource task force, and routinely rewarded his direct reports for their successes. Now he began to say such things as "This is actually fun." Ownership had arrived. The people in his department began to see him as one of the best executives in the company. This process took over 2 years.

The manager did not have a personality transformation. He did not change his opinion of some of his superiors, and he was right in pointing out that they didn't follow their own rules. But he did change his behavior because he gradually discovered meaningful incentives to do so.

Contracts. The use of contracts to structure the helping process itself was discussed in Chapter 3. Self-contracts—that is, contracts that clients make with themselves—can also help clients commit themselves to new courses of action. Although contracts are promises clients make to themselves to behave in certain ways and to attain certain goals, they are also ways of making goals more focused. It is not only the expressed or implied promise that helps but also the explicitness of the commitment. Consider the following example, in which one of Dora's sons disappears without a trace.

About a month after one of Dora's two young sons disappeared, she began to grow listless and depressed. She was separated from her husband at the time the boy disappeared. By the time she saw a counselor a few months later, a pattern of depressed behavior was quite pronounced. Although her conversations with the counselor helped ease her feelings of guilt—for instance, she stopped engaging in self-blaming rituals—she remained listless. She shunned relatives and friends, kept to herself at work, and even distanced herself emotionally from her other son. She resisted developing images of a better future, because the only better future she would allow herself to imagine was one in which her son had returned.

Some strong challenging from Dora's sister-in-law, who visited her from time to time, helped jar her loose from her preoccupation with her own misery. "You're trying to solve one hurt, the loss of Bobby, by hurting Timmy and hurting yourself. I can't imagine in a thousand years that this is what Bobby would want!" her sister-in-law screamed at her one night. Afterward, Dora and the counselor discussed a "recommitment" to Timmy, to herself, to the extended family, and to their home. Through a series of contracts, she began to reintroduce patterns of behavior that had been characteristic of her before the tragedy. For instance, she contracted to opening her life up to relatives and friends once more, creating a much more positive atmosphere at home, encouraging Timmy to have his friends over, and so forth. Contracts worked for Dora because, as she said to the counselor, "I'm a person of my word."

When Dora first began implementing these goals, she felt she was just going through the motions. However, what she was really doing was acting herself into a new mode of thinking. Contracts helped Dora in both her initial commitment to a goal and her movement to action. In counseling, contracts are not legal documents but human instruments to be used if they are helpful. They often provide both the structure and the incentives some clients need.

Even self-contracts have a shadow side. There is no such thing as a perfect contract. Most people don't think through the consequences of all the provisions of a contract, whether it be marriage, employment, or self-contracts designed to enhance a client's commitment to goals. And even people of goodwill unknowingly add covert codicils to contracts they make with themselves and others—"I'll pursue this goal—until it begins to hurt" or "I won't be abusive—unless she pushes me to the wall." The codicils are buried deep in the decision-making process and only gradually make their way to the surface.

Help clients deal with competing agendas. Clients often set goals and formulate programs for constructive change without taking into account competing agendas— other things in their lives that soak up time and energy, such as job, family, and leisure pursuits. The world is filled with distractions. For instance, one manager wanted to begin developing computer and Internet-related skills, but the daily push of business and a divorce set up competing agendas and sapped his resources. Not one of the goals of his self-development agenda was accomplished.

Programs for constructive change often involve a rearrangement of priorities. If a client is to be a full partner in the reinvention of his marriage, then he cannot spend as much time "with the boys." Or the underemployed blue-collar worker might have to put aside some parts of her social life if she wants a more fulfilling job. She eventually discovers a compromise. A friend introduces her to the job search possibilities on the Internet. She discovers that she can work full time to support herself, do a better job looking for new employment on the Internet than by using traditional methods, and still have some time for a reasonable social life.

This is not to suggest that all competing agendas are frivolous. Sometimes clients have to choose between right and right. The woman who wants to expand her horizons by getting involved in social settings outside the home still has to figure out how to handle the tasks at home. This is a question of balance, not frivolity. The single parent who wants a promotion at work needs to balance her new responsibilities with involvement with her children. A counselor who had worked with a two-career couple as they made a decision to have a child helped them think of competing agendas once the pregnancy started. A year after the baby was born, they saw the counselor again for a couple of sessions to work on some issues that had come up. However, they started the session by saying, "Are we glad that you talked about competing agendas when we were struggling with the decision to become parents! After the baby was born, we went back time and time again and reviewed what we said about managing competing and conflicting priorities. It helped stabilize us for the last 2 years."

Box 12-3 indicates the kinds of questions you can help clients ask themselves about their commitment to their change agendas.

Great Expectations: Encourage Client Self-Efficacy—"I Can, I Will"

The role of expectations in life is being explored more broadly, more deeply, and more practically (Kirsch, 1999). Clients need to find the motivation to seize their goals and run with them. The more they find their motivation within themselves the better. "Self-regulation" is the ideal. The counselor's role in client self-regulation is to help clients choose goals, develop commitment to them, and develop a sense of agency and assertiveness (Galassi & Bruch, 1992). Expectations, whether "great" or not, also play part in self-regulation. Here we look at client expectations through the lens of "self-efficacy" (Bandura, 1986, 1989, 1991, 1995, 1997, 2001; Cervone, 2000; Cervone & Scott, 1995; Lightsey, 1996; Locke & Latham, 1990; Maddux, 1995; Schwarzer, 1992). Self-efficacy is an extremely useful concept when it comes to constructive change. It is impossible to do justice to it here. What follows will, hopefully, pique your interest and help you relate self-efficacy to helping. You can feast on the vast self-efficacy literature later.

Box 12-3 Questions for Evaluating Clients' Commitment to Goals

Here are the kinds of questions you can help clients ask themselves in order to test their commitment to goals they are setting.

- What is my state of readiness for change in this area at this time?
- How badly do I want what I say I want?
- How hard am I willing to work?
- To what degree am I choosing this goal freely?
- How highly do I rate the personal appeal of this goal?
- How do I know I have the courage to work on this?
- What's pushing me to choose this goal?
- What incentives do I have for pursuing this change agenda?
- What rewards can I expect if I work on this agenda?
- If this goal is in any way being imposed by others, what am I doing to make it my own?
- What difficulties am I experiencing in committing myself to this goal?
- In what way is it possible that my commitment is not a true commitment?
- What can I do to get rid of the disincentives and overcome the obstacles?
- What can I do to increase my commitment?
- In what ways can the goal be reformulated to make it more appealing?
- To what degree is the timing for pursuing this goal poor?
- What do I have to do to stay committed?
- What resources can help me? What kind of support do I need?

The nature of self-efficacy. As Bandura (1995, p. 2) notes, "Perceived self-efficacy refers to beliefs in one's capabilities to organize and execute the courses of action required to manage prospective situations. Efficacy beliefs influence how people think, feel, motivate themselves, and act." People's expectations of themselves and can-do beliefs have a great deal to do with their willingness to put forth effort to cope with difficulties, the amount of effort they will expend, and their persistence in the face of obstacles. Clients with higher self-efficacy will make bolder choices, moving from adaptation toward stretch goals. Clients tend to take action when two conditions are fulfilled:

1. **Outcome expectations.** Clients tend to act when they see that their actions will most likely lead to certain desirable results or accomplishments. "I will end up with a better relationship with Sophie."

2. **Self-efficacy beliefs.** People tend to act when they are reasonably sure that they have the wherewithal—for instance, working knowledge, skills, time, stamina, guts and other resources—to successfully engage in the kind of behavior that will lead to the desired outcomes. "I have the ability to deal with the conflicts Sophie and I have. I can do this. I'm going to do this."

Now let's see these two factors operating together in a few examples. Yolanda, who has had a stroke, not only believes that participation in a rather painful and demanding physical rehabilitation program will literally help her get on her feet again (an outcome expectation), but she also believes that she has what it takes to inch her way through the program (a self-efficacy belief). She therefore enters the program with a very positive attitude and makes good progress.

Yves, on the other hand, is not convinced that an aggressive drug rehabilitation program will lead to a more fulfilling life (a negative outcome expectation), even though he knows he could "get through" the program (a self-efficacy belief). So he says no to the therapist. Even though the therapist has "promised" him a "drug-free" life, Yves keeps saying to himself, "Drug-free for what?" He sees being drug-free as an instrumental goal. But he has not yet come up with an attractive ultimate goal.

Xavier is convinced that a series of radiation and chemotherapy treatments would help him (a positive outcome expectation), but he does not feel that he has the stamina and courage to go through with them (a negative self-efficacy expectation). He, too, refuses the treatment.

Outcome expectations and self-efficacy beliefs are factors, not just in helping, but in everyday life. Do an Internet search on that term and you will find a rich body of literature covering all facets of life—for instance, applications to education (Lopez, Lent, Brown, & Gore, 1997; Multon, Brown, & Lent, 1991; Smith & Fouad, 1999; Zimmerman, 1995, 1996), health care (O'Leary, 1985; Schwarzer & Fuchs, 1995; Schwarzer & Renner, 2000), physical rehabilitation (Altmaier, Russell, Kao, Lehmann, & Weinstein, 1993), and work (Donnay & Borgen, 1999).

Helping clients develop self-efficacy. People's sense of self-efficacy can be strengthened in a variety of ways (see Mager, 1992). Lest self-efficacy be seen as a paradigm that applies only to the weak, let's take the case of a very strong manager, let's call him Nick, who wanted to change his abrasive supervisory style but was doubtful that he could do so. "After all these years, I am what I am," he would say. It would have been silly to merely tell him, "Nick, you can do it; just believe in yourself." It was necessary to help him do a number of things to help strengthen his sense of self-efficacy in supervision.

Skills: Make sure that clients have the skills they need to perform desired tasks. Self-efficacy is based on ability and the conviction that you can use this ability to get a task done. Nick first read about and then attended some skill-building sessions on such "soft" skills as listening, responding with empathic highlights, giving feedback that is softer on the person but harder on the problem, and constructive challenging. In truth, he had many of these skills, but they lay dormant. These short training experiences put him back in touch with some things he could do but didn't do. Please note, however, that merely acquiring skills does not by itself increase clients' self-efficacy. The way they acquire them must give them a sense of their competence. "I now have these skills and I am positive that I can use them to get this task done."

Corrective feedback: Provide feedback that is based on deficiencies in performance, not on deficiencies in the client's personality. Corrective feedback can help clients develop a sense of self-efficacy because it helps clear away barriers to the use of resources. Because I attended many meetings with Nick, I routinely described the ups and downs of his performance. I'd say such things as, "Nick, in yesterday's meeting you listened to and responded to everyone's ideas. Let me make a suggestion. You don't have to respond, as you did, in a positive way to every suggestion. Crap is still crap. Do some sorting as you listen and respond. Show why good ideas are good and why lousy ideas are bad. Then, whether the ideas are good or bad, everyone learns something."

When corrective feedback sounds like an attack on clients' personalities, their sense of self-efficacy will suffer. My feedback helped Nick's self-efficacy belief because it pointed out that he could be decent and listen well and still use his excellent critical abilities. People would leave the room enlightened, not angry. When you give feedback, you would do well to ask yourself, "In what ways will this feedback help increase the client's sense of self-efficacy?"

Positive feedback: Provide positive feedback and make it as specific as corrective feedback. Positive feedback strengthens clients' self-efficacy by emphasizing their strengths and reinforcing what they do well. This is especially true when it is specific. Too often negative feedback is very detailed, whereas positive feedback is perfunctory—"Nice job." This and other throwaway phrases probably sound like clichés. Here's one bit of feedback I gave Nick: "Yesterday, you interrupted Jeff, who was engaging in another one of his monologues. You summarized his main ideas. Then, with a few questions you showed him why only part of his plan was viable. The others were glad you took Jeff on. He learned something. And you saved us all a lot of time."

The formula for giving specific positive feedback goes something like this. "Here's what you did. Here's the positive outcome it had. And here's the wider upbeat impact." Helping Nick see the value of this pattern of behavior helped him engage in it more frequently and increased his sense of self-efficacy. "I can combine the hard stuff and the soft stuff." Clients need to see feedback as information they need to accomplish a task.

Using success as a reinforcer: Challenge clients to engage in actions that produce positive results. Even small successes can increase a client's sense of self efficacy. Success is reinforcing. Often success in a small endeavor will give clients the courage to try something more difficult. "I can do even more." Nick began delegating a few minor tasks to some of his direct reports. They handled their assignments very well. When I commented, "They seem to be doing pretty well," Nick replied, "I think that I can safely begin to put more on their plate. They like it, and I like seeing them succeed." Successful delegation increased Nick's sense of supervisory self-efficacy. He could say to himself more assuredly, "I can delegate without worrying whether or not it's going to get done." Make sure, however, that the link between success and increased self-confidence is forged. A series of successes on its own does not necessarily increase the strength of a client's self-efficacy beliefs. Success has to be linked to a sense of increased competence.

Models: Help clients increase their own sense of self-efficacy by learning from others. I asked Nick to name the best manager in the division. He mentioned a name. "What's he like?" I asked. Of course Nick talked about how competent this guy was, how effective he was in getting results, and how tough he was. Tongue in cheek, I remarked, "But I suppose that he's not very good with people." Nick exploded. "Of course he's fair. He's as good at all of this soft stuff as anyone else." He went on to name ways in which the guy was "good with people." Then suddenly he stopped, looked at me, and smiled. "Caught me, didn't you?" Learning makes clients more competent and increases their self-efficacy. Learning from models is, as we have seen, a bit tricky. Nick had too much pride to think that he could learn very much from others.

Providing encouragement: Support clients' self-efficacy beliefs without being patronizing. We took a brief look at motivational interviewing and encouragement in Chapter 8. However, if your support is to increase clients' sense of self-efficacy, it must be real, and what you support in them must be real. Encouragement and support must be tailored to each client in each instance. A supportive remark to one client might sound patronizing to another. Had I patronized Nick—"Give it a try, Nick, I know that you can do it"—I would have failed. My encouragement was, let's say, more subtle and indirect.

Reducing fear and anxiety: Help clients overcome their fears. Fear blocks clients' sense of self-efficacy. If clients fear that they will fail, they will be reluctant to act. Therefore, procedures that reduce fear and anxiety help heighten their sense of self-efficacy. Deep down, Nick was fearful of two things regarding changing his supervisory style—messing up the business and making a fool of himself. As he tentatively changed some of his supervisory practices, business results held steady. He even noticed that two of his team members seemed to become more productive.

Helping him allay his fear of making a fool of himself by being too soft was a bit trickier. His behavior outside the office came to the rescue. Although he was often an ogre in the office, Nick was very upbeat when we visited teams out in the field. He was as good at "rallying the troops" as anyone I have ever seen. And he was real. Discussions about his two different styles helped him get rid of fears that he would make a fool of himself with his direct reports by engaging them instead of driving them.

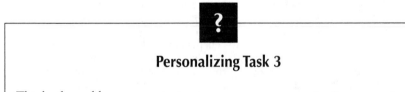

Personalizing Task 3

Think of a problem you are trying to manage or an opportunity you are trying to develop. Once you have set some goals, use the following questions to assess your commitment.

- What do I need to do to commit myself to a better future?
- What do I do to make sure that the goals I am setting are really mine and not someone else's?
- How effectively do I do a cost-benefit analysis of the goals I am choosing?
- To what degree do I focus on the appealing dimensions of the goals I am setting?
- How effectively do I perceive and deal with my misgivings about the goals I am formulating?
- What self-contracts would help me commit myself to my goals more effectively?
- What am I doing to identify, explore, and manage competing agendas?
- How do I move to initial goal-accomplishing action?
- What can I do to help increase my sense of self-efficacy?
- To what degree do I use the tasks of Stage II as opportunities to engage in actions, large or small, that help me move forward?

STAGE II AND ACTION

The work of Task 1—developing possibilities for a better future—is just what some clients need. It frees them from thinking solely about problem situations and unused resources and enables them to begin fashioning a better future. Once they identify some of their wants and needs and consider a few possible goals, they move into action.

Francine is depressed because her aging and debilitated father has been picking on her even though she has put off marriage in order to take care of him. Some of the things he says to her are quite hurtful. The situation has begun to affect her productivity at work. A counselor suggests to her that the hurtful things her father says to her is not her father but his illness speaking. This gives her a whole new perspective and frees her to think about other possibilities. Once she spends a bit of time brainstorming answers to the counselor's question—"What do you want for both yourself and your father?"—she says things like, "I'd like both of us to go through this with our dignity intact" and "I'd like to be living the kind of life he would want me to have if his mind wasn't so clouded." Once she brainstorms some possibilities for a better arrangement with her father, she needs little further help. Her usual resourcefulness returns. She gets on with life.

For other clients, Task 2 is the trigger for action. Shaping goals helps them see the future in a very different way. Once they have a clear idea of just what they want or need, they go for it.

While driving under the influence of alcohol, Nero, a man in his early 20s, had a car accident that took his wife's life. Strangely, Nero is filled with self-pity rather than remorse. The counselor, at her wit's end, confronts him about remaining wrapped up in himself. She says, "Who's the most decent person you know?" After fudging around a bit, he names Saul, an uncle. "Describe his lifestyle to me," she urges, "What makes him so decent?" With some prodding, he describes the lifestyle of this decent man. Then she says, "Do the description again, but instead of saying 'Saul' say 'Nero.'" Nero sweats, but the session has enormous impact on him. The picture of the contrast between his uncle's lifestyle and his own haunts him for days after. But he begins to stop feeling so sorry for himself; he visits his wife's parents and begs their forgiveness. He begins to see that there are other people in the world besides Nero.

For still other clients, Task 3—the search for incentives for commitment—is the trigger for action. Once they see what's in it "for me"—a kind of upbeat and productive selfishness, if you will—they move into action.

Callahan is seeing a consultant because he is very distressed. He owns and runs a small business. A few of his employees have gotten together and filed a workplace discrimination suit against him. The "troublemakers" he calls them, meaning a few women, a couple of Hispanics, and three African Americans. The consultant finds out that Callahan believes that they are "decent workers." In fact, they are more than decent. Callahan tells the counselor that he is paying them "scale," but actually he underpays them. Callahan also says that he doesn't expect his supervisors to "bend over backwards to become their friends." The truth is that some supervisors—all but two are white males—are sometimes abusive.
     The consultant convinces Callahan to attend an excellent program on diversity "before some court orders you to." A couple of weeks after returning from the program, he has a session with the consultant. He says that he had never even once considered the advantages of diversity in the workplace. All the term had meant to him was "a bunch of politicians looking for votes." Now that he saw the business reasons for diversity, he knew there were a few things he could do, but he still needed the

consultant's help and guidance. "I don't want to look like a soft jerk." Callahan's newly acquired "human touch" is far from being soft. He remains a rather rough-and-tough business guy.

Callahan didn't change his stripes overnight, but finding a package of incentives certainly helped him move toward much needed action. Who knows, the whole situation might have even made a dent in his deeply ingrained prejudices.

THE SHADOW SIDE OF GOAL SETTING

Despite the advantages of goal setting outlined in this chapter, some helpers and clients seem to conspire to avoid goal setting as an explicit process. It is puzzling to see counselors helping clients explore problem situations and unused opportunities and then stopping short of asking them what they want and helping them set goals. As Bandura (1990, p. xii) put it, "Despite this unprecedented level of empirical support [for the advantages of goal setting], goal theory has not been accorded the prominence it deserves in mainstream psychology." Years ago, the same concern was expressed differently. A U.S. developmental psychologist was talking to a Russian developmental psychologist. The Russian said, "It seems to me that American researchers are constantly seeking to explain how a child came to be what he is. We in the USSR are striving to discover how he can become what he not yet is" (see Bronfenbrenner, 1977, p. 258). One of the main reasons that counselors do not help clients develop realistic life-enhancing goals is that they are not trained to do so.

There are other reasons. First, some clients see goal setting as very rational, perhaps too rational. Their lives are so messy and goal setting seems so sterile. Both helpers and clients object to this overly rational approach. There is a dilemma. On the one hand, many clients need or would benefit by a rigorous application of the problem-management process, including goal setting. On the other, they resist its rationality and discipline. They find it alien. Second, goal setting means that clients have to move out of the relatively safe harbor of discussing problem situations and of exploring the possible roots of those problems in the past and move into the uncharted waters of the future. This may be uncomfortable for client and helper alike. Third, clients who set goals and commit themselves to them move beyond the victim-of-my-problems game. Victimhood and self-responsibility make poor bedfellows.

Fourth, goal setting involves clients' placing demands on themselves, making decisions, committing themselves, and moving to action. If I say, "This is what I want," then, at least logically, I must also say, "Here is what I am going to do to get it. I know the price and I'm willing to pay it." Because this demands work and pain, clients will not always be grateful for this kind of "help." Fifth, though goals are liberating in many respects, they also hem clients in. If a woman chooses a career, then it might not be possible for her to have the kind of marriage she would like. If a man commits himself to one woman, then he can no longer play the field.

There is some truth in the ironic statement "There is only one thing worse than not getting what you want, and that's getting what you want." The responsibilities accompanying getting what you want—a drug-free life, a renewed marriage, custody of the children, a promotion, the peace and quiet of retirement, freedom from an abusing husband—often open up a new set of problems. Even good solutions create new problems. It is one thing for parents to decide to give their children more freedom; it is another to watch them use that freedom. Finally, there is a phenomenon called post-decisional depression. Once choices are made, clients begin to have second thoughts that often keep them from acting on their decisions.

As for action, some clients move into action too quickly. The focus on the future liberates them from the past, and the first few possibilities are very attractive. They fail to get the kind of focus and direction provided by Task 2. So they go off half-cocked. Failing to weigh alternatives and shape goals often means that they have to do the process all over again.

Culture plays an important role in goal setting. Wosket (2006) points out that goal setting has cultural implications that helpers too easily overlook.

> Objections are sometimes legitimately raised about the endorsement of the pursuit of individual over collective goals that is explicit or implicit in most Eurocentric and westernized approaches to counseling—the skilled helper model included. So here the counselor has to be careful not to contaminate the client's process with his own conscious or unconscious bias toward the reinforcement of predominant cultural norms attached to goal setting. The process of goal setting can still be usefully applied to communal or collective contexts, for instance where the client's allegiance to family or cultural expectations prevails over individual preferences or objectives. Committing to a course of action that honors a sense of duty is a legitimate goal. For instance, the goal of keeping the family together may be a high priority for an Irish Catholic woman and one that, if accomplished, might give her more of a sense of achievement and fulfillment than pursuing the individual goal of leaving an unsatisfying relationship.

Effective helpers know what lurks in the shadows of goal setting both for themselves and for their clients and are prepared to manage their own part of it and help clients manage theirs. The answer to all of this lies in helpers' being trained in the entire problem-management process and in their sharing a picture of the entire process with the client. Then goal setting, described in the client's language, will be a natural part of the process. Artful helpers weave goal setting, under whatever name, into the flow of helping. They do so by moving easily back and forth among the stages and tasks of the helping process even in brief therapy.

CHAPTER

13

STAGE III: HELP CLIENTS DEVELOP STRATEGIES AND PLANS TO ACCOMPLISH THEIR GOALS

INTRODUCTION TO STAGE III

In its broadest sense, planning includes all the steps of Stages II and III—that is, it deals with solutions with a Big S and a small s. In a narrower sense, planning deals with identifying, choosing, and organizing the strategies needed to accomplish goals. Whereas Stage II is about outcomes—goals or accomplishments "powerfully imagined"—Stage III is about the activities or the work needed to produce those outcomes.

When helped to explore what is going wrong in their lives, clients often ask, "Well, what should I do about it?" That is, they focus on actions they need to take in order to "solve" things. But, as we shall see, action—though essential—is valuable only to the degree that it leads to problem-managing and opportunity-developing *outcomes*. Of course, outcomes are valuable only to the degree that they have a con-structive impact on the life of the client. The distinction between action, outcomes, and impact is seen in the following example.

Lacy, a 40-year-old single woman, is making a great deal of progress in controlling her drinking through her involvement with an AA program. She attends AA meetings, follows the 12 steps, stays away from situations that would tempt her to drink, and calls fellow AA members when she feels depressed or when the temptation to drink is pushing her hard. The outcome is that she has stayed sober for over 7 months. She feels that this is quite an accomplishment. The impact of all this is very rewarding. She feels better about herself, and she has had both the energy and the enthusiasm to do things that she has not done in years—developing a circle of friends, getting interested in church activities, and doing a bit of travel.

But Lacy is also struggling with a troubled relationship with a man. In fact, her drinking was, in part, an ineffective way of avoiding the problems in the relationship. She knows that she no longer wants to tolerate the psychological abuse she has been getting from her male friend, but she's afraid of the vacuum she will create by cutting off the relationship. She is, therefore, trying to determine what she wants, almost fearing that ending the relationship might turn out to be the best option.

She has attempted to manage the relationship in a number of ways. For instance, she has become much more assertive with her friend. She now cuts off contact whenever he becomes abusive. And she no longer lets him make all the decisions about what they are going to do together. But the relationship remains troubled. Even though she is doing many things, there is no satisfactory outcome. She has not yet determined what the outcome should be; that is, she has not determined what kind of relationship she would like and if it is possible to have such a relationship with this man. Nor has she determined to end the relationship.

Finally, after one seriously abusive episode, she tells him that she is ending the relationship. She does what she has to do to sever all ties with him (action), and the outcome is that the relationship ends permanently. The impact is that she feels liberated but lonely. The helping process needs to be recycled to help her with this new problem.

Stage III has three interrelated tasks. They are all aimed at action on the part of the client.

- **Task 1: Strategies.** Help clients develop possible strategies for accomplishing their goals. "What kind of actions will help me get what I need and want?"

- **Task 2: Best-fit strategies.** Help clients choose strategies that are effective, efficient, and tailored to their preferences and resources. "What actions are best for me?"

- **Task 3: Plans.** Help clients turn strategies into a realistic plan. "What should my campaign for constructive change look like? What do I need to do first? Second? Third?"

Stage III, highlighted in Figure 13-1, adds the final pieces to a client's planning a program for constructive change. Stage III deals with the "game plan." However, these three tasks constitute *planning* for action and should not be confused with action itself. Without action, a program for constructive change is nothing more than a wish list. The implementation of plans is discussed in the next chapter.

TASK 1: HELP CLIENTS DEVELOP STRATEGIES FOR ACCOMPLISHING THEIR GOALS

Strategies are actions that help clients accomplish their goals. Task 1, developing a range of possible strategies to accomplish goals, is a powerful exercise. Clients who feel hemmed in by their problems and unsure of the viability of their goals are liberated

The Skilled-Helper Model

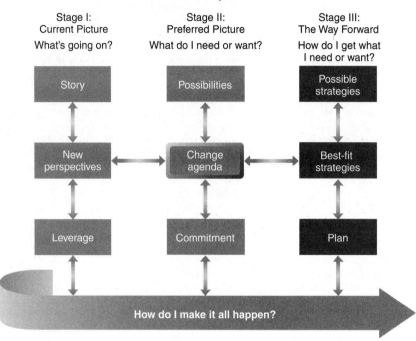

Stage I: Current Picture What's going on?	Stage II: Preferred Picture What do I need or want?	Stage III: The Way Forward How do I get what I need or want?
Story	Possibilities	Possible strategies
New perspectives	Change agenda	Best-fit strategies
Leverage	Commitment	Plan

How do I make it all happen?

FIGURE 13-1
The Helping Model—Stage III

through this process. Clients who see clear pathways to their goals have a greater sense of self-efficacy. "I can do this."

Strategy is the art of identifying and choosing realistic courses of action for achieving goals and doing so under adverse conditions, such as war. The problem situations in which clients are immersed constitute adverse conditions; clients often are at war with themselves and the world around them. Helping clients develop strategies to achieve goals can be a most thoughtful, humane, and fruitful way of being with them. This step in the counseling process is another that helpers sometimes avoid because it is too jargon-laden and mechanical. They do their clients a disservice. Clients with goals but no clear idea of how to accomplish them are still at sea. Once more it is a question of helping clients stimulate their imaginations and engage in divergent thinking. Most clients do not instinctively seek different routes to goals and then choose the ones that make most sense.

Use Brainstorming to Stimulate Clients' Thinking

Brainstorming, discussed in Chapter 12, plays an important part in strategy development. The more routes to the achievement of a goal, the better. Consider the case of Karen, who came to realize that heavy drinking is ruining her life. Her goal was to stop drinking. She felt that it simply would not be enough to cut down; she had to stop. To her, the way forward seemed simple enough: Whereas before she drank, now she wouldn't. Because of the novelty of not drinking, she was successful for a few days; then she fell off the wagon. This happened a number of times until she finally realized that she could use some help. Stopping drinking, at least for her, was not as simple as it seemed.

A counselor at a city alcohol and drug treatment center helps her explore a number of techniques that could be used in an alcohol-management program. Together they come up with the following possibilities:

- Just stop cold turkey and get on with life.
- Join Alcoholics Anonymous.
- Move someplace declared "dry" by local government.
- Take a drug that causes nausea if followed by alcohol.
- Replace drinking with other rewarding behaviors.
- Join some self-help group other than Alcoholics Anonymous.
- Get rid of all liquor in the house.
- Take the "pledge" not to drink; to make it more binding, take it in front of a minister.
- Join a residential hospital detoxification program.
- Avoid friends who drink heavily.
- Change other social patterns; for instance, find places other than bars and cocktail lounges to socialize.
- Try hypnosis to reduce the urge to drink.
- Use behavior modification techniques to develop an aversion for alcohol; for instance, pair painful but safe electric shocks with drinking or even thoughts about drinking.
- Change self-defeating patterns of self-talk, such as "I have to have a drink" or "One drink won't hurt me."
- Become a volunteer to help others stop drinking.
- Read books and view films on the dangers of alcohol.
- Stay in counseling as a way of getting support and challenge for stopping.
- Share intentions to stop drinking with family and close friends.
- Spend a week with an acquaintance who does a great deal of work in the city with alcoholics, and go with him on his rounds.
- Walk around skid row meditatively.
- Have a discussion with members of the family about the impact drinking has on them.
- Eat foods such as sweets that can help reduce the craving for alcohol.
- Get a hobby or an avocation that demands time and energy.
- Substitute a range of self-enhancing activities such as exercise or surfing the web for drinking.

This list contained many more items than Karen would have thought of had she not been stimulated by the counselor to take a census of possible strategies. One of the reasons that clients are clients is that they are not very creative in looking for ways of getting what they want. Once goals are established, getting them accomplished is not just a matter of hard work. It is also a matter of imagination.

If a client is having a difficult time coming up with strategies, the helper can "prime the pump" by offering a few suggestions. Driscoll (1984, p. 167) put it well.

> Alternatives are best sought cooperatively, by inviting our clients to puzzle through with us what is or is not a more practical way to do things. But we must be willing to introduce the more practical alternatives ourselves, for clients are often unable to do so on their own. Clients who could see for themselves the more effective alternatives would be well on their way to using them. That clients do not act more expediently already is in itself a good indication that they do not know how to do so.

Although the helper may need to suggest alternatives, he or she can do so in such a way that the principal responsibility for evaluating and choosing possible strategies stays with the client. For instance, there is the "prompt and fade" technique. The counselor can say, "Here are some possibilities. . . . Let's review them and see whether any of them make sense to you or suggest further possibilities." Or, "Here are some of the things that people with this kind of problem situation have tried. . . . How do they sound to you?" The "fade" part of this technique keeps it from being advice giving. It remains clear that the client must think these strategies over, choose the right ones, and commit to them.

> Elton, a graduate student in counseling psychology, is plagued with perfectionism. Although he is an excellent student, he worries about getting things right. After he writes a paper or practices counseling, he agonizes over what he could have done better. The kind of behavior puts him on edge when he practices counseling with his fellow trainees. They tell him that his "edge" makes them uncomfortable and interferes with the flow of the helping process. One student says to him, "You make me feel as if I'm not doing the right things as a client."
>
> Elton realizes that "less is more"—that is, becoming less preoccupied with the details of helping will make him a more effective helper. His goal is to become more relaxed in the helping sessions, free his mind of the "imperatives" to be perfect, and learn from mistakes rather than expending an excessive amount of effort trying to avoid them. He and his supervisor talk about ways he can free himself of these inhibiting imperatives.

SUPERVISOR: What kinds of things can you do to become more relaxed?

ELTON: I need to focus my attention on the client and the client's goals instead of being preoccupied with myself.

SUPERVISOR: So a basic shift in your orientation right from the beginning will help.

ELTON: Right. . . . And this means getting rid of a few inhibiting beliefs.

SUPERVISOR: Such as . . .

ELTON: That technical perfection in the helping model is more important than the relationship with the client. I get lost in the details of the model and have forgotten that I'm a human being with another human being.

SUPERVISOR: So "rehumanizing" the helping process in your own mind will help. . . . Any other internal behaviors need changing?

ELTON: Another belief is that I have to be the best in the class. That's my history, at least in academic subjects. Being as effective as I can be in helping a client has nothing to do with competing with my fellow students. Competing is a distraction. I know it's in my bones. It might have been all right in high school, but. . . .

SUPERVISOR: Okay, so the academic-game mentality doesn't work here . . .

ELTON (interrupting): That's precisely it. Even the practicing we do with one another is real life, not a game. You know that a lot of us talk about real issues when we practice.

SUPERVISOR: You've been talking about getting your attitudes right and the impact that can have on helping sessions. Are there any external behaviors that might also help?

ELTON (pauses): I'm hesitating because it strikes me how I'm in my head too much, always figuring me out. . . . On a much more practical basis, I like what Jerry and Philomena do. Before each session with their "clients" in their practice sessions, they spend 5 or 10 minutes reviewing just where the client is in the overall helping process and determining what they might do in the next session to add value and move things forward. That puts the focus where it belongs, on the client.

SUPERVISOR: So a mini-prep for each session can help you get out of your world and into the client's.

ELTON: Also in debriefing the training videos we make each week. . . . I now see that I always start by looking at my behavior instead of what's happening with the client. . . . Oh, there's another thing I can do. I can share just what we've been discussing here with my training partner. She can help me refocus myself.

SUPERVISOR: I'm not sure whether you bring up the perfectionism issues when you're the "client" in the practice sessions or in the weekly lifestyle group meetings.

ELTON (hesitating): Well, not really. I'm just coming to realize how pervasive it is in my life. . . . To tell you the truth I think I haven't brought it up because I'd rather have my fellow trainees see me as competent, not perfectionistic. . . . Well, the cat is out of the bag with you, so I guess it makes sense to put it on my lifestyle group agenda.

This dialogue, which includes empathy, probes, and challenge from the supervisor, produces a number of strategies that Elton can use to develop a more client-focused mentality. He ends by saying that all these can be reinforced through his interactions with his training partner.

Use Frameworks for Stimulating Clients' Thinking About Strategies

Helpers can use simple frameworks to formulate questions or probes that help clients develop a range of strategies. Simple frameworks can help. Consider the following case.

> Jackson has terminal cancer. He has been in and out of the hospital several times over the past few months, and he knows that he probably will not live more than a year. He would like the year to be as full as possible, and yet he wants to be realistic. He hates being in the hospital, especially a large hospital, where it is so easy to be anonymous. One of his goals is to die outside the hospital. He would like to die as peacefully as possible and retain possession of his faculties as long as possible. How is he to achieve these goals?

You can use probes and prompts to help clients discover possible strategies by helping them investigate resources in their lives, including people, models, communities, places, things, organizations, programs, and personal resources.

Individuals. What individuals might help clients achieve their goals? Jackson gets the name of a local doctor who specializes in the treatment of chronic, cancer-related pain. The doctor teaches people how to use a variety of techniques to manage pain. Jackson says that perhaps his wife and daughter can learn how to give simple injections to help him control the pain. A friend of his has mentioned that his father got excellent hospice care and died at home. Also, he thinks that talking every once in a while with a friend whose wife died of cancer, a man he respects and trusts, will help him find the courage he needs.

Models and exemplars. Who is presently doing what the client wants to do? One of Jackson's fellow workers died of cancer at home. Jackson visited him there a couple of times. That's what gave him the idea of dying at home, or at least outside the hospital. He noticed that his friend never allowed himself to engage in poor-me talk. He refused to see dying as anything but part of living. This touched Jackson deeply at the time, and now reflecting on that experience may help him develop the same kind of upbeat attitudes.

Communities. What communities of people are there through which clients might identify strategies for implementing their goals? Even though Jackson has not been a regular churchgoer, he does know that the parish within which he resides has some resources for helping those in need. A brief investigation reveals that the parish has developed a relatively sophisticated approach to providing various services for the sick. He also does an Internet search and discovers that there are a number of self-help groups for people like himself.

Places. Are there particular places that might help? Jackson immediately thinks of Lourdes, the shrine to which Catholic believers flock with all sorts of human problems. He doesn't expect miracles, but he feels that he might experience life more deeply there. It's a bit wild, but why not a pilgrimage? He still has the time and also enough money to do it. He also finds a high-tech place—an Internet chat room for cancer patients *and* their caregivers. This helps him get out of himself and, at times, become a helper instead of a client.

Things. What things can help clients achieve their goals? Jackson has read about the use of combinations of drugs to help stave off pain and the side effects of chemotherapy. He has heard that certain kinds of electric stimulation can ward off chronic pain. He explores all these possibilities with his doctor and even arranges for second opinions.

Organizations. Jackson runs across an organization that helps young cancer patients get their wishes. He volunteers. In his role as helper, he finds he receives as much help and motivation and solace as he gives.

Programs. Are there any ready-made programs for people in the client's position? He learns that a new hospice in his part of town has three programs. One helps people who are terminally ill stay in the community as long as they can. A second makes provision for part-time residents. The third provides a residential program for those who can spend little or no time in the community. The goals of these programs are practically the same as Jackson's. Box 13-1 outlines some questions that you can help clients ask themselves to develop strategies for accomplishing goals.

> **Box 13-1   Questions for Developing Strategies**
>
> Here are the kinds of questions you can help clients ask themselves
> in their search for ways of accomplishing their goals.
>
> - Now that I know what I want, what do I need to do?
> - Now that I know my destination, what are the different routes for getting
>   there?
> - What actions will get me to where I want to go?
> - Now that I know the gaps between what I have and what I want and need,
>   what do I need to do to bridge those gaps?
> - How many ways are there to accomplish my goals?
> - How do I get started?
> - What can I do right away?
> - What do I need to do later?

Help Clients Find Social Support in Their Efforts to Change

Task 1 includes helping clients identify the resources, both internal and environmental, they need to pursue goals. One of the most important resources is social support (Barker & Pistrang, 2002), though, surprisingly enough, research to verify the
usefulness of social support is skimpy (Hogan, Linden, & Najarian, 2002). Of course,
helpers themselves can provide a great deal of support (Alford & Beck, 1997;
Arkowitz, 1997; Castonguay, 1997; Yalom & Bugental, 1997). But if clients are to
pursue goals "out there" in their real lives, they also need support there.
Unfortunately, such support may not always be easy to find (Putnam, 2000). In North
American society, the supply of "social capital"—both informal social connectedness
and formal civic engagement—has fallen dangerously low. We belong to fewer
organizations that conduct meetings, know our neighbors less, meet with friends less
frequently, and even socialize with our families less often. Yet this is the environment in which clients must do the work of constructive change.

However, social support is a key element in change.

> Social support has . . . been examined as a predictor of the course of mental
> illness. In about 75% of studies with clinically depressed patients, social-
> support factors increased the initial success of treatment and helped patients
> maintain their treatment gains. Similarly, studies of people with schizophre-
> nia or alcoholism revealed that higher levels of social support are correlated
> with fewer relapses, less frequent hospitalizations, and success and mainte-
> nance of treatment gains. (Basic Behavioral Science Task Force of the
> National Advisory Mental Health Council, 1996, p. 628)

In a study on weight loss and maintaining the loss (Wing & Jeffery, 1999), clients
who enlisted the help of friends were much more successful than clients who took the
solo path. This is called "social facilitation" and is quite different from dependence.

Social facilitation, a positive psychology approach, is energizing, whereas dependence is often depressing. Therefore, a culture of social isolation does not bode well for clients. Of course, all of this reinforces what we already know through common sense. Who among us has not been helped through difficult times by family and friends?

When it comes to social support, there are two categories of clients. First, there are those who lead an impoverished social life. The objective with this group is to help them find social resources, to get back into community in some productive way. But as Putnam points out, even when clients, at least on paper, have a social system, they may not use it very effectively. This provides counselors with a different challenge—that is, helping clients tap into those human resources in a way that helps them manage problem situations more effectively.

Indeed, the National Advisory Mental Health Council study just mentioned showed that people who are highly distressed and therefore most in need of social support may be the least likely to receive it because their expressions of distress drive away potential supporters. Who among us has not avoided at one time or another a distressed friend or colleague? Therefore, distressed clients can be helped to learn how to modulate their expressions of distress. Who wants to help whiners? On the other hand, potential supporters can learn how to deal with distressed friends and colleagues, even when these friends and colleagues let themselves become whiners.

The Task Force study suggested two general strategies for fostering social support—helping clients mobilize or increase support from existing social networks and "grafting" new ties onto impoverished social networks. Both of these come into play in the following case.

> Casey, a bachelor whose job involved frequent travel literally around the world, fell ill. He had many friends, but they were spread around the world. Because he was neither married nor in a marriage-like relationship, he had no primary caregiver in his life. He received excellent medical care, but his psyche fared poorly.
>
> Once out of the hospital, he recuperated slowly, mainly because he was not getting the social support he needed. In desperation, he had a few sessions with a counselor, sessions that proved to be quite helpful. The counselor challenged him to "ask for help" from his local friends. He had underplayed his illness with them because he didn't want to be a "burden." He discovered that his friends were more than ready to help. But because their time was limited, he, with some hesitancy, "grafted" onto his rather sparse hometown social network some very caring people from the local church. He was fearful that he would be deluged with piety, but instead he found people like himself. Moreover, they were, in the main, socially intelligent. They knew how much or how little care to give. In fact, most of the time their care was simple friendship. Finally, he hired a couple of students from a local university to do word processing and run errands for him from time to time. They also provided some social support.

As the Task Force authors note, it's important not only that people be available to provide support but also that those needing support perceive that it is available. This may mean, as in Casey's case, working with the client's attitudes and openness to receive support.

Eventually, all clients have to make it without the help of a counselor. Therefore, effective helpers right from the beginning try to help them explore the social-support dimensions of problem situations. At the Action Arrow stage, questions like the following are appropriate: Who might help you do this? Who's going to challenge you when you want to give up? With whom can you share these kinds of concerns? Who's going to give you a pat on the back when you accomplish your goal?

Although social support is often key, it is not the only resource clients need to pursue their goals. Effective helpers build some kind of resource census into the helping process.

Help Clients Determine the Kind of Working Knowledge and Skills They Will Need to Move Forward

People often get into trouble or fail to get out of it because they lack the needed life skills or coping skills to deal with problem situations. When this is the case, helping clients find ways of learning the life skills they need to cope more effectively is an important broad strategy. Indeed, the use of skills training as part of therapy—what Carkhuff years ago (1971) called "training as treatment"—might be essential for some clients. Challenging clients to engage in activities for which they don't have the skills compounds, rather than solves, the problem. What kinds of working knowledge and skills does this client need to get where he or she wants to go? Consider the following case.

Jerzy and Zelda fell in love. They married and enjoyed a relatively trouble-free honeymoon period of about 2 years. Eventually, however, the problems that inevitably arise from living closely together asserted themselves. For instance, they found that they counted too heavily on romantic fervor to help them overcome—or, more often, ignore—difficulties. Once that fervor had cooled, they began fighting about finances, sex, and values. They lacked certain critical interpersonal communication skills. Furthermore, their 9-year age difference became problematic because they lacked understanding of each other's developmental needs. Jerzy had little working knowledge of the developmental demands of a 20-year-old woman; Zelda had little working knowledge of the kinds of cultural blueprints that were operative in the lifestyle of her 29-year-old husband. The relationship began to deteriorate. Because they had few problem-solving skills, they didn't know how to handle their situation.

Jerzy and Zelda needed skills. This is hardly surprising. Lack of requisite interpersonal communication and other life skills is often at the heart of relationship breakdowns.

One marriage counselor I know does marriage counseling in groups of four couples. Training in communication skills is part of the process. He separates men from women and trains them in tuning in, active listening, and responding with empathy. For skills practice, he begins by pairing a woman with a woman and a man with a man. Next he pairs a man and a woman, but not spouses. Finally, spouses are paired, taught a simple version of the problem-management process outlined in this book, and then helped to use the skills they have learned to engage in problem solving with each other. In sum, he equips them in two sets of life skills—interpersonal communication and problem solving. This is his way of pursuing the second goal of helping outlined in Chapter 1.

Help Clients Link Possible Strategies to Action

Although all the steps of the helping process can and should stimulate client action, this is especially true of Task 1. Many clients, once they begin to see what they can do to get what they want, begin acting immediately. They don't need a formal plan. Here are a couple of examples of clients who, once they were helped to identify strategies for implementing their goals, acted on them.

Jeff had been in the army for about 10 months. He found himself both overworked and, perhaps not paradoxically, bored. He had a couple of sessions with one of the educational counselors on the base. During these sessions, Jeff began to see quite clearly that not having a high school diploma was working against him. The counselor mentioned that he could finish high school while in the army. Jeff realized that this possibility had been pointed out to him during the orientation talks, but he hadn't paid any attention to it. He had joined the army because he wasn't interested in school and, being unskilled, couldn't find a decent job. Now he decided that he would get a high school diploma as soon as possible.

Jeff obtained the authorization needed from his company commander to go to school. He found out what courses he needed and enrolled in time for the next school session. It didn't take him long to finish. Once he received his high school degree, he felt better about himself and found that opportunities for more interesting jobs opened up for him in the army. Achieving his goal of getting a high school degree helped him manage the problem situation.

Jeff was one of those fortunate ones who, with a little help, quickly set a goal (the "what") and identified and implemented the strategies (the "how") to accomplish it. Note, too, that his goal of getting a diploma was also a stepping stone to other goals—feeling good about himself and getting better jobs in the army.

Grace's road to problem management was quite different from Jeff's. She needed much more help.

As long as she could remember, Grace had been a fearful person. She was especially afraid of being rejected and of being a failure. As a result, she had an impoverished social life. She had held a series of jobs that were safe but boring. She became so depressed that she made a half-hearted suicide attempt, probably more an expression of anguish and a cry for help than a serious attempt to get rid of her problems by getting rid of herself.

During her stay in the hospital, Grace had a few therapy sessions with one of the staff psychiatrists. The psychiatrist was supportive and helped her handle both the guilt she felt because of the suicide attempt and the depression that had led to the attempt. Just talking to someone about things she usually kept to herself seemed to help. She began to see her depression as a case of "learned helplessness." She saw quite clearly how she had let her choices be dictated by her fears. She also began to realize that she had a number of underused resources. For instance, she was intelligent and, though not what most people would consider good-looking, attractive in other ways. She had a fairly good sense of humor, though she seldom gave herself the opportunity to use it. She was also sensitive to others and basically caring.

After Grace was discharged from the hospital, she returned for a few outpatient sessions. She got to the point where she wanted to do something about her general fearfulness and her passivity, especially the passivity in her social life. A psychiatric social worker taught her relaxation and thought-control techniques that helped her reduce her anxiety. As she became less anxious, she was in a better position to do something about establishing some social relationships. With the social worker's help, she set goals of acquiring a couple of friends and becoming a member of some social groups. However, she was at a loss as to how to proceed, because she thought that friendship and a fuller social life were things that should happen "naturally." She soon came to realize that many people had to work at acquiring a more satisfying social life, that for some people there was nothing automatic about it at all.

The social worker helped Grace identify various kinds of social groups that she might join. She was then helped to see which of these would best meet her needs without placing too much stress on her. She finally chose to join an arts and crafts group at a local YMCA. The group gave her an opportunity to begin developing some of her talents and to meet people without having to face demands for intimate social contact. It also allowed her to take a look at other, more socially oriented programs sponsored by the Y. In the arts and crafts program, she met a couple of people she liked and who seemed to like her. She began having coffee with them once in a while and then an occasional dinner.

Grace still needed support and encouragement from her helper, but she was gradually becoming less anxious and feeling less isolated. Once in a while she would let her anxiety get the better of her. She would skip a meeting at the Y and then lie about having attended. However, as she began to let herself trust her helper more, she revealed this self-defeating game. The social worker helped her develop coping strategies for those times when anxiety seemed to be higher.

Grace's problems were more severe than Jeff's, and she did not have as many immediate resources. Therefore, she needed both more time and more attention to develop goals and strategies.

Personalizing Task 1

Consider some change you would like to make in your life. Then use the following questions to view how well you are engaging in Task 1 of the planning process. How effectively are you doing the following?

- Probing, prompting, and challenging yourself to identify possible goal-accomplishing strategies.
- Engaging in divergent thinking with respect to strategies.
- Brainstorming as many ways as possible to accomplish your goals.
- Using some kind of framework to help yourself become more creative in identifying strategies.
- Identifying and beginning to gather the resources you need to accomplish your goals.
- Identifying and developing the skills you need to accomplish your goals.
- Using the search for strategies as a way of identifying actions that will move you toward your goal.

TASK 2: HELP CLIENTS CHOOSE BEST-FIT STRATEGIES

In Task 2, clients are in decision-making mode once more. After brainstorming strategies for accomplishing goals, they need to choose a strategy or a "package" of strategies that best fits their situation, resources, personality, and preferences and turn them into some kind of plan for constructive change. Whether these tasks are done with the kind of formality outlined here is not the point. Counselors, understanding what planning is and what makes it work, can add value by helping clients find ways of accomplishing goals (getting what they need and want) in a systematic, flexible, personalized, and cost-effective way.

Once they are helped to develop a range of strategies to implement goals, some clients move forward on their own; that is, they choose the best strategies, put together an action plan, and implement it. Others, however, need help in choosing strategies that best fit their situation, and so we add Task 2 to the helping process. It is useless to have clients brainstorm if they don't know what to do with all the action strategies they generate.

Bud's Amazing Odyssey

Consider the case of Bud, a man who was helped to discover two best-fit strategies for achieving emotional stability in his life. With these, he achieved outcomes that surpassed anyone's wildest expectations.

One morning, Bud, then 18 years old, woke up unable to speak or move. He was taken to a hospital, where catatonic schizophrenia was diagnosed. After repeated admissions to hospitals, where he underwent both drug and electroconvulsive therapy (ECT), his diagnosis was changed to paranoid schizophrenia. He was considered incurable.

A quick overview of Bud's earlier years suggests that much of his emotional distress was caused by unmanaged life problems and the lack of human support. He was separated from his mother for 4 years when he was young. They were reunited in a city new to both of them, and there he suffered a great deal of harassment at school because of his "ethnic" looks and accent. There was simply too much stress and change in his life. He protected himself by withdrawing. He was flooded with feelings of loss, fear, rage, and abandonment. Even small changes became intolerable. His catatonic attack occurred in the autumn on the day of the change from daylight saving to standard time. It was the last straw.

In the hospital, Bud became convinced that he and many of his fellow patients could do something about their illnesses. They did not have to be victims of their own helplessness or of the institutions designed to help them. Reflecting on his hospital stays and the drug and ECT treatments, he later said he found his "help" so debilitating that it was no wonder that he got crazier. Somehow Bud, using his own inner resources, managed to get out of the hospital. Eventually, he got a job, found a partner, and got married.

One day, after a series of problems with his family and at work, Bud felt himself becoming agitated and thought he was choking to death. His doctor sent him to the hospital "for more treatment." There Bud had the good fortune to meet Sandra, a psychiatric social worker who was convinced that many of the hospital's patients were there because of lack of support before, during, and after their bouts of illness. She helped him see his need for social support, especially during times of stress. She also discovered in her inpatient counseling groups that Bud had a knack for helping others. Bud's broad goal was still emotional stability, and he wanted to do whatever was necessary to achieve it. Finding support and helping others cope with their problems, instrumental goals, were his best strategies for achieving the stability he wanted.

Once discharged from the hospital, Bud got to work. He enrolled Sandra to coach his wife on the best ways to provide support for him at times of stress. He then started a self-help group for ex-patients like himself. In the group, he was a full-fledged participant. But because he also had a deep desire to help others like himself, he developed the self-help group into a network of self-help groups for ex-patients.

This is an amazing example of a client who focused on one broad goal, emotional stability; translated it into a number of immediate, practical goals; discovered two broad strategies—finding ongoing emotional support and helping others—for accomplishing those goals; translated the strategies into practical applications; and by doing all that found the emotional stability he was looking for.

Criteria for Choosing Goal-Accomplishing Strategies

The criteria for choosing goal-accomplishing strategies are somewhat like the criteria for choosing goals outlined in Stage II. These criteria are reviewed briefly here through a number of examples. Strategies to achieve goals should be, like goals themselves, specific, robust, prudent, realistic, sustainable, flexible, cost-effective, and in keeping with the client's values. Let's take a look at a few of these criteria as applied to choosing strategies.

Specific strategies. Strategies for achieving goals should be specific enough to drive behavior. In the preceding example, Bud's two broad strategies for achieving emotional stability, tapping into human support and helping others, were translated into quite specific strategies—keeping in touch with Sandra, getting help from his wife, participating in a self-help group, starting a self-help group, and founding and running a self-help organization. Contrast Bud's case with Stacy's.

> Stacy was admitted to a mental hospital because she had been exhibiting bizarre behavior in her neighborhood. She dressed in a slovenly way and went around admonishing the residents of the community for their "sins." Her condition was diagnosed as schizophrenia, simple type. She had been living alone for about 5 years, since the death of her husband. It seems that she had become more and more alienated from herself and others. In the hospital, medication helped control some of her symptoms. She stopped admonishing others and took reasonable care of herself, but she was still quite withdrawn. She was assigned to "milieu" therapy, a euphemism meaning that she was helped to follow the more or less general routine of the hospital—a bit of work, a bit of exercise, some programmed opportunities for socializing. She remained withdrawn and usually seemed moderately depressed. No therapeutic goals had been set, and the nonspecific program to which she was assigned was totally inadequate.

So-called milieu therapy did nothing for Stacy because in no way was it specific to her needs. It was a general program that was only marginally better than drug-focused standard care. Bud's strategies, on the other hand, proved to be powerful. They not only helped him gain stability but also gave him a new perspective on life.

Substantive strategies. Strategies are substantive to the degree that they challenge the client's resources and, when implemented, actually achieve the goal. Not only was Stacy's program too general, but it also lacked bite. Bud's strategies, on the other hand, were substantive, especially the strategy of starting and running a self-help organization. What could be done for Stacy?

> A newly hired psychiatrist, who had been influenced by Corrigan's (1995) notion of "champions of psychiatric rehabilitation," saw immediately that Stacy needed more than either standard psychiatric or milieu-centered care. He involved her in a new comprehensive social-learning program, which included cognitive restructuring, social-skills training, and behavioral-change interventions based on incentives, shaping, modeling, and rewards. Despite a few ups and downs, Stacy responded very well to the new rather intensive program. She was discharged within 6 months and, with the help of an outpatient extension of the program, remained in the community.

For Stacy this program proved to be specific, substantive, prudent, realistic, sustainable, flexible, cost-effective, and in keeping with her values. It was cost-effective in two ways. First, it was the best use of Stacy's time, energy, and psychological resources. Second, it helped her and others like her to get back into the community and stay there. It was in keeping with her values because, even though some staff members at the hospital had concluded that all she wanted was "to be left alone," Stacy did value human companionship and freedom. She did better in a community setting.

Realistic strategies. When clients choose strategies that are beyond their resources, they are doing themselves in. Strategies are realistic when they can be carried out with the resources the client has, are under the client's control, and are unencumbered by obstacles. Bud's strategies would have appeared unrealistic to most clients and helpers. But this highlights an important point. Just as we should help clients set stretch goals whenever possible, so we should not underestimate what clients are capable of doing. In the following case, Desmond moves from unrealistic to realistic strategies for getting what he wants.

> Desmond was in a halfway house after leaving a state mental hospital. From time to time he still had bouts of depression that would incapacitate him for a few days. He wanted to get a job because he thought that it would help him feel better about himself, become more independent, and manage his depression better. He answered job advertisements in a rather random way and was constantly turned down after being interviewed. He simply did not yet have the kinds of resources needed to put himself in a favorable light in job interviews. Moreover, he was not yet ready for a regular, full-time job.

On his own, Desmond does not do well in choosing strategies to achieve even modest goals. But here's what happened next.

> A local university received funds to provide outreach services to halfway houses in the metropolitan area. The university program included finding companies that were willing, on a win-win basis, to work with halfway-house residents. A counselor from the program helped Desmond to get in contact with companies that had specific programs to help people with psychiatric problems. He found two that he thought would fit his needs. Some of the best workers in these companies had a variety of disabilities, including psychiatric problems. After a few interviews, Desmond got a job in one of these companies that fitted his situation and capabilities. The entire work culture was designed to provide the kind of support he needed.

Of course, there is a difference between realism and allowing clients to sell themselves short. Substantive strategies that make clients stretch for a valued goal can be most rewarding. Bud's case is an exceptional example of that.

Strategies in keeping with the client's values. Make sure that the strategies chosen are consistent with the client's values. Let's return to the case of the priest who had been unjustly accused of child molestation.

> In preparing for the court case, the priest and his lawyer had a number of discussions. The lawyer wanted to do everything possible to destroy the accusers' credibility. He had dug into their pasts and dredged up some dirt. The priest objected to these tactics. "If I let you do this," he said, "I descend to their level. I can't do that." The priest discussed this with his counselor, his superiors, and another lawyer. He stuck to his guns. They prepared a strong case but without any sleaze.
> After the trial was over and he was acquitted, the priest said that his discussion about the lawyer's preferred tactics was one of the most difficult issues he had to face. Something in him said that because he was innocent, any means to prove his innocence was allowed. Something else told him that this was not right. The counselor helped him clarify and challenge his values but made no attempt to impose either his own or the lawyer's values on his client.

Box 13-2 outlines the kinds of questions you can help clients ask themselves in order to choose best-fit strategies.

Strategy Sampling

Some clients find it easier to choose strategies when they first sample some of the possibilities. Consider this case.

> Two business partners were in conflict over ownership of the firm's assets. Their goals were to see justice done, to preserve the business, and, if possible, to preserve their

Box 13-2 Questions on Best-Fit Strategies

Here are some questions you can help clients ask themselves to determine which strategies will best fit their situation.

- Which strategies will be most useful in helping me get what I need and want?
- Which strategies are best for this situation?
- Which strategies best fit my resources?
- Which strategies will be most economic in the use of resources?
- Which strategies are most powerful?
- Which strategies best fit my preferred way of acting?
- Which strategies best fit my values?
- Which strategies will have the fewest unwanted consequences?

relationship. A colleague helped them sample some possibilities. Under her guidance, they discussed with a lawyer the process and consequences of bringing their dispute to the courts, they had a meeting with a consultant-counselor who specialized in these kinds of disputes, and they visited an arbitration firm.

In this case, the sampling procedure had the added effect of giving them time to let their emotions simmer down. They agreed to go the consultant-counselor route.

Karen, the woman who brainstormed a wide range of strategies for disengaging from alcohol with the help of her counselor, decided to sample some of the possibilities.

Surprised by the number of program possibilities to achieve the goal of getting liquor out of her life, Karen decided to sample some of them. She went to an open meeting of Alcoholics Anonymous, attended a meeting of a women's lifestyle-issues group, visited a hospital that had a residential treatment program, and joined a 2-week trial physical fitness program at a YMCA. She did not engage in these activities frantically. Rather, she tried them out and then discussed them with her counselor. Her search for the programs that were best for her did occupy her energies and strengthened her resolve to do something about her alcoholism.

Of course, some clients could use strategy sampling as a way of putting off action. That was certainly not the case with Bud. His attending the meeting of a self-help group after leaving the hospital was a form of strategy sampling. Although he was impressed by the group, he thought that he could start a group limited to ex-patients that would focus more directly on the kinds of issues he and other ex-patients were facing.

A Balance-Sheet Method for Choosing Strategies

Some form of balance sheet can be used to help clients make decisions in general. The methodology could be used for any key decision related to the helping process—whether to get help in the first place, to work on one problem rather than another, or to choose this rather than that goal. Balance sheets deal with the acceptability and unacceptability of both benefits and costs. A balance-sheet approach, applied to choosing strategies for achieving goals, poses questions such as the following:

- What are the benefits of choosing this strategy? for myself? for significant others?
- To what degree are these benefits acceptable? to me? to significant others?
- In what ways are these benefits unacceptable? to me? to significant others?
- What are the costs of choosing this strategy? for myself? for significant others?
- To what degree are these costs acceptable? to me? to significant others?
- In what ways are these costs unacceptable? to me? to significant others?

Let's return to Karen. She used the balance-sheet method to assess the viability, not of a goal, but of strategies to achieve a goal. Karen's goal was to stop drinking. One possible strategy for accomplishing that goal was to spend a month as an inpatient at an alcoholic treatment center. This possibility appealed to her. However, because choosing this strategy would be a serious decision, the counselor, Joan, helped Karen use the balance sheet to weigh possible costs and benefits. After filling it out, Karen and Joan discussed Karen's findings. She chose to consider the pluses and minuses for herself and for her husband and children.

Benefits of Choosing the Residential Program	
For me	**For significant others**
It would help me because it would be a dramatic sign that I want to do something to change my life. It's a clean break, as it were. It would also give me time just for myself. I'd get away from all my commitments to family, relatives, friends, and work. I see it as an opportunity to do some planning. I'd have to figure out how I would act as a sober person.	I'm thinking mainly of my family here. It would give them a breather, a month without an alcoholic wife and mother around the house. I'm not saying that to put myself down. I think it would give them time to reassess family life and make some decisions about any changes they'd like to make. I think something dramatic like my going away would give them hope. They've had very little reason to hope for the last 5 years.

Acceptability of Benefits	
For me	**For significant others**
I feel torn here. But looking at it just from the viewpoint of acceptability, I feel kind enough toward myself to give myself a month off. Also something in me longs for a new start in life. And it's not just time off. The program is a demanding one.	I think that my family would have no problem letting me take a month off. I'm sure that they'd see it as a positive step from which all of us would benefit.

Unacceptability of Benefits	
For me	**For significant others**
Going away for a month seems such a luxury, so self-indulgent. Also, even though taking such a dramatic step would give me an opportunity to change my current lifestyle, it would also place demands on me. My fear is that I would do fine while in the program but that I would come out and fall on my face. I guess I'm saying it would give me another chance at life, but I have misgivings about having another chance. I need some help here.	The kids are young enough to readjust to a new me. But I'm not sure how my husband would take this "benefit." He has more or less worked out a lifestyle that copes with my being drunk a lot. Though I have never left him and he has never left me, still I wonder whether he wants me back sober. Maybe this belongs under the "cost" part of this exercise. I need some help here. And, of course, I need to talk to my husband about all this. I also notice that some of my misgivings relate not to a residential program as such but to a return to a lifestyle free of alcohol. Doing this exercise helped me see that more clearly.

Costs of Choosing the Residential Program	
For me	**For significant others**
Well, there's the money. I don't mean the money just for the program, but I would be losing 4 weeks' wages. But I've lost a lot of wages through drinking. The major cost seems to be the commitment I have to make about a lifestyle change. And I know the residential program won't be all fun. I don't know exactly what they do there, but some of it must be demanding. Probably a lot of it.	It's a private program, and it's going to cost the family a lot of money. The services I have been providing at home will be missing for a month. It could be that I'll learn things about myself that will make it harder to live with me—though living with a drunken spouse and mother is no joke. What if I come back more demanding of them—I mean, in good ways? I need to talk this through more thoroughly.

Acceptability of Costs	
For me	**For significant others**
I have no problem at all with the money or with whatever the residential program demands of me physically or psychologically. I'm willing to pay. What about the costs of the demands the program will place on me for substantial lifestyle changes? Well, in principle I'm willing to pay what that costs. But I'm not sure what these are. I need some help here.	They will have to make financial sacrifices, but I have no reason to think that they would be unwilling. Still, I can't be making decisions for them. I see much more clearly the need to have a counseling session with my husband and children present. I think they're also willing to have a "new" person around the house, even if it means making adjustments and changing their lifestyle a bit. I want to check this out with them, but I think it would be helpful to do this with the counselor. I think they will be willing to come.

Unacceptability of Costs	
For me	**For significant others**
Although I'm ready to change my lifestyle, I hate to think that I will have to accept some dumb, dull life. I think I've been drinking, at least in part, to get away from dullness; I've been living in a fantasy world, a play world a lot of the time. A stupid way of doing it, perhaps, but it's true. I have to do some life planning of some sort. I need some help here.	It strikes me that my family might have problems with a sober me if it means that I will strike out in new directions. I wonder if they want the traditional homebody wife and mother. I don't think I could stand that. All this should come out in the meeting with the counselor.

Karen concludes, "All in all, it seems like the residential program is a good idea. There is something much more substantial about it than an outpatient program. But that's also what scares me."

Karen's use of the balance sheet helps her make an initial program choice, but it also enables her to discover issues that she has not yet worked out completely. By using the balance sheet, she returns to the counselor with work to do. This highlights the usefulness of exercises and other forms of structure that help clients take more responsibility for what happens both in the helping sessions and outside.

Realism in using the balance sheet. Now, let's look at a more practical and flexible approach to using the balance sheet. It is not to be used with every client to work out the pros and cons of every course of action. Tailor the balance sheet to the needs of the client. Choose the parts of the balance sheet that will add most value with *this* client pursuing *this* goal or set of goals. In fact, one of the best uses of the balance sheet is not to use it directly at all. Keep it in the back of your mind whenever clients are making decisions. Use it as a filter to listen to clients. Then turn relevant parts of it into probes to help clients focus on issues they may be overlooking. "How will this decision affect the significant people in your life?" is a probe that originates in the balance sheet. "Is there any downside to that strategy?" might help a client who is being a bit too optimistic. No formula.

Help Clients Link Best-Fit Strategies to Action

Some clients are filled with great ideas for getting things done but never seem to do anything. They lack the discipline to evaluate their ideas, choose the best, and turn them into action. Often this kind of work seems too tedious to them, even though it is precisely what they need. Consider the following case.

Clint came away from the doctor feeling depressed. He was told that he was in the high-risk category for heart disease and that he needed to change his lifestyle. He was cynical, very quick to anger, and did not readily trust others. Venting his suspicions and hostility did not make them go away, however; it only intensified them. Therefore, one critical lifestyle change was to change this pattern and develop the ability to trust others.

He developed three broad goals: reducing mistrust of others' motives, reducing the frequency and intensity of such emotions as rage, anger, and irritation, and learning how to treat others with consideration. Clint read through the strategies suggested to help people pursue these broad goals (see Williams, 1989). They included the following:

- Keeping a hostility log to discover the patterns of cynicism and irritation in one's life
- Finding someone to talk to about the problem, someone to trust
- "Thought stopping," catching oneself in the act of indulging in hostile thoughts or in thoughts that lead to hostile feelings
- Talking sense to oneself when tempted to put others down
- Developing empathic thought patterns—that is, walking in the other person's shoes
- Learning to laugh at one's own silliness
- Using a variety of relaxation techniques, especially to counter negative thoughts
- Finding ways of practicing trust
- Developing active listening skills
- Substituting assertive for aggressive behavior
- Getting perspective, seeing each day as one's last
- Practicing forgiving others without being patronizing or condescending

> Clint prided himself on his rationality (though his "rationality" was one of the things that got him into trouble). So, as he read down the list, he chose strategies that could form an "experiment," as he put it. He decided to talk to a counselor (for the sake of objectivity), keep a hostility log (data gathering), and use the tactics of thought stopping and talking sense to himself whenever he felt that he was letting others get under his skin. The counselor noted to himself that none of these necessarily involved changing Clint's attitudes toward others. However, he did not challenge Clint at this point. His best bet was that through "strategy sampling" Clint would learn more about his problem, that he would find that it went deeper than he thought. Clint set himself to his experiment with vigor.

Clint chose strategies that fit his values. The problem was that the values themselves needed reviewing. But Clint did act, and action gave him the opportunity to learn.

The Shadow Side of Selecting Strategies

The shadow side of decision making, discussed in Chapter 11, is certainly at work in clients' choosing strategies to implement goals. Goslin (1985, pp. 7, 9) put it well:

> In defining a problem, people dislike thinking about unpleasant eventualities, have difficulty in assigning . . . values to alternative courses of action, have a tendency toward premature closure, overlook or undervalue long-range consequences, and are unduly influenced by the first formulation of the problem. In evaluating the consequences of alternatives, they attach extra weight to those risks that can be known with certainty. They are more subject to manipulation . . . when their own values are poorly thought through. . . . A major problem . . . for . . . individuals is knowing when to search for additional information relevant to decisions.

In choosing a course of action, clients often fail to evaluate the risks involved and determine whether the risk is balanced by the probability of success. Gelatt, Varenhorst, and Carey (1972) suggested four ways in which clients may try to deal with the factors of risk and probability: wishful thinking, playing it safe, avoiding the worst outcome, and achieving some kind of balance. The first three are often pursued without reflection and therefore lie in the "shadows."

Wishful thinking. In this case, clients choose a course of action that might (they hope) lead to the accomplishment of a goal regardless of risk, cost, or probability. For instance, Jenny wants her ex-husband to increase the amount of support he is paying for the children. She tries to accomplish this by constantly nagging him and trying to make him feel guilty. She doesn't consider the risk (he might get angry and stop giving her anything), the cost (she spends a great deal of time and emotional energy arguing with him), or the probability of success (he does not react favorably to nagging). The wishful-thinking client operates blindly, engaging in some course of action without taking into account its usefulness. At its worst, this is a reckless approach. Clients who "work hard" and still "get nowhere" may be engaged in wishful thinking, persevering in using means they prefer but that are of doubtful efficacy. Effective helpers find ways of challenging wishful thinking. "Jenny, let's review what you've been doing to get Tom to pay up and how successful you've been."

Playing it safe. In this case, the client chooses only safe courses of action, ones that have little risk and a high degree of probability of producing at least limited success. For instance, Liam, a manager in his early 40s, is very dissatisfied with the way his boss treats him at work. His ideas are ignored, the delegation he is supposed to have is preempted, and his boss does not respond to his attempts to discuss career development. His goals center around his career. He wants to let his boss know about his dissatisfaction and he wants to learn what his boss thinks about him and his career possibilities. These are instrumental goals, of course, because his overall goal is to carve out a career path. However, he fails to bring these issues up when his boss is "out of sorts." On the other hand, when things are going well, Liam doesn't want to "upset the applecart." He drops hints about his dissatisfaction, even joking about them at times. He tells others in hopes that word will filter back to his boss. During formal appraisal sessions he allows himself to be intimidated by his boss. However, in his own mind, he is doing whatever could be expected of a "reasonable" man. He does not know how safe he is playing it. The helper says, "Liam, you're playing pretty safe with your boss. And, although it's true that you haven't upset him, you're still in the dark about your career prospects."

Avoiding the worst outcome. In this case, clients choose means that are likely to help them avoid the worst possible result. They try to minimize the maximum danger, often without identifying what that danger is. Crissy, dissatisfied with her marriage, sets a goal to be "more assertive." However, even though she has never said this either to herself or to her counselor, the maximum danger for her is to lose her partner. Therefore, her "assertiveness" is her usual pattern of compliance, with some frills. For instance, every once in a while she tells her husband that she is going out with friends and will not be around for supper. Without her knowing it, he actually enjoys these breaks. At some level of her being, she realizes that her absences are not putting him under any pressure, but she continues to be assertive in this way. She never sits down with her husband to review where they stand with each other because that might be the beginning of the end. At the beginning of one session, the counselor says, "What if some good friend were to say to you, 'Bill has you just where he wants you.' How would you react?" Crissy is startled, but she comes away from the session with a much more realistic view of her situation.

Striking a balance. In the ideal case, clients choose strategies for achieving goals that balance risks against the probability of success. This "combination" approach is the most difficult to apply, for it involves the right kind of analysis of problem situations and opportunities, choosing goals with the right edge, being clear about one's values, ranking a variety of strategies according to these values, and estimating how effective any given course of action might be. Even more to the point, it demands challenging the blind spots that might distort these activities. Because some clients have neither the skills nor the will for this combination approach, it is essential that their counselors help them engage in the kind of dialogue that will help them face up to this impasse.

?

Personalizing Task 2

Review the strategies you brainstormed in Task 1. Then use the following guidelines to help yourself choose the best-fit strategy or strategies for accomplishing your goals effectively and efficiently.

- Choose strategies that are clear and specific, that best fit your capabilities, that are linked to goals, that have power, and that are suited to your style and values.
- If it is hard to choose, try some strategy sampling.
- Use the balance sheet as a way of choosing strategies by outlining the principal benefits and costs for yourself, others, and relevant social settings.
- Manage the shadow side of selecting courses of action—that is, wishful thinking, playing it too safe, focusing on avoiding the worst possible outcome rather than on getting what you want, and wasting time by trying to spell out a perfectly balanced set of strategies.
- Use the act of choosing strategies to stimulate problem-managing action.

TASK 3: HELP CLIENTS MAKE PLANS

After identifying and choosing strategies to accomplish goals, clients need to organize these strategies into a plan. In this task, counselors help clients come up with the plan itself, the sequence of actions—What should I do first, second, and third?—that will get them what they want, their goals.

The Case of Frank: No Plan of Action
The lack of a plan—that is, a clear step-by-step process to accomplish a goal—keeps some clients mired in their problem situations. Consider the case of Frank, a vice president of a large West Coast corporation.

Frank was a go-getter. He was very astute about business and had risen quickly through the ranks. Vince, the president of the company, was in the process of working out his own retirement plans. From a business point of view, Frank was the heir apparent. But there was a glitch. Vince was far more than a good manager; he was a leader. He had a vision of what the company should look like 5 to 10 years down the line. Early on, he saw the power of the Internet and used it wisely to give the business a competitive edge.

Though tough, Vince related well to people. People constituted the human capital of the company. He knew that products *and* people kept customers happy. He also took to heart the results of a millennium survey of some two million employees in the United States. One of the sentences in the summary of the survey results haunted him—"People join companies but leave supervisors." In the "war for talent," he couldn't afford supervisors who alienated their team members.

Frank was quite different. He was a "hands-on" manager, meaning, in his case, that he was slow to delegate tasks to others, however competent they might be. He kept second-guessing others when he did delegate, reversed their decisions in a way that made them feel put down, listened poorly, and took a fairly short-term view of the business—"What were last week's figures like?" He was not a leader but an "operations" man. His direct reports called him a micromanager.

One day, Vince sat down with Frank and told him that he was considering him as his successor down the line, but that he had some concerns. "Frank, if it were just a question of business acumen, you could take over today. But my job, at least in my mind, demands a leader." Vince went on to explain what he meant by a leader and to point out the things in Frank's style that had to change.

So Frank did something that he never thought we would do. He began seeing a coach. Roseanne had been an executive with another company in the same industry but had opted to be a consultant for family reasons. Frank chose her because he trusted her business acumen. That's what meant most to him. They worked together for over a year, often over lunch and in hurried meetings early in the morning or late in the evening. And, indeed, he valued their dialogues about the business.

Frank's ultimate aim was to become president. If getting the job meant that he had to try to become the kind of leader his boss had outlined, so be it. Because he was very bright, he came up with some inventive strategies for moving in that direction. But he could never be pinned down to an overall program with specific milestones by which he could evaluate his progress. Roseanne pushed him, but Frank was always "too busy" or would say that a formal program was "too stifling." That was odd, because formal planning was one of his strengths in the business world.

Frank remained as astute as ever in his business dealings. But he merely dabbled in the strategies meant to help him become the kind of leader Vince wanted him to be. Frank had the opportunity of not just correcting some mistakes, but also developing and expanding his managerial style. But he blew it. At the end of 2 years, Vince appointed someone else president of the company.

Frank never got his act together. He never put together the kind of change program needed to become the kind of leader Vince wanted as president. Why? Frank had two significant blind spots that the consultant did not help him overcome. First, he never really took Vince's notion of leadership seriously. So he wasn't really ready for a change program. He thought the president's job was his, that business acumen alone would win out in the end. Second, he thought he could change his management style at the margins, when more substantial changes were called for.

Roseanne never challenged Frank as he kept "trying things" that never led anywhere. Maybe things would have been different if she had said something like, "Come on, Frank, you know you don't really buy Vince's notion of leadership. But you can't just give lip service to it. Vince will see right through it. We're just messing around. You don't want a program because you don't believe in the goal. Let's do something or call these meetings off." In a way she was a coconspirator because she, too, relished their business discussions. When Frank didn't get the job, he left the company, leaving Roseanne to ponder her success as a consultant but her failure as a coach.

How Plans Add Value to Clients' Change Programs

Some clients, once they know what they want and some of the things they have to do to get what they want, get their act together, develop a plan, and move forward. Other clients need help. Because some clients (and some helpers) fail to appreciate the power of a plan, it is useful to start by reviewing the advantages of planning.

Not all plans are formal. "Little plans," whether called such or not, are formulated and executed throughout the helping process. Tess, an alcoholic who wants to stop drinking, feels the need for some support. She contacts Lou, a friend who has shaken a drug habit, tells him of her plight, and enlists his help. He readily agrees. Objective accomplished. This "little plan" is part of her overall change program. Change programs are filled with setting "little objectives" and developing and executing "little plans" to achieve them.

Formal planning usually focuses on the sequence of the "big steps" clients must take in order to get what they need or want. Clients are helped answer the question, "What do I need to do first, second, and third?" A formal plan in its most formal version takes strategies for accomplishing goals, divides them into workable steps, puts the steps in order, and assigns a timetable for the accomplishment of each step. Formal planning, provided that it is adapted to the needs of individual clients, has a number of advantages.

Plans help clients develop needed discipline. Many clients get into trouble in the first place because they lack discipline. Planning places reasonable demands on clients to develop discipline. Desmond, the halfway-house resident discussed earlier in this chapter, needed discipline and benefitted greatly from a formal job-seeking program. Indeed, ready-made programs such as the 12-step program of Alcoholics Anonymous are in themselves plans that demand or at least encourage self-discipline.

Plans keep clients from being overwhelmed. Plans help clients see goals as doable. They keep the steps toward the accomplishment of a goal "bite-size." Amazing things can be accomplished by taking bite-size steps toward substantial goals. Bud, the ex-psychiatric patient who ended up creating a network of self-help groups for ex-patients, started with the bite-size step of participating in one of those groups himself. He did not become a self-help entrepreneur overnight. It was a step-by-step process.

Formulating plans helps clients search for more useful ways of accomplishing goals—that is, even better strategies. Sy Johnson was an alcoholic. When Mr. Johnson's wife and children, working with a counselor, began to formulate a plan for coping with their reactions to his alcoholism, they realized that the strategies they had been trying were hit-or-miss. With the help of an Al-Anon self-help group, they went back to the drawing board. Mr. Johnson's drinking had introduced a great deal of disorder into the family. Planning would help them restore order.

Plans provide an opportunity to evaluate the realism and adequacy of goals. This is an example of the "dialogue" that should take place among the stages of the helping process. When Walter, a middle manager who had many problems in the workplace, began tracing out a plan to cope with the loss of his job and with a lawsuit filed against him by his former employer, he realized that his initial goals—getting his job back and filing and winning a countersuit—were unrealistic. His revised goals included getting his former employer to withdraw the suit and getting into better shape to search for a job by participating in a self-help group of managers who had lost their jobs.

Plans make clients aware of the resources they will need to implement their strategies. When Dora was helped by a counselor to formulate a plan to pull her life together after the disappearance of her younger son, she realized that she lacked the social support needed to carry out the plan. She had retreated from friends and even relatives, but now she knew she had to get back into community. Normalizing life demanded ongoing social involvement and support. A goal of finding the support needed to get back into community was added to her constructive-change program.

Formulating plans helps clients uncover unanticipated obstacles to the accomplishment of goals. Ernesto, a U.S. soldier who had accidentally killed an innocent bystander during his stint in Kosovo, was seeing a counselor because of the difficulty he was having returning to civilian life. Only when he began pulling together and trying out plans for normalizing his social life did he realize how ashamed he was of what had happened to him in the military. He felt so flawed because of if that it was almost impossible to involve himself intimately with others. Helping him deal with his shame became one of the most important parts of the healing process.

Formulating plans will not solve all our clients' problems, but it is one way of making time an ally instead of an enemy. Many clients engage in aimless activity in their efforts to cope with problem situations. Plans help clients make the best use of their time. Finally, planning itself has a hefty shadow side. For a good review of the shadow side of planning, see Dorner (1996, pp. 153–183).

Shaping the Plan: Three Cases

Plans need "shape" to drive action. A formal plan identifies the activities or actions needed to accomplish a goal or a subgoal, puts those activities into a logical but flexible order, and sets a time frame for the accomplishment of each key step. Therefore, a plan should include the answers to these three simple questions.

- What are the concrete things that need to be done to accomplish the goal or the subgoal?

- In what sequence should these be done? What should be done first, what second, what third?

- What is the time frame? What should be done today, tomorrow, next month?

If clients choose goals that are complex or difficult, then it is useful to help them establish subgoals as a way of moving step-by-step toward the ultimate goal. For instance, once Bud decided to start an organization of self-help groups composed of ex-patients from mental hospitals, there were a number of subgoals he needed to accomplish before the organization would become a reality. His first step was to set up a test group. This instrumental goal provided the experience needed for further planning. A later step was to establish some kind of charter for the organization. "Charter in place" was one of the subgoals leading to his main goal.

In general, the simpler the plan the better. However, simplicity is not an end in itself. The question is not whether a plan or program is complicated but whether it is well shaped and designed to produce results. If complicated plans are broken down into subgoals and the strategies or activities needed to accomplish them, they are as capable of being achieved, if the time frame is realistic, as simpler ones. In schematic form, shaping looks like this:

Subprogram 1 (a set of activities) leads to subgoal 1 (usually an instrumental goal).

Subprogram 2 leads to subgoal 2.

Subprogram n (the last in the sequence) leads to the accomplishment of the ultimate goal.

The case of Wanda. Take the case of Wanda, a client who set a number of goals in order to manage a complex problem situation. One of her goals was finding a job. The plan leading to this goal had a number of steps, each of which led to the accomplishment of a subgoal. The following subgoals were part of Wanda's job-finding program. They are stated as accomplishments (the outcome or results approach).

Subgoal 1: Résumé written.

Subgoal 2: Kind of job wanted determined.

Subgoal 3: Job possibilities canvassed.

Subgoal 4: Best job prospects identified.

Subgoal 5: Job interviews arranged.

Subgoal 6: Job interviews completed.

Subgoal 7: Offers evaluated.

The accomplishment of these subgoals leads to the accomplishment of the overall goal of Wanda's plan—that is, getting the kind of job she wants.

Wanda also had to set up a step-by-step process or program to accomplish each of these subgoals. For instance, the process for accomplishing the subgoal "job possibilities canvassed" included such things as doing an Internet search on one or more of the many of the job-search sites, reading the "Help Wanted" sections of the local papers, contacting friends or acquaintances who could provide leads, visiting employment agencies, reading the bulletin boards at school, and talking with someone in the job placement office. Sometimes the sequencing of activities is important, sometimes not. In Wanda's case, it's important for her to have her résumé completed before she begins to canvass job possibilities, but when it came to using different methods for identifying job possibilities, the sequence does not make any difference.

The case of Harriet: The economics of planning. Harriet, an undergraduate student at a small state college, wants to become a counselor. Although the college offers no formal program in counseling psychology, with the help of an advisor she identifies several undergraduate courses that would provide some of the foundation for a degree in counseling. One is called Social Problem-Solving Skills; a second, Effective Interpersonal Communication Skills; a third, Developmental Psychology: The Developmental Tasks of Late Adolescence and Early Adulthood. Harriet takes the courses as they came up. Unfortunately, Social Problem-Solving Skills is the first course. The good news is that it includes a great deal of practice in the skills. The bad news is that it assumes competence in interpersonal communication skills. Too late she realizes that she is taking the courses out of optimal sequence. She would be getting much more from the courses had she taken the communication skills course first.

Harriet also volunteers for the dormitory peer-helper program run by the Center for Student Services. The Center's counselors are very careful in choosing people for the program, but they don't offer much training. It is a learn-as-you-go approach. Harriet realizes that the developmental psychology course would have helped her enormously in this program. It would have helped her understand both herself and her peers better. She finally realizes that she needs a better plan. In the next semester, she drops out of the peer counselor program. She sits down with one of the Center's psychologists, reviews the schools offerings with him, determines which courses will help her most, and determines the proper sequencing of these courses. He also suggests a couple of courses she could take in a local community college. Harriet's opportunity-development program would have been much more efficient had it been better shaped in the first place.

The case of Frank revisited. Let's see what planning might have done for Frank, the vice president who needed leadership skills. In this fantasy, Frank, like Scrooge, gets a second chance.

What does Frank need to do? To become a leader, Frank decides to reset his managerial style with his subordinates by involving them more in decision making. He wants to listen more, set work objectives through dialogue, ask subordinates for suggestions, and delegate more. He knows he should coach his direct reports in keeping with their individual needs, give them feedback on the quality of their work, recognize their contributions, and reward them for achieving results beyond their objectives.

In what sequence should Frank do these things? Frank decides that the first thing he will do is call in each subordinate and ask, "What do you need from me to get your job done? How can I add value to your work? And what management style on my part would help you most?" Their dialogue around these issues will help him tailor his supervisory interventions to the needs of each team member. The second step is also clear. The planning cycle for the business year is about to begin, and each team member needs to know what his or her objectives are. It is a perfect time to begin setting objectives through dialogue rather than simply assigning them. Frank therefore sends a memo to each of his direct reports, asking them to review the company's strategy and business plan and the strategy and plan for each of their functions, and to write down what they think their key managerial objectives for the coming year should be. He asks them to include "stretch" goals.

What is Frank's time frame? Frank calls in each of his subordinates immediately to discuss what they need from him. He completes his objective-setting sessions with them within 3 weeks. He puts off further action on delegation until he gets a better reading on their performance. This is a rough idea of what a plan for Frank might have looked like and how it might have improved his chaotic and abortive effort to change his managerial style—on the condition, of course, that he was convinced that a different approach to management and supervision made personal

Box 13-3   Questions on Planning

Here are some questions you can help clients ask themselves in order to come up with a viable plan for constructive change.

- Which sequence of actions will get me to my goal?
- Which actions are most critical?
- How important is the order in which these actions take place?
- What is the best time frame for each action?
- Which step of the program needs substeps?
- How can I build informality and flexibility into my plan?
- How do I gather the resources, including social support, needed to implement the plan?

and business sense. Box 13-3 is a list of questions you can use to help clients think systematically about crafting a plan to get what they need and want.

Humanizing the Mechanics of Constructive Change

Some years ago I lent a friend of mine an excellent, though somewhat detailed, book on self-development. About 2 weeks later he came back, threw the book on my desk, and said, "Who would go through all of that!" I retorted, "Anyone really interested in self-development." That was the righteous, not the realistic, response. Planning in the real world seldom looks like planning in textbooks. Textbooks do provide useful frameworks, principles, and processes, but they are seldom implemented exactly as they are outlined. Most people are too impatient to do the kind of planning just outlined. One of the reasons for the dismal track record of discretionary change mentioned earlier is that even when clients do set realistic goals, they lack the discipline to develop reasonable plans. The detailed work of planning is too burdensome.

Therefore, Stages II and III of the helping process together with their six tasks need a human face. If helpers skip the goal-setting and planning steps clients need, they shortchange them. On the other hand, if they are pedantic, mechanistic, or awkward in their attempts to help clients engage in these steps—failing to give these processes a human face—they run the risk of alienating the people they are trying to serve. Clients might well say, "I'm getting a lot of boring garbage from him." Here, then, are some principles to guide the constructive-change process.

Build a planning mentality into the helping process right from the start. A constructive-change mind-set should permeate the helping process right from the beginning. This is part of the hologram metaphor—the whole model should be found in each of its parts—mentioned in Chapter 2. Helpers need to see clients as self-healing agents capable of changing their lives, not just as individuals mired in problem situations. Even while listening to a client's story, the helper needs to begin thinking of how the situation can be remedied and through probes find out what approaches to change clients are thinking about—no matter how tentative these ideas might be. As mentioned earlier, helping clients act in their real world right from the beginning of the helping process helps them develop some kind of initial planning mentality. If helping is to be solution-focused, thinking about strategies and plans must be introduced early. When a client tells of some problem, the helper can ask early on, "What kinds of things have you done so far to try to cope with the problem?"

> Cora, a battered spouse, did not want to leave her husband because of the kids. Right from the beginning, the helper saw Cora's problem situation from the point of view of the whole helping process. While she listened to Cora's story, without distorting it, she saw possible goals and strategies. Within the helping sessions, the counselor helped Cora learn a great deal about how battered women typically respond to their plight and how dysfunctional some of those responses are. She also learned how to stop blaming herself for the violence and to overcome her fears of developing more active coping strategies. At home she confronted her husband and stopped submitting to the violence in a vain attempt to avoid further abuse. She also joined a local self-help group for battered women. There she found social support and learned how to invoke both police protection and recourse through the courts. Further sessions with the counselor helped her gradually change her identity from battered woman to survivor and, eventually, to doer. She moved from simply facing problems to developing opportunities.

Constructive-change scenarios like this must be in the helper's mind from the start, not as preset programs to be imposed on clients but as part of a constructive-change mentality.

Adapt the constructive-change process to the style of the client. Setting goals, devising strategies, and making and implementing plans can be done formally or informally. There is a continuum. Some clients actually like the detailed work of devising plans; it fits their style.

> Gitta sought counseling as she entered the "empty nest" period of her life. Although there were no specific problems, she saw too much emptiness as she looked into the future. The counselor helped her see this period of life as a normal experience rather than a psychological problem. It was a developmental opportunity and challenge (see Raup & Myers, 1989). It was an opportunity to reset her life. After spending a bit of time discussing some of the maladaptive responses to this transitional phase of life, they embarked on a review of possible scenarios. Gitta loved brainstorming, getting into the details of the scenarios, weighing choices, setting strategies, and making formal plans. She had been running her household this way for years. So the process was familiar even though the content was new.

Here is another case.

> Connor, rebuilding his life after a serious automobile accident, very deliberately planned both a rehabilitation program and a career change. Keeping to a schedule of carefully planned actions not only helped him keep his spirits up but also helped him accomplish a succession of goals. These small triumphs buoyed his spirits and moved him, however slowly, along the rehabilitation path.

Both Gitta and Connor readily embraced the positive psychology approach embedded in constructive change programs. They thrived on both the work and the discipline to develop plans and execute them. However, many, if not most, people, are not like Gitta and Connor. The distribution is skewed toward the "I hate all this detail and won't do it" end of the continuum.

Kirschenbaum (1985) challenged the notion that planning should always provide an exact blueprint for specific actions, their sequencing, and the time frame. He suggested that helpers consider these three questions:

1. How specific do the activities have to be?
2. How rigid does the order have to be?
3. How soon does each activity have to be carried out?

Kirschenbaum (p. 492) suggested that, at least in some cases, being less specific and rigid about actions, sequencing, and deadlines can "encourage people to pursue their goals by continually and flexibly choosing their activities." That is, flexibility in planning can help clients become more self-reliant and proactive. Rigid planning strategies can lead to frequent failure to achieve short-term goals. Consider the case of Yousef.

> Yousef was a single parent with a mentally retarded son. He was challenged one day by a colleague at work. "You've let your son become a ball and chain and that's not good for you or him!" his friend said. Yousef smarted from the remark, but eventually—and reluctantly—sought counseling. He never discussed any kind of extensive change program with his helper, but with a little stimulation from her he began doing little things differently at home. When he came home from work especially tired and frustrated, he had a friend in the apartment building stop by. This helped him to refrain from taking his frustrations out on his son. Then, instead of staying cooped up over the weekend, he found simple things to do that eased tensions, such as going to the zoo and to the art museum with a woman friend and his son. He discovered that his son enjoyed these pastimes immensely despite his limitations. In short, he discovered little ways of blending caring for his son with a better social life. His counselor had a constructive-change mentality right from the beginning but did not try to engage Yousef in overly formal planning activities.

On the other hand, a slipshod approach to planning—"I will have to pull myself together one of these days"—is also self-defeating. We need only look at our own experience to see that such an approach is fatal. Overall, counselors should help clients embrace the kind of rigor in planning that make sense for them in their situations. There are no formulas; there are only client needs and common sense. Some things need to be done now, some later. Some clients need more slack than others. Sometimes it helps to spell out the actions that need to be done in quite specific terms; at other times it is necessary only to help clients outline them in broad terms and leave the rest to their own sound judgment. If therapy is to be brief, help clients start doing things that lead to their goals. Then, in a later session, help them review what they have been doing, drop what is not working, continue what is working, add more effective strategies, and put more organization in their programs. If you have a limited number of sessions with a client, you can't engage in extensive goal setting and planning. "What can I do that will add most value?" is the ongoing challenge in brief therapy.

Devise a plan for the client and then help the client tailor it to his or her needs. The more experienced helpers become, the more they learn about the elements of program development and the more they come to know what kinds of programs work for different clients. They build up a stockpile of useful programs and know how to stitch pieces of different programs together to create new programs. And so they can use their knowledge and experience to fashion a plan for any clients who lack the skills or the temperament to pull together a plan for themselves. Of course, their objective is not to foster dependence but to help clients grow in self-determination. For instance, they can first offer a plan as a sketch or in outline form rather than as a detailed program. Helpers then work with clients to fill out the sketch and adapt it to their needs and style. Consider the following case.

> Katrina, a woman who dropped out of high school but managed to get a high school equivalency diploma, was overweight and reclusive. Over the years she had restricted her activities because of her weight. Sporadic attempts at dieting had left her even heavier. Because she was chronically depressed and had little imagination, she was not able to come up with any kind of coherent plan. Once her counselor understood the dimensions of Katrina's problem situation, she pulled together an outline of a change program that included such things as blame reduction, the redefinition of beauty, decreasing self-imposed social restrictions, cognitive restructuring activities aimed at lessening depression (see Robinson & Bacon, 1996). She also provided information about obesity and suggestions for dealing with it drawn from health-care sources. She presented these in a simple format, adding detail only for the sake of clarity. She added further detail as Katrina got involved in the planning process and in making choices.

Although this counselor pulled together elements of a range of already existing programs, counselors are, of course, free to make up their own programs based on their expertise and experience. The point is to give clients something to work with, something to get involved in. The elaboration of the plan emerges through dialogue with the client and in the kind of detail the client can handle.

The ultimate test of the effectiveness of plans lies in the problem-managing and opportunity-developing action clients engage in to get what they need and want. There is no such thing as a good plan in and of itself. Results, not planning or hard work, are the final arbiter. The next and final chapter deals with turning planning into accomplishments.

Tailor ready-made programs to clients' needs. Counselors add value by helping clients adapt "set" programs to their particular needs. There are many ready-made programs for clients with particular problems. They are often tried-and-true constructive-change programs. The 12-step approach of Alcoholics Anonymous is one of the most well known. It has been adapted to other forms of substance abuse and addiction. Systematic desensitization, a behavioral approach, has been used to treat clients with posttraumatic stress disorder (PTSD; Frueh, de Arellano, & Turner, 1997). This program includes sessions in muscle relaxation, the development of a fear hierarchy, and, finally, weekly sessions in the systematic desensitization of these fears. The program helps alleviate such debilitating symptoms as intrusive thoughts, panic attacks, and episodic depression. Here are a couple of other ready-made programs.

A prevention program for pedophilia. Although there are many treatment programs for pedophilic clients *after* the fact, prevention programs are much scarcer. Consider this case.

> After a couple of rather aimless sessions, the helper said to Ahmed, "We've talked about a lot of things, but I'm still not sure why you came in the first place." This challenged Ahmed to reveal the central issue, though he needed a great deal of help to do so. It turned out that Ahmed was sexually attracted to prepubescent children of both sexes. Although he had never engaged in pedophilic behavior, the temptation to do so was growing.

The counselor adapted a New Zealand program called Kia Marama (Hudson, Marshall, Ward, Johnston, et al., 1995), a comprehensive cognitive-behavioral program for incarcerated child molesters, to Ahmed's situation. The original program includes intensive work in challenging distorted attitudes, reviewing a wide range of sexual issues, seeing the world from the victim's point of view, developing problem-solving and interpersonal-relationship skills, stress management, and relapse-prevention training. Counselor and client spent some time assessing which parts of the program might be of most help before embarking on an intensive tailored program.

The economics of prevention far outweigh the economics of rehabilitation. Not only did Ahmed stay out of trouble, but much of what he learned from the program—for instance, stress management—also applied to other areas of his life.

A program for helping people on welfare be successful at work. One community-based mental health center worked extensively with people on welfare. When new legislation was passed forcing welfare recipients to get work, they searched for programs that helped people on welfare get and keep jobs. They learned a great deal from one program sponsored by a major hotel chain (see Milbank, 1996). The hotel targeted welfare recipients because it made both economic and social sense. Because of the problems with this particular population, however, the hotel's recruiters, trainers, and supervisors had to become paraprofessional helpers, though they never used that term. The people they recruited—battered women, ex-convicts, addicts, homeless people, including those who had been thrown out of shelters, and so forth—had all sorts of problems. In the beginning, the hotel's staff did many things *for* the trainees.

> They drive welfare trainees to work, arrange their day care, negotiate with their landlords, bicker with their case workers, buy them clothes, visit them at home, coach them in everything from banking skills to self-respect and promise those who stick with it full-time jobs. (Milbank, 1996, p. A1)

But the trainers also challenged their "clients'" mind-set that they were not responsible for what happened to them, enforced the hotel's code of behavior with equity, and persevered. The hotel program was far from perfect, but it did help many of the participants develop much-needed self-discipline and find a new life both at work and outside.

The counselors from a local mental health center who acted as consultants to the program learned that some of the new employees benefitted greatly from wholesale upfront involvement of trainers and supervisors in their lives. It kick-started a constructive-change process. They also saw that the recruiters, trainers, and supervisors also benefitted. So they started a volunteer program at the mental health center, looking for people willing to do the kinds of things that the hotel trainers and supervisors did. They knew that both the clients and the volunteers would benefit.

Help clients develop contingency plans. If the future is uncertain, it pays to have a broad range of options open. There is no use investing a great deal of time and energy in a goal or in a program to accomplish a goal that will have to be changed because the client's world changes. Therefore, help clients choose one or more backup goals to take care of such eventualities. In this way, clients have direction, but they also have contingency. If the world changes, then the client can choose the goal that best fits the circumstances at the time. So choosing a goal or a program to achieve a goal is not necessarily a once-and-forever decision. The client says to herself, "I'll stick with this goal until I see that it is no longer viable. Or until a better goal emerges."

Having viable options helps you kill, or at least put on the back burner, an option that is no longer working. Backup plans provide freedom and flexibility. They also keep clients from falling into the status quo decision-making trap outlined earlier. Consider Linda's case.

> Linda (not her real name) is a young woman working for a computer firm in Mexico. She was born and raised in Iraq. She has made a tortuous journey through South and Central America as an illegal immigrant. Her journey included prostitution and a range of harrowing, even life-threatening, experiences. The upside of all this is that she has learned to live by her wits. After returning from an illegal trip to the United States, she has one goal—to live there permanently. She takes counsel with a friend of hers, a lawyer in Mexico, telling him of her plan to live as an illegal in the United States. Both intelligent and socially savvy, she feels that she can pull it off.
>
> Her lawyer friend, knowing that her ultimate goal is to live permanently in the United States, helps her review a range of instrumental goals—goals in themselves but steps toward helping her achieve her ultimate goal. They discuss possibilities. Options other than living by her wits as an illegal immigrant include obtaining political refugee status, becoming a green card holder, marrying a U.S. citizen, marrying a foreigner who is most likely to get a green card, and being included in the quota of immigrants allowed permanent resident status because they have essential skills such as those needed in booming technology industries. A plan would be needed to pursue each of these. Linda's future is certainly filled with risk and uncertainty. She has to choose an instrumental goal that she thinks offers the best possibility for achieving her ultimate goal, but after her discussion she has a range of fall-back options.

If at times goals need to be changed—"If I don't get into medical school, then the nurse-practitioner route is still attractive"—it is also true that strategies for achieving goals might have to change. Contingency plans answer the question, "What will I do if the plan of action I choose is not working?" They help make clients more effective tacticians. We make contingency plans because we live in an imperfect world.

> Jackson, the man dying of cancer, decided to become a resident in the hospice he had visited. The hospice had an entire program in place for helping patients like Jackson die with dignity. Once there, however, he had second thoughts. He felt incarcerated. Fortunately, he had worked out alternative scenarios with his helper. One was living at the home of an aunt he loved and who loved him dearly, with some outreach services from the hospice. He moved out of the hospice into his aunt's home and then spent his final days at the hospice.

Contingency plans are needed especially when clients choose a high-risk program to achieve a critical goal. Having backup plans also helps clients develop more responsibility. If they see that a plan is not working, then they have to decide whether to try the contingency plan. Backup plans need not be complicated. A counselor might merely ask, "If that doesn't work, then what will you do?" As in the case of Jackson, clients can be helped to specify a contingency plan further once it is clear that the first choice is not working out. Box 13-3 (on page 325) lists the kinds of questions you can help clients ask themselves about making plans.

General Well-Being Programs: Exercise and Diet

Some programs that contribute to general well-being can be used as adjuncts to all approaches to helping. Take diet. In many ways we are what we eat. But it is not just a question of obesity. Many of us could achieve significant health benefits by losing just a few pounds (Martin, 2002). The incidence of cancer could be cut in half if we were to avoid toxins in the environment and eat well. As helpers, we cannot afford to remain ignorant of the role of a good diet in *mental* health and of programs that promote healthy eating. The best place to start is with ourselves.

Exercise programs are probably one of the most underused adjuncts to helping (Burks & Keeley, 1989; DeAngelis, 2002; Servan-Schreiber, 2004). Because self-efficacy often improves with exercise (McAuley, Mihalko, & Bane, 1997), exercise deserves a place in many action programs. The self-discipline developed through exercise programs can be a stimulus to increased self-regulation in other areas of life. There is evidence showing that exercise programs can help in the treatment of schizophrenia and alcohol dependence (Read & Brown, 2003; Tkachuk & Martin, 1999). Such programs also help more directly to reduce depression, manage chronic pain, and control anxiety. Kate Hays (1999) has done a comprehensive review of the positive psychology possibilities of exercise in *Working It Out: Using Exercise in Psychotherapy*. I'll end on a personal note regarding exercise. Once I got my gear together and started out to get some exercise. When I hesitated, I asked myself: "Have you *ever* regretted exercising?" I answered, "Never," and headed out the door.

Treatment Manuals

Manualized treatment programs are also examples of ready-made programs. Manuals have been around for some time. For instance, Donald Meichenbaum (1994) published a comprehensive handbook, including a practical manual, for dealing with PTSD. However, the evidence-based practice movement has accelerated their development enormously. There are hundreds of efficacy studies (Luborsky, 1993; Seligman, 1995), many of them very well designed, that demonstrate the efficacy of a particular therapy for a particular psychological disorder and for particular populations. Over the years, these studies have led to "empirically supported treatments" (ESTs) or "empirically validated treatments" (EVTs; Nathan, 1998; Waehler, Kalodner, Wampold, & Lichtenberg, 2000).

Treatments that have proved to be substantive have been "manualized." Manuals outline step-by-step processes and programs for helping clients achieve a specific goal, such as reducing the debilitating symptoms of PTSD. Manuals describe how to deliver treatment for a particular kind of disorder—for instance, agoraphobia (Barlow, 2004; Rothbaum, 2006). Some are written for practitioners (Barlow, 2004; Zuercher-White, 1998), some as self-help guides for clients (Pollard & Zuercher, 2003), and some for both helpers and clients (Andrews, Creamer, Crino, Hunt, Lampe, & Page, 2002).

Manuals have been developed for a wide range of human problems (Nathan & Gorman, 2002), including anxiety, phobias, depression, personality disorders, post-traumatic stress disorder, substance abuse, panic, and borderline personality disorder. New manuals for different psychological disorders are being created continually. Indeed, merely reading through Chambless's (2005) compendium of empirically supported therapies and Lambert et al.'s (1995) compendium of psychotherapy treatment manuals is daunting. One publisher, Hogrefe & Huber, in cooperation with the Society of Clinical Psychology (APA Division 12) has begun to publish manuals in its *Advances in Psychotherapy—Evidence-Based Practice* series. Each book in the series is a compact "how-to" reference on a particular disorder. Topics to be available by the end of 2006 include bipolar disorder, heart disease, obsessive-compulsive disorder, childhood maltreatment, schizophrenia, and prob-lem and pathological gambling.

Of course, when it comes to treatment manuals, the debate on evidence-based practice referred to in Chapter 1 spills over and even intensifies (Weisz, Weersing, & Henggeler, 2005; Westen, Novotny, & Thompson-Brenner, 2004, 2005). According to Waehler and his associates (2000), the forces driving the creation of manualized treatments include managed care's push to maximize the value of all treatments, including mental health treatments, the challenge coming from biological psychia-try to justify empirically the efficacy of psychosocial approaches to helping, and the attempt of psychology itself to answer the question the nagging question: "What treatment, by whom, is most effective for this individual with that specific problem under which set of circumstances?" (Paul, 1967).

The debate is highly polarized—pro and con. For instance, on the "pro" side is a study of 47 therapists' views of treatment manuals (Najavits, Weiss, Shaw, & Dierberger, 2000). Findings indicated that helpers had a very positive view of man-uals, used them extensively, and had few concerns with them. However, 4 years later, the findings from a study from the same research setting (Najavits et al., 2004) were more nuanced: "Therapists were highly positive about the treatments. However, their likelihood of using them in the future without modification was low, and they viewed them as too short. Supervision was seen as more important than manuals and taping of sessions more important than adherence scales. It took therapists an aver-age of 8 months to feel comfortable with the treatments" (p. 26).

Many psychologists see a bright future for manualized treatments (Nathan, 1998; Wilson, 1998). Two journals provide, in special sections, a range of articles on empirically supported therapies and manualized treatments. Kazdin (1996) intro-duces a series of articles on empirically validated treatments in *Clinical Psychology: Science and Practice*. And Kendall (1998) introduces a special section of the *Journal of Consulting and Clinical Psychology* on empirically supported therapies.

On the down side, Garfield (1996) responded to guidelines on the use of empir-ically validated treatments issued by the Division of Clinical Psychology of the American Psychological Association (Division 12 Task Force, 1995). He expressed a range of concerns that have since been echoed and amplified by other researchers and practitioners. Some common concerns are that the language of the Task Force report was too strong, some of its recommendations were premature, manuals often idealize and thus distort psychotherapy, the research base underpinning some manu-als is questionable, clients are messy and don't come to therapy with neatly catego-rized problems, studies don't factor in the competence (or the lack thereof) of the therapists involved, manuals ignore the role of the therapist as model of adult living, the place of art and clinical judgment (see Soldz & McCullough, 2000; Waddington, 1997) is demeaned or ignored, and manual treatments are often highly specialized, cumbersome, time consuming, expensive, and not user friendly.

One of the major annoyances for many, mentioned in Chapter 1, is that the evi-dence-based practice movement requires buying into the medical model of psycho-logical treatment (Barlow, 2004). In an historical examination of evidence-based practice, Wampold and Bhati (2004) found that evidence-based treatments overem-phasize treatments and treatment differences and ignore aspects of therapy demon-strated to be related to positive outcomes, such as variations among psychologists, the helping relationship, the subjective experience of the client, and other common factors. These themes have been repeated year after year in a host of articles.

Many clients have problem situations rather than specific problems such as phobias or PTSD symptoms. Or they have both. Many of the cases reviewed in this book focus on clients with complex problems in living that are not amenable to manualized treatments. Manuals are used when the "diagnosis" is clear and the desired outcomes are clear. Because anxiety and stress are part of most problem situations, manualized treatments for them can play an important role (Lebow, 2006). Furthermore, manuals focus on problems—not on the development of opportunities, which should be one of the principal forms of helping.

More common sense and balance are needed in this debate (Deegear & Lawson, 2003). Many psychologists see manuals as just another tool and fold them into their practice whenever they are deemed useful (Weisz, Sandler, Durlak, & Anton, 2005). Others see the need to tailor such programs to the individual (Nicholson, Anderson, Fox, & Brenner, 2002; Schmidt & Taylor, 2002). Still others want more research into manualized treatments to start and end with real clients and based on case studies (Edwards, Dattilio, & Bromley, 2004; Messer, 2004). If you intend to become a professional helper, you will have to come to terms with the evidence-based practice controversy and treatment manuals. Current literature on these topics is rich, thought provoking, political, and confusing. Treatment manuals are not going to go away (Foxhall, 2000), nor is the controversy they have precipitated.

Finally, not all useful ready-made programs are found in sophisticated evidence-based manuals. Many are found in the best of the self-help literature. Books like *Thoughts and Feelings* (McKay, Davis, & Fanning, 1997) are filled with systematic strategies for the treatment of a wide variety of psychological problems. The best are realistic, practical, and translations of some of the best thinking in the field or a "manualization" of common sense. John Norcross and his colleagues (2003) have published a very useful "authoritative guide" to self-help resources in mental health.

You need to study both the problems and potential associated with evidence-based practice manuals. Your major criterion should be usefulness for your clients. The January 2006 issue of *American Psychologist* (Volume 61, Number 1) is devoted in large part to the concept of "arbitrary metrics." It is well worth reading because it challenges the way we "prove" things in psychology. Kazdin (2006, p. 42) applies the concept of arbitrary metrics to evidence-based treatments: "Research designed to establish the empirical underpinnings of psychotherapy relies heavily on arbitrary metrics, and researchers often do not know if clients receiving an evidence-based treatment have improved in everyday life or changed in a way that makes a difference, apart from the changes the arbitrary metrics may have shown." The points he makes do not constitute an attack on evidence-based treatments. Rather, they are a plea for linking measurements to life-enhancing outcomes.

On the other hand, Lebow outlines what he calls the "incontestable track record" (2006, p. 77) evidence-based treatments have in helping clients deal with the anxiety related to panic disorders, obsessive-compulsive behavior, simple phobias, and generalized anxiety disorder. He claims that the evidence for the effectiveness of such treatments is so strong that, "as responsible therapists, we need to know how to practice these techniques or be prepared to refer these clients to therapists who do" (p. 79). Such referrals are common practice at the clinic where he works. In the end, theory, research, and ideology need to serve or give way to the needs of clients.

Once more, it is not a question of accepting or rejecting evidence-based treatments but of using a critical client-focused clinical eye to integrate selective treatments into your practice.

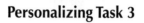

Personalizing Task 3

In facing your own problems and unused opportunities, think of a concrete situation: How effective a planner are you? Ask yourself the following questions.

- To what degree do I prize and practice planning in my own life?
- How effectively do I get into a planning mind-set right from the beginning of the problem management process?
- How quickly do I move to the action planning stage once I realize that's what I need?
- What do I do to overcome my resistance to planning? How effectively do I identify the incentives for and the payoff of planning for myself?
- How well do I formulate subgoals that lead to the accomplishment of overall preferred-picture goals?
- How practical am I in identifying the actions needed to accomplish subgoals, sequencing those actions, and establishing realistic time frames for them?
- In managing my own problems, how easily do I move back and forth among the different stages and steps of the helping model as the need arises?
- When I work with others, how human is the technology of constructive change in my hands?
- How well do I adapt the helping process to the needs and style of people I work with?
- What generic or ready-made change programs have I tried? How well did I adapt them to my specific needs?
- How well do I understand the evidence-based practice and manualized treatment movements and the debate surrounding them? What are my reactions to these movements?

THE ACTION ARROW: MAKING IT ALL HAPPEN

The Action Arrow of the helping model in Figure 14-1 (p. 338) highlights the difference between planning and action. Stages I, II, III, and their nine tasks all revolve around planning for change, not change itself. However, the need to incorporate action into planning and planning into action has been emphasized throughout the book. That is, the "little actions" needed to get the change process moving right from the start have been noted and illustrated. We now take a more formal look at results-producing action and how to help clients identify obstacles to action and find ways to move beyond them.

GETTING THERE: HELP CLIENTS IMPLEMENT THEIR CHANGE PROGRAMS—"HOW DO I MAKE IT ALL HAPPEN?"

In a book called *True Success*, Tom Morris (1994) lays down the conditions for achieving success. They include the following:

- determining what you want—that is, a goal or a set of goals "powerfully imagined"
- focus and concentration in preparation and planning
- the confidence or belief in oneself to see the goal through—that is, self-efficacy
- a commitment of emotional energy
- being consistent and persistent in the pursuit of the goal
- the kind of integrity that inspires trust and gets people pulling for you
- a capacity to enjoy the process of getting there

The role of the counselor is to help clients engage in these kinds of internal and external behaviors in the interest of goal accomplishment.

Some clients, once they have a clear idea of what to do to handle a problem situation or develop some opportunity, go ahead and do it, whether or not they have a formal plan. They need little or no further support and challenge from their helpers. They either find the resources they need within themselves or get support and challenge from the significant others in the social settings of their lives. However, other clients choose goals and come up with strategies for implementing them, but are, for whatever reason, stymied when it comes to action. Most clients fall between these two extremes.

Discipline and self-control—an important part of social-emotional intelligence—play an important role in implementing change programs. Kirschenbaum (1987) found that many things can contribute to not getting started or to giving up: low initial commitment to change, weak self-efficacy, poor outcome expectations, the use of self-punishment rather than self-reward, depressive thinking, failure to cope with emotional stress, lack of consistent self-monitoring, failure to use effective habit-change techniques, giving in to social pressure, failure to cope with initial relapse, and paying attention to the wrong things—for instance, focusing on the difficulty of the change process rather than the attractiveness of the outcome.

We have seen that self-determination and self-control are essential for action. Kanfer and Schefft (1988, p. 58) differentiated between two kinds of self-control. In *decisional self-control*, a single choice terminates a conflict. For instance, a couple makes the decision to get a divorce and goes through with it. In *protracted self-control*, continued resistance to temptation is required. For instance, it is not enough for a client to decide that she has to keep her anger under control when disagreements with others arise. Each time a conflict arises she has to renew her resolve. It helps enormously if she develops the attitude that conflicts are learning opportunities and not just interpersonal struggles. This is a positive way of staying on guard.

Most clients need both kinds of self-control to manage their lives better. A client's choice to give up alcohol completely (decisional self-control) needs to be complemented by the ability to handle inevitable longer-term temptations. Protracted self-control calls for a preventive mentality and a certain degree of street

The Skilled-Helper Model

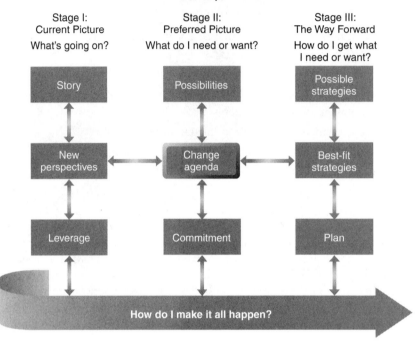

FIGURE 14-1
The Helping Model Complementing Planning With Action

smarts. It is easier for the client who has given up alcohol to turn down an invitation to go to a bar in the first place than to sit in a bar all evening with friends and refrain from drinking. Figure 14-1 adds the Action Arrow to the helping model.

HELP CLIENTS MOVE FROM PLANNING TO ACTION

In the implementation phase, strategies for accomplishing goals need to be complemented by tactics and logistics. A strategy is a practical plan to accomplish some objective. Tactics is the art of *adapting* a plan to the immediate situation. This includes being able to change the plan on the spot to handle unforeseen complications. Logistics is the art of being able to provide the *resources* needed for the implementation of a plan *in a timely way*.

> During the summer, Rebecca wanted to take an evening course in statistics so that the first semester of the following school year would be lighter. Having more time would enable her to act in one of the school plays, a high priority for her. But she didn't have the money to pay for the course, and at the university she planned to attend prepayment for summer courses was the rule. Rebecca had counted on paying for the course from her summer earnings, but she would not have the money until later. Consequently, she did some quick shopping around and found that the same course was being offered by a community college not too far from where she lived. Her tuition there was minimal, because she was a resident of the area the college served.

In this example, Rebecca keeps to her overall plan (strategy). However, in light of an unforeseen circumstance, the demand for prepayment, she adapts her plan (a tactic) by locating another resource (logistics).

Principles of Effective Implementation

Because many well-meaning and motivated clients are simply not good tacticians and are not good at finding the resources they need (logistics), counselors can add value by using the following principles to help them engage in focused and sustained goal-accomplishing action.

Help clients develop "implementation intentions." Commitment to goals (see Chapter 12) must be followed by commitment to courses of action. Gollwitzer (1999) has researched a simple way to help clients cope with the common problems associated with translating goals into action—failing to get started, becoming distracted, reverting to bad habits, and so forth. Strong commitment to goals is not enough. Equally strong commitment to specific actions to accomplish goals is required. Good intentions, Gollwitzer points out, don't deserve their poor reputation. Strong intentions—"I strongly intend to study for an hour every weekday before dinner"—are "reliably observed to be realized more often than weak intentions" (p. 493).

> Implementation intentions are subordinate to goal intentions and specify the when, where, and how of responses leading to goal attainment. They have the structure of "When situation x arises, I will perform response y!" and thus link anticipated opportunities with goal-directed responses. (p. 494)

So Gwendolyn, an aide in a nursing home, may say, "When Enid [a patient] becomes abusive, I will not respond immediately. I'll tell myself that it's her illness that's talking. Then I'll respond with patience and kindness." Her ongoing goal is to control her anger and other negative responses to patients. However, Gwendolyn keeps pursuing this goal by continually refreshing her strong implementation intentions. Because Enid has been a particularly difficult patient, Gwendolyn needs to refresh her intentions frequently. However, her initial strong intention to substitute anger and impatience with kindness and equanimity means that in most cases her responses are more or less automatic. The environmental cue—patient anger, abuse, lack of consideration, and whatever—"triggers" the appropriate response in Gwendolyn. In a way, poor patient behavior become "opportunities" for her responses. You can help clients enunciate to themselves strong specific intentions that will help them "automatically" handle many of the obstacles to goal implementation.

Help clients avoid imprudent action. For some clients, the problem is not that they refuse to act but that they act imprudently. Rushing off to try the first "strategy" that comes to mind is often imprudent.

> Elmer injured his back and underwent a couple of operations. After the second operation he felt a little better, but then his back began troubling him again. When the doctor told him that further operations would not help, Elmer was faced with the problem of handling chronic pain. It soon became clear that his psychological state affected the level of pain. When he was anxious or depressed, the pain always seemed much worse.

Elmer was talking this through with a counselor. One day he read about a pain clinic located in a western state. Without consulting anyone, he signed up for a 6-week program. Within 10 days he was back, feeling more depressed than ever. He had gone to the program with extremely high expectations because his needs were so great. The program was a holistic one that helped the participants develop a more realistic lifestyle. It included programs dealing with nutrition, stress management, problem solving, and quality of interpersonal life. Group counseling was part of the program, and training was part of the group experience. For instance, the participants were trained in behavioral approaches to the management of pain.

The trouble was that Elmer had arrived at the clinic, which was located on a converted farm, with unrealistic expectations. He had bought a "packaged" program without studying the package carefully. Because he had expected to find marvels of modern medicine that would magically help him, he was extremely disappointed when he found that the program focused mainly on reducing and managing rather than eliminating pain.

Elmer's goal was to be completely free of pain, but he failed to explore the realism of his goal. A more realistic goal would have centered on the reduction and management of pain. Elmer's counselor failed to help him avoid two mistakes—setting an unrealistic goal and, in desperation, acting on the first strategy that came along. Obviously, action cannot be prudent if it is based on flawed assumptions—in this case, Elmer's assumption that he could be pain free.

Help clients overcome procrastination. At the other end of the spectrum are clients who keep putting action off. There are many reasons for procrastination. Take the case of Eula.

Eula, disappointed with her relationship with her father in the family business, decided that she wanted to start her own. She thought that she could capitalize on the business skills she had picked up in school and in the family business. Her goal, then, was to establish a small software firm that created products for the family-business market.

But a year went by and she still did not have any products ready for market. A counselor helped her see two things. First, her activities—researching the field, learning more about family dynamics, going to information technology seminars, getting involved for short periods with professionals such as accountants and lawyers who did a great deal of business with family-owned firms, drawing up and redrafting business plans, and creating a brochure—were helpful, but they did not produce products. The counselor helped Eula see that at some level of her being she was afraid of starting a new business. She had a lot of half-finished products. Overpreparation and half-finished products were signs of that fear. So she plowed ahead, finished a product, and brought it to market on the Internet. To her surprise, it was successful. Not a roaring success, but it meant that the cork was out of the bottle. Once she got one product to market, she had little problem developing and marketing others.

Eula certainly was not lazy. She was very active. She did all sorts of useful things. But she avoided the most critical actions—creating and marketing products.

Help clients identify possible obstacles to and resources for implementing plans. Years ago Kurt Lewin (1969) codified common sense by developing what he called "force-field analysis." In ordinary language, this is simply a review by the client of the major obstacles to and the major facilitating forces for implementing action plans. The slogan is "forewarned is forearmed."

Obstacles. The identification of possible obstacles or restraining forces to the implementation of a program helps forewarn clients.

Raul and Maria were a childless couple living in a large midwestern city. They had been married for about 5 years and had not been able to have children. They finally decided that they would like to adopt a child, so they consulted a counselor familiar with adoptions. The counselor helped them work out a plan of action that included helping them examine their motivation, reviewing their suitability to be adoptive parents, contacting an agency, and preparing themselves for an interview. After the plan of action had been worked out, Raul and Maria, with the help of the counselor, identified two possible obstacles or pitfalls: the negative feelings that often arise on the part of prospective parents when they are being scrutinized by an adoption agency and the feelings of helplessness and frustration caused by the length of time and uncertainty involved in the process.

The assumption here is that if clients are aware of some of the "wrinkles" that can accompany any given course of action, they will be less disoriented when they encounter them. Identifying possible obstacles is, at its best, a straightforward census of likely pitfalls rather than a self-defeating search for every possible thing that could go wrong.

Obstacles can come from within the clients themselves, from others, from the social settings of their lives, and from larger environmental forces. Once an obstacle is spotted, counselors can help clients identify ways of coping with it. Sometimes simply being aware of a pitfall is enough to help clients mobilize their resources to handle it. At other times a more explicit coping strategy is needed. For instance, the counselor arranged a couple of role-playing sessions with Raul and Maria in which she assumed the role of the examiner at the adoption agency and took a "hard line" in her questioning. These rehearsals helped them stay calm during the actual interviews. The counselor also helped them locate a mutual-help group of parents working their way through the adoption process. The members of the group shared their hopes and frustrations and provided support for one another. In short, Raul and Maria were trained to cope with the restraining forces they might encounter on the road toward their goal.

Facilitating forces. In a more positive vein, counselors can help their clients identify unused resources that facilitate action.

Nora found it extremely depressing to go to her weekly dialysis sessions. She knew that without them she would die, but she wondered whether it was worth living if she had to depend on a machine. The counselor helped her see that she was making life more difficult for herself by letting herself think such discouraging thoughts. He helped her learn how to think thoughts that would broaden her vision of the world instead of narrowing it down to herself, her discomfort, and the machine. Nora was a religious person and found in the Bible a rich source of positive thinking. She initiated a new routine: The day before she visited the clinic, she began to prepare herself psychologically by reading from the Bible. Then, as she traveled to the clinic and underwent treatment, she meditated slowly on what she had read.

In this case, the client substituted positive thinking, an underused resource, for poorme thinking. Brainstorming resources that can counter obstacles to action can be very helpful for some clients. Helping clients brainstorm facilitating forces raises the probability that they will act in their own interests. They can be simple things. George was avoiding an invasive diagnostic procedure. After a brainstorming session, he decided to get a friend to go with him. This meant two things. Once he asked for his friend's help, he "had to go through with it." Second, his friend's very presence distracted him from his fears.

Or consider Lucy. She had a history of letting her temper get the better of her. This was especially the case when she returned home after experiencing crises at work. Her mother-in-law and children became the target of her wrath. After a counseling session she took two photographs with her to work. One was a wedding-day picture that included her mother-in-law. The second was a recent picture of her three children. When she parked the car at work, she placed the pictures on the driver's seat. Then, when she got in the car in the evening, the first thing she saw was two photographs of her life at its best. This made her reflect on the drive home on how she wanted to enter the house.

Help clients find incentives and rewards for sustained action. Clients avoid engaging in action programs when the incentives and the rewards for not engaging in the program outweigh the incentives and the rewards for doing so.

Miguel, a police officer on trial for use of excessive force with a young offender, had a number of sessions with a counselor from an HMO that handled police health insurance. During the sessions, the counselor learned that although this was the first time Miguel had run afoul of the law, it was in no way the first expression of a brutal streak within him. He was a bully on the beat and a despot at home, and had run-ins with strangers when he visited bars with his friends. During the trial witnesses recalled instances of these behaviors.

Up to the time of his arrest, he had gotten away with all of this, even though his friends had often warned him to be more cautious. His badge had become a license to do whatever he wanted. His arrest and now the trial shocked him. Before, he had seen himself as invulnerable; now, he felt very vulnerable. The thought of being an ex-cop in prison understandably horrified him. He was found guilty, was suspended from the force for several months, and received probation on the condition that he continue to see the counselor.

Beginning with his arrest, Miguel had modified his aggressive behavior a great deal, even at home. Of course, fear of the consequences of his aggressive behavior was a strong incentive to change. The next time, the courts would show no sympathy. The counselor took a tough approach to this tough cop. He confronted Miguel for "remaining an adolescent" and for "hiding behind his badge." He called the power Miguel exercised over others "cheap power." He challenged the "decent person" to "come out from behind the screen." He told Miguel point-blank that the fear he was experiencing was probably not enough to keep him out of trouble in the future. After probation, the fear would fade and Miguel could easily fall back into his old ways. Even worse, fear was a "weak man's" crutch.

On a more positive note, the counselor saw in Miguel's expressions of vulnerability the possibility of a much more decent human being, one "hiding" under the tough exterior. The real incentives, he suggested, came from the "decent guy" buried inside. He had Miguel paint a picture of a "tough but decent" cop, family man, and friend. He had Miguel come up with "experiments in decency"—at home, on the beat, with his buddies—to get first-hand experience of the rewards associated with decency.

The counselor was not trying to change Miguel's personality. Indeed, he didn't believe in personality transformations. But he pushed him hard to find and bring to the surface a different, more constructive set of incentives to guide his dealings with people. The new incentives had to drive out the old.

The incentives and rewards that help a client get going on a program of constructive change in the first place may not be the ones that keep the client going.

Dwight, a man in his early 30s who was recovering from an accident at work that had left him partially paralyzed, had begun an arduous physical rehabilitation program with great commitment. Now, months later, he was ready to give up. The counselor asked him to visit the children's ward. Dwight was both shaken by the experience and amazed at the courage of many of the kids. He was especially struck by one teenager who was undergoing chemotherapy. "He seems so positive about everything," Dwight said. The counselor told him that the boy was tempted to give up, too. Dwight and the boy saw each other frequently. Dwight put up with the pain. The boy hung in there. Three months later, the boy died. Dwight's response, besides grief, was, "I can't give up now; that would really be letting him down."

Dwight's partnership with the teenager proved to be an excellent incentive. It helped him renew his resolve. Although the counselor joined with Dwight in celebrating his newfound commitment, he also worked with Dwight to find "backup" incentives for those times when current incentives seem to lose their power. One was the possibility of marrying and starting a family despite residual limitations due to the accident.

Constructive-change activities that are not rewarded tend over time to lose their vigor, decrease, and even disappear. This process is called extinction. It was happening with Luigi.

Luigi, a middle-aged man, had been in and out of mental hospitals a number of times. He discovered that one of the best ways of staying out was to use some of his excess energy helping others. He had not returned to the hospital once during the 3 years he worked at a soup kitchen. However, finding himself becoming more and more manic over the past 6 months and fearing that he would be rehospitalized, he sought the help of a counselor.

Luigi's discussions with the counselor led to some interesting findings. He discovered that, whereas in the beginning he had worked at the soup kitchen because he wanted to, he was now working there because he thought he should. He felt guilty about leaving and also thought that doing so would lead to a relapse. In sum, he had not lost his interest in helping others, but his current work was no longer interesting or challenging. As a result of his sessions with the counselor, Luigi began to work for a group that provided housing for the homeless and the elderly. He poured his energy into his new work and no longer felt manic.

The lesson here is that incentives cannot be put in place and then be taken for granted. They need tending.

Help clients develop action-focused self-contracts and agreements. Earlier we discussed self-contracts as a way of helping clients commit themselves to what they want—that is, their goals. Self-contracts are also useful in helping them both initiate and sustain problem-managing action and the work involved in developing opportunities. For instance, Feller (1984) developed a "job-search agreement" to help job seekers persist in their search. The following agreement—clients were to respond "true" to all the following statements and then act on these "truths"—requires clients to commit themselves not only to job-seeking behavior but also to sound psychological practices that promote the right mentality for such behavior.

I agree that no matter how many times I enter the job market, or the level of skills, experiences, or academic success I have, the following appear TRUE:

1. It takes only one YES to get a job; the number of no's does not affect my next interview.
2. The open market lists about 20% of the jobs presently open to me.
3. About 80% of the job openings are located by talking to people.
4. The more people who know my skills and know that I'm looking for a job, the more I increase the probability that they'll tell me about a job lead.

5. The more specifically I can tell people about the problems I can solve or out-comes I can attain, rather than describe the jobs I've had, the more jobs they may think I qualify for.

I agree that regardless of how much I need a job, the following appear TRUE:

6. If I cut expenses and do more things for myself, I reduce my money problems.
7. The more I remain positive, the more people will be interested in me and my job skills.
8. If I relax and exercise daily, my attitude and health will appear attractive to potential employers.
9. The more I do positive things and the more I talk with enthusiastic people, the more I will gain the attention of new contacts and potential employers.
10. Even if things don't go as I would like them to, I choose my own thoughts, feel-ings, and behaviors each day.

It is easy to see how similar "agreements" could act as drivers of action in many dif-ferent kinds of problem-managing and opportunity-developing situations. Self-contracts and agreements with others focus clients' energies. Here is an example. In this case, several parties had to commit themselves to the provisions of the contract.

> A boy in the seventh grade was causing a great deal of disturbance by his outbursts in class, which included verbal jousting with his friends and profanity. The apparent pur-pose of the disruptions was to position himself among his friends. He seemed to want to cultivate a reputation for being unafraid of the teachers and principal. The punish-ments handed down by them were dwarfed by what he saw as the admiration of his friends. After the teacher discussed the situation with the school counselor, the coun-selor called a meeting of all the stakeholders—the boy, his parents, the teacher, and the principal. The counselor offered a simple contract. When the boy disrupted the class, he would spend the next day working by himself under the direction of a teacher's aide. This would take him away from his friends. The day after, he would return to the classroom. There would be no further punishment. Concurrently, the counselor would work with him on what "leadership" behavior in the classroom might look like.
>
> The first month, the boy spent a fair number of days with the teacher's aide. The second month, however, he missed only 2 days, and the third month only 1. The truth is that he really wanted to be in school with his classmates. That's where the action was. And so he paid the price of self-control to get what he wanted. He also began to discover more constructive forms of leadership behavior. For instance, on occasion he challenged what the teacher was saying about a particular topic. Sometimes this led to a very lively classroom debate. Even the teacher liked this.

The counselor had suspected that the boy found socializing with his classmates rewarding. But now he had to pay for the privilege of socializing. Reasonable behav-ior in the classroom was not too high a price.

GETTING ALONG WITHOUT A HELPER: DEVELOPING SOCIAL NETWORKS FOR SUPPORT AND CHALLENGE

In most cases, helping is a relatively short-term process. But even in longer-term therapy, clients must eventually get on with life without their helpers. Ideally, the counseling process not only helps clients deal with specific problem situations and unused opportunities, but also, as outlined in Chapter 1, equips them with the working knowledge and skills needed to manage those situations more effectively on their own.

Because adherence to constructive-change programs is often difficult (Bosworth, Oddone, & Weinberger, 2005), social support and challenge in clients' everyday lives can help them move to action, persevere in action programs, and both consolidate and maintain gains. When it comes to social support and challenge, there are a number of possible scenarios at the implementation stage and beyond.

- Counselor helps clients with their plans for constructive change and then clients, using their own initiative and resources, take responsibility for the plans and pursue them on their own.
- Clients continue to see a helper regularly in the implementation phase.
- Clients see a helper occasionally, either "on demand" or in scheduled stop-and-check sessions.
- Clients join some kind of self-help group concurrently with one-to-one counseling sessions, which are eventually eliminated.
- Clients develop social relationships that provide both ongoing support and challenge for the changes they are making.

Support was discussed earlier, and, because the literature tends to focus on caring support, a few words about caring challenge in everyday life are in order.

Challenging relationships. As suggested earlier, support without challenge can be hollow and that challenge without support can be abrasive. Ideally, the people in the lives of clients provide a judicious mixture of support and challenge.

> Harry, a man in his early 50s, was suddenly stricken with a disease that called for immediate and drastic surgery. He came through the operation quite well, even getting out of the hospital in record time. For the first few weeks he seemed, within reason, to be his old self. However, he had problems with the drugs he had to take following the operation. He became quite sick and took on many of the mannerisms of a chronic invalid. Even after the right mix of drugs was found, he persisted in invalid-like behavior. Whereas right after the operation he had "walked tall," he now began to shuffle. He also talked constantly about his symptoms and generally used his "state" to excuse himself from normal activities.
>
> At first Harry's friends were in a quandary. They realized the seriousness of the operation and tried to put themselves in his place. They provided all sorts of support. But gradually they realized that he was adopting a style that would alienate others and keep him out of the mainstream of life. Support was essential, but it was not enough. They used a variety of ways to challenge his behavior, mocking his "invalid" movements, engaging in serious one-to-one talks, turning a deaf ear to his discussion of symptoms, and routinely including him in their plans.

Harry did not always react graciously to his friends' challenges, but in his better moments he admitted that he was fortunate to have such friends. Counselors can help clients find people willing to provide a judicious mixture of support and challenge.

Feedback from significant others. Gilbert (1978, p. 175), in his book on human competence, claimed that "improved information has more potential than anything else I can think of for creating more competence in the day-to-day management of performance." Feedback is certainly one way of providing both support and challenge. If clients are to be successful in implementing their action plans, they need adequate information about how well they are performing. Sometimes they know themselves; other times they need a more objective view. The purpose of feedback is not to pass judgment on clients' performance but rather to provide guidance, support, and challenge. There are two kinds of feedback.

1. **Confirmatory feedback.** Through confirmatory feedback, significant others—such as helpers, relatives, friends, and colleagues—let clients know that they are on course; that is, moving successfully through the steps of an action program toward a goal.

2. **Corrective feedback.** Through corrective feedback, significant others let clients know that they have wandered off course and specify what they need to do to get back on.

Corrective feedback, whether from helpers or people in the client's everyday life, should incorporate the following principles:

- Give feedback in the spirit of caring.
- Remember that mistakes are opportunities for growth.
- Use a mix of both confirmatory and corrective feedback.
- Be concrete, specific, brief, and to the point.
- Focus on the client's behaviors rather than on more elusive personality characteristics.
- Tie behavior to goals.
- Explore the impact and implications of the behavior.
- Avoid name-calling.
- Provide feedback in moderate doses. Overwhelming the client defeats the purpose of the entire exercise.
- Engage the client in dialogue. Invite the client not only to comment on the feedback but also to expand on it. Lectures don't usually help.
- Help the client discover alternative ways of doing things. If necessary, prime the pump.
- Explore the implications of changing over not changing.

The spirit of these "rules" should also govern confirmatory feedback. Very often people give very detailed corrective feedback and then just say "nice job" when a person does something well. All feedback provides an opportunity for learning. Consider the following statement from a father talking to his son, who stood up for the rights of a friend who was being bullied by some of his high school classmates:

"Jeb, I'm proud of you. You stood your ground even when they turned on you. They were mean. You weren't. You gave your opinion calmly, but forcefully. You didn't apologize for what you were saying. You were ready to take the consequences. It's easier now that a couple of them have apologized to you, but at the time you didn't know they would. You were honest to yourself. And now the best of them appreciate it. It made me think of myself. I'm not sure that I would have stood my ground the way you did. . . . But I'm more likely to do so now."

Although brief, this is much more powerful than "I'm proud of you, son." Being specific about behavior and pointing out the impact of the behavior turn positive feedback into a learning experience.

Of course, one of the main problems with feedback is finding people in the client's day-to-day life who see the client in action enough to make it meaningful, who care enough to give it, and who have the skills to provide it constructively.

An amazing case of getting along without a helper. As indicated earlier, many client problems are coped with and managed, not solved. Consider the following, very real case of a woman who certainly did not choose not to change. Quite the contrary. Her case is a good example of a no-formula approach to developing and implementing a program for constructive change.

> Vicky readily admits that she has never fully "conquered" her illness. Some 20 years ago, she was diagnosed with manic-depressive disorder (now called "bipolar disorder"). The picture looked something like this: She would spend about 6 weeks on a high; then the crash would come, and for about 6 weeks she'd be in the pits. After that she'd be normal for about 8 weeks. This cycle meant many trips to the hospital. Some 7 years into her illness, during a period in which she was in and out of the hospital, she made a decision. "I'm not going back into the hospital again. I will so manage my life that hospitalization will never be necessary." This nonnegotiable goal was her manifesto.
>
> Starting with this declaration of intent, Vicky moved on, in terms of Task 2 of Stage II, to spell out what she wanted: (1) She would channel the energy of her "highs"; (2) she would consistently manage or at least endure the depression and agony of her "lows"; (3) she would not disrupt the lives of others by her behavior; (4) she would not make important decisions when either high or low. Vicky, with some help from a rather nontraditional counselor, began to do things to turn those goals into reality. She used her broad goals to provide direction for everything she did.
>
> Vicky learned as much as she could about her illness, including cues about crisis times and how to deal with both highs and lows. To manage her highs, she learned to channel her excess energy into useful—or at least nondestructive—activity. Some of her strategies for controlling her highs centered on the telephone. She knew instinctively that controlling her illness meant not just managing problems but also developing opportunities. During her free time, she would spend long hours on the phone with a host of friends, being careful not to overburden any one person. Phone marathons became part of her lifestyle. She made the point that a big phone bill was infinitely better than a stay in the hospital. She called the telephone her "safety valve." She even set up her own phone-answering business and worked very hard to make it a success.
>
> She would also do whatever she had to do to tire herself out and get some sleep, for she had learned that sleep was essential if she was to stay out of the hospital. This included working longer shifts at the business. She developed a cadre of supportive people, including her husband. She took special care not to overburden him. She made occasional use of a drop-in crisis center but preferred avoiding any course of action that reminded her of the hospital.

It must be noted that the central driving force in this case was Vicky's decision to stay out of the hospital. Her determination drove everything else. This case also exemplifies the spirit of action that ideally characterizes the implementation stage of the helping process. Here is a woman who, with occasional help from a counselor, took charge of her life. She set some simple goals and devised a set of simple strategies for accomplishing them. She never looked back. And she was never hospitalized again.

Some will say that she was not "cured" by this process. But her goal was not to be cured but to lead as normal a life as possible in the real world. Some would say

Box 14-1   Questions on Implementing Plans

Here are the kinds of questions you can help clients ask themselves
in their search for ways of implementing their plans.

- Now that I have a plan, how do I move into action?
- What kind of self-starter am I? How can I improve?
- What obstacles lie in my way? Which are critical?
- How can I manage these obstacles?
- How do I keep my efforts from flagging?
- What do I do when I feel like giving up?
- What kind of support will help me to keep going?

that her approach lacked elegance. But it certainly did not lack results. Box 14-1
outlines the kinds of questions you can help clients ask themselves as they implement change programs.

THE SHADOW SIDE OF IMPLEMENTING CHANGE

There are many reasons why clients fail to act in their own behalf. Three are discussed here: helpers who do not have an action mentality, client inertia, and client
entropy. As you read about these common phenomena, recall what was said about
"implementation intentions" earlier. They can play an important role in managing
the shadow side obstacles outlined here.

Helpers as Agents
Driscoll (1984, pp. 91–97) discussed the temptation of helpers to respond to the passivity of their clients with a kind of passivity of their own, a "sorry, it's up to you"
stance. This, he claimed, is a mistake.

> A client who refuses to accept responsibility thereby invites the therapist to
> take over. In remaining passive, the therapist foils the invitation, thus forcing
> the client to take some initiative or to endure the silence. A passive stance is
> therefore a means to avoid accepting the wrong sorts of responsibility. It is
> generally ineffective, however, as a long-run approach. Passivity by a thera-
> pist leaves the client feeling unsupported and thus further impairs the already
> fragile therapeutic alliance. Troubled clients, furthermore, are not merely
> unwilling but generally and in important ways unable to take appropriate
> responsibility. A passive countermove is therefore counterproductive, for nei-
> ther therapist nor client generates solutions, and both are stranded together
> in a muddle of entangling inactivity. (p. 91)

To help others act, helpers must be agents and doers in the helping process, not mere
listeners and responders. The best helpers are active in the helping sessions. They

keep looking for ways to enter the worlds of their clients, to get them to become more active in the sessions, to get them to own more of the helping process, to help them see the need for action—action in their heads and action outside their heads in their everyday lives. And they do all this while espousing the client-centered values outlined in Chapter 3. Although they don't push reluctant clients too hard, thus turning reluctance into resistance, neither do they sit around waiting for reluctant clients to act.

Client Inertia: Reluctance to Get Started

Inertia is the human tendency to put off problem-managing action. I sometimes say to clients who I suspect are prone to inertia something like this, "The action program you've come up with seems to be a sound one. The main reason that sound action programs don't work, however, is that they are never tried. Don't be surprised if you feel reluctant to act or are tempted to put off the first steps. This is quite natural. Ask yourself what you can do to get by that initial barrier." The sources of inertia are many, ranging from pure sloth to paralyzing fear. Understanding what inertia is like is easy. We need only look at our own behavior. The list of ways in which we avoid taking responsibility is endless. We'll examine several of them here: passivity, learned helplessness, disabling self-talk, and getting trapped in vicious circles.

Passivity. One of the most important ingredients in the generation and perpetuation of the "psychopathology of the average" is passivity, the failure of people to take responsibility for themselves in one or more developmental areas of life or in various life situations that call for action. Passivity takes many forms: doing nothing—that is, not responding to problems and options; uncritically accepting the goals and solutions suggested by others; acting aimlessly; and becoming paralyzed—that is, shutting down or becoming violent, blowing up (see Schiff, 1975).

> When Zelda and Jerzy first noticed small signs that things were not going right in their relationship, they did nothing. They noticed certain incidents, mused on them for a while, and then forgot about them. They lacked the communication skills to engage each other immediately and to explore what was happening. Zelda and Jerzy had both learned to remain passive before the little crises of life, not realizing how much their passivity would ultimately contribute to their downfall. Endless unmanaged problems led to major blowups until they decided to end their marriage.

Passivity in dealing with little things can prove very costly. Little things have a way of turning into big things.

Learned helplessness. Seligman's (1975, 1991) concept of "learned helplessness" and its relationship to depression is an important one (Garber & Seligman, 1980; Peterson, Maier, & Seligman, 1995). Some clients learn to believe from an early age that there is nothing they can do about certain life situations. There are degrees in feelings of helplessness—from mild forms of "I'm not up to this" to feelings of total helplessness coupled with deep depression. Learned helplessness, then, is a step beyond mere passivity.

Bennett and Bennett (1984) saw the positive side of helplessness. If clients' problems are indeed out of their control, then it is not helpful for them to have an illusory sense of control, unjustly assign themselves responsibility, and indulge in excessive expectations. Somewhat paradoxically, they found that challenging clients' tendency to blame themselves for everything actually fostered realistic hope and change.

The trick is helping clients learn what is and what is not in their control. A man with a physical disability may not be able to do anything about the disability itself, but he does have some control over how he views his disability and his power to pursue certain life goals despite it. The opposite of helplessness is "learned optimism" (Seligman, 1998) and resourcefulness. If helplessness can be learned, so can resourcefulness. Indeed, increased resourcefulness is one of the principal goals of successful helping. Optimism, however, is not an unmixed blessing; nor is pessimism always a disaster (Chang, 2001). Although optimists do live longer and enjoy greater success than pessimists, pessimists are better predictors of what is likely to happen. The price of optimism is being wrong a lot of the time. Perhaps we should help our clients be hopeful realists rather than optimists or pessimists.

Disabling self-talk. Challenging clients' dysfunctional self-talk was discussed earlier. Clients often talk themselves out of things, thus talking themselves into passivity. They say to themselves such things as "I can't do it," "I can't cope," "I don't have what it takes to engage in that program; it's too hard," and "It won't work." Such self-defeating conversations with themselves get people into trouble in the first place and then prevent them from getting out. Helpers can add great value by helping clients challenge the kind of self-talk that interferes with action.

Vicious circles. Pyszczynski and Greenberg (1987) developed a theory about self-defeating behavior and depression. They said that people whose actions fail to get them what they want can easily lose a sense of self-worth and become mired in a vicious circle of guilt and depression.

> Consequently, the individual falls into a pattern of virtually constant self-focus, resulting in intensified negative affect, self-derogation, further negative outcomes, and a depressive self-focusing style. Eventually, these factors lead to a negative self-image, which may take on value by providing an explanation for the individual's plight and by helping the individual avoid further disappointments. The depressive self-focusing style then maintains and exacerbates the depressive disorder. (p. 122)

It does sound depressing. One client, Amanda, fits this theory perfectly. She had aspirations of moving up the career ladder where she worked. She was very enthusiastic and dedicated, but she was unaware of the "gentleman's club" politics of the company in which she worked and didn't know how to "work the system." She kept doing the things that she thought should get her ahead. They didn't. Finally, she got down on herself, began making mistakes in the things that she usually did well, and made things worse by constantly talking about how she "was stuck," thus alienating her friends. By the time she saw a counselor, she felt defeated and depressed. She was about to give up. The counselor focused on the entire "circle"—low self-esteem producing passivity producing even lower self-esteem—and not just the self-esteem part. Instead of just trying to help her change her inner world of disabling self-talk, he also helped her intervene in her life to become a better problem solver. Small successes in problem solving led to the start of a "benign" circle—success producing greater self-esteem leading to greater efforts to succeed.

Disorganization. Tico lived out of his car. No one knew exactly where he spent the night. The car was chaos, and so was his life. He was always going to get his career, family relations, and love life in order, but he never did. Living in disorganization was his way of putting off life decisions. Ferguson (1987, p. 46) painted a picture that may well remind us of ourselves, at least at times.

> When we saddle ourselves with innumerable little hassles and problems, they distract us from considering the possibility that we may have chosen the wrong job, the wrong profession, or the wrong mate. If we are drowning in unfinished housework, it becomes much easier to ignore the fact that we have become estranged from family life. Putting off an important project—painting a picture, writing a book, drawing up a business plan—is a way of protecting ourselves from the possibility that the result may not be quite as successful as we had hoped. Setting up our lives to insure a significant level of disorganization allows us to continue to think of ourselves as inadequate or partially-adequate people who don't have to take on the real challenges of adult behavior.

Many things can be behind this unwillingness to get our lives in order, like defending ourselves against a fear of succeeding.

Driscoll (1984, pp. 112–117) has provided us with a great deal of insight into this problem. He described inertia as a form of control. He says that if we tell some clients to jump into the driver's seat, they will compliantly do so—at least until the journey gets too rough. The most effective strategy, he claimed, is to show clients that they have been in the driver's seat all along: "Our task as therapists is not to talk our clients into taking control of their lives, but to confirm the fact that they already are and always will be." That is, inertia, in the form of staying disorganized, is itself a form of control. The client is actually successful, sometimes against great odds, at remaining disorganized and thus preserving inertia. Once clients recognize their power, then we can help them redirect it.

Entropy: The Tendency of Things to Fall Apart

Entropy is the tendency to give up action that has been initiated. Kirschenbaum (1987), in a review of the research literature, uses the term "self-regulatory failure." Programs for constructive change, even those that start strong, often dwindle and disappear. All of us have experienced problems trying to implement programs. We make plans, and they seem realistic to us. We start the steps of a program with a good deal of enthusiasm. However, we soon run into tedium, obstacles, and complications. What seemed so easy in the planning stage now seems quite difficult. We become discouraged, flounder, recover, flounder again, and finally give up, rationalizing to ourselves that we did not want to accomplish those goals anyway.

Under the rubric of "false hopes of self-change," Polivy and Herman (2002) suggest that this scenario occurs all too frequently. Perhaps it is even the norm in self-change programs such as dieting, which they use as their point of reference. At the center of the false-hope syndrome, they say, are the clients' unrealistic expectations. They refer to things like New Year's Eve resolutions. Most of us can immediately think of many of our own resolutions that fell by the wayside. Fletcher (2003), Lowe (2003), and Snyder and Rand (2003) all quite vigorously challenge Polivy and Herman's findings and even the concept of "false hopes." They say that the authors paint an overly pessimistic picture of self-change programs, especially dieting. That makes a priori sense if we consider outcome research in helping. When it comes to counseling, if we start with the premise that helping does help (see Chapter 1), then the kind of pessimism that Polivy and Herman suggest must be wrong. Right? Well, let's take a look.

Substantive change is hard work. Helpers see some counseling through to a successful conclusion. But many clients start with helpers only to end with the task of implementing self-change programs. The helping literature, for obvious reasons, tends to focus on successes, not failures. It is probably safe to say that there is some "true hope" and some "false hope" out there, but how much of each we don't know. Even if the work of Polivy and Herman is as flawed as its critics say—and by the way, how could something so flawed end up in the *American Psychologist?*—there is something about it that rings true.

The track record of *discretionary* change—change that is not forced in one way or another—on the part of both individuals and institutions is poor. This is my read of individuals (including myself), companies, institutions, and countries. If they don't have to change, then often they do not, even though they say they want to. In my work with institutions, I talk about the Okavango-Kalahari phenomenon. When the waters from the highlands in the north flood into the Okavango Delta in Botswana, it becomes an ecological wonderland. But somehow those waters disappear into the Kalahari Desert, though hydrologists don't know exactly how. I ask the managers, "Does this sound like any of your change programs?" They laugh. "Where's that management development program you started so vigorously 2 years ago?" "In the Kalahari!" shouts one. I'm not sure that I have the Okavango-Kalahari hydrology right, but something tells me that the "false hopes" debate is not over. To say that entropy is here to stay is not a defeatist statement. If in helping there are false hopes based on unrealistic expectations, then we helpers need to face up to it. "Forewarned is forearmed" is realism, not pessimism.

Phillips (1987, p. 650) identified what he called the "ubiquitous decay curve" in both helping and in medical-delivery situations. Attrition, noncompliance, and relapse are the name of the game. A married couple trying to reinvent their marriage might eventually say to themselves, "We had no idea that it would be so hard to change ingrained ways of interacting with each other. Is it worth the effort?" Their motivation is on the wane. Wise helpers know that the decay curve is part of life and help clients deal with it. With respect to entropy, a helper might say, "Even sound action programs begun with the best of intentions tend to fall apart over time, so don't be surprised when your initial enthusiasm seems to wane a bit. That's only natural. Rather, ask yourself what you need to do to keep yourself at the task."

Brownell and her associates (1986) provided a useful caution. They drew a fine line between preparing clients for mistakes and giving them "permission" to make mistakes by implying that mistakes are inevitable. They also made a distinction between "lapse" and "relapse." A slip or a mistake in an action program (a lapse) need not lead to a relapse—that is, giving up the program entirely. Consider Graham, a man who has been trying to change what others see as his "angry interpersonal style." Using a variety of self-monitoring and self-control techniques, he has made great progress in changing his style. On occasion, he loses his temper, but never in any extreme way. He makes mistakes, but he does not let an occasional lapse end up in relapse.

RESILIENCE REVISITED: PEOPLE'S ABILITY TO HOLD THEMSELVES TOGETHER, BOUNCE BACK, AND GROW

There are a number of different definitions of resilience and a fair amount of debate about just what it entails and where it comes from (Bonanno, 2004, 2005; T. Kelley, 2005; Linley & Joseph, 2005; Litz, 2005; Maddi, 2005; Roisman, 2005). Unlike some debates, this one sheds a good deal of light on the concept. The term *resilience* has a literary flair to it. Turning it into a concept that is amenable to research is not easy. As to the reality of resilience, let me offer a case.

> Juliana was a headstrong girl right from the start. High school was a bit tumultuous, but the seeds of really significant problems were sown during her college years. There she learned both the pleasures and the misery of drugs. Because her friends were "druggies," she choose boyfriends who were also playing with drugs and who could accommodate her headstrong nature. Predictably, boyfriends came and went. The drugs were a constant. After college she stuck to this lifestyle. Now boyfriends and jobs came and went.
>
> Her parents loved her dearly and did everything they could to help her. She played them like a harp. At times she would give hints that she was tired of her lifestyle, and their hopes were raised. Then she plunged back into her old ways with a vengeance. Her parents were always there for her. In a sense, this was part of the problem situation. Their financial and social support allowed her to continue in her unhealthy ways. After some sessions with a counselor, her parents decided to cut her loose. She had to leave the house and make her own way financially. This ripped them apart but they did not know what else to do.
>
> Off Juliana went. The ensuing picture was not pretty at all. It almost seemed that she was going to get even worse as a way of getting back at her parents. To their horror, she ended up in jail. But jail seemed to be the turning point. After getting out, she all but quit taking drugs. She got a number of jobs, but her headstrong nature still asserted itself. She moved from job to job either because she could not stand the people at work or they could no longer put up with her. She was really shaken up when her father lost his job. By then she had reestablished some kind of minimal relationship with her family. She began to see a counselor again. He soon recognized a number of Juliana's inner resources. Headstrong, yes, but a woman with guts, too. A woman who could not seem to establish reasonable relationships with others—especially men— but also a woman with a keen understanding of human nature. In spite of all the messy years she had not lost her spark, and her rather substantial inner resources were hidden but intact.
>
> Juliana finally nailed down a job she liked, was promoted, was headhunted by another company, but stayed with the original one because she got a better position with better pay. She established a home for her two children who had been farmed out to relatives. She subscribed to and promoted the values, such as self-discipline and decency in interpersonal relationships, that her children had learned in their foster homes. She reestablished full contact with her family. She became a caring companion to the man with whom she lived. In short, she successfully rejoined the community she had abandoned.

The counselor played a catalytic role in this process of renewal, but Juliana did the work. He encouraged her to tap into the wellspring of resilience or hardiness (Khoshaba & Maddi, 2004; Maddi, 2002) or growth-through-adversity (Jeseph & Linley, 2005) inside her, and she did. She moved from self-loathing to a nonindulgent form of self-esteem. Success in one area of life (work) spread to others (interpersonal relationships). Like many converts, she even became a crusader with her children, relatives, and friends for hard work, self-discipline, and the total avoidance

Box 14-2   Guidelines for the Action Arrow

Here are some guidelines for helping clients develop a bias for
problem-managing and opportunity-developing action as they strive
to implement their plans.

- Understand how widespread both inertia and entropy are and how they are
  affecting this client.
- Help clients become effective tacticians.
- Help clients form "implementation intentions" especially when obstacles to
  goal attainment are foreseen.
- Help clients avoid both procrastination and imprudent action.
- Help clients develop contingency plans.
- Help clients discover and manage obstacles to action.
- Help clients discover resources that will enable them to begin acting, to
  persist, and to accomplish their goals.
- Help clients find the incentives and the rewards they need to persevere in
  action.
- Help clients acquire the skills they need to act and to sustain goal-accom-
  plishing action.
- Help clients develop a social support and challenge system in their day-to-day
  lives.
- Prepare clients to get along without a helper.
- Come to grips with the fact that helpers need to become agents of change
  in their own lives.
- Face up to the fact that not every client wants to change.

of self-pity. The point is that even in the most difficult cases, there is probably some
residue of resilience. Tapping into it is the challenge.

Will Juliana stay the course? No one can say for sure, but her actions and
accomplishments so far have a sense of permanency to them, which is not always
the case. So there is, as you might expect, a literature on helping clients stay the
course (Bosworth & Oddone, 2005; Christensen, 2004; Cockell, Zaitsoff, &
Geller, 2004; Colan, 2003; Meichenbaum & Turk, 1987; Witkiewitz & Marlatt,
2005). The literature on terminating psychotherapy also has some helpful guidelines. Juliana made do with her own revived inner strengths and with the social
support she received, mainly from family members. She was lucky because she
came from a family of good communicators. Too many people who would like to
provide support are ineffective interpersonal communicators. They either do it
poorly or their support remains buried inside (Cohen, Underwood, & Gotlieb,
2000; Goldsmith, 2004).

Choosing Not to Change

Some clients who seem to do well in analyzing problems, developing goals, and even identifying reasonable strategies and plans end up by saying—in effect, if not directly—something like this: "Even though I've explored my problems and understand why things are going wrong—that is, I understand myself and my behavior better, and I realize what I need to do to change—right now I don't want to pay the price called for by action. The price of more effective living is too high."

The question of human motivation seems almost as enigmatic now as it must have been at the dawning of the history of the human race. So often we seem to choose our own misery. Worse, we choose to stew in it rather than endure the relatively short-lived pain of behavioral change. Helpers can and should challenge clients to search for incentives and rewards for managing their lives more effectively. They should also help clients understand the consequences of not changing. But in the end, it is the client's choice.

The shadow side of change stands in stark contrast to the case of Vicky. Savvy helpers are not magicians, but they do understand the shadow side of change, learn to see signs of it in each individual case, and, in keeping with the values outlined in Chapter 3, do whatever they can to challenge clients to deal with the shadow side of themselves and the world around them.

ABRAMSON, P. R., Cloud, M. Y., Keese, N., & Keese, R. (1994, Spring). How much is too much? Dependency in a psychotherapeutic relationship. *American Journal of Psychotherapy, 48,* 294–301.

ACKERMAN, S. J., & Hilsenroth, M. J. (2003). A review of therapist characteristics and techniques positively impacting the therapeutic alliance. *Clinical Psychology Review, 23,* 1–33.

ACKOFF, R. (1974). *Redesigning the future.* New York: Wiley.

ADLER, R. B., Proctor, R. F., & Towne, N. (2005). *Looking out, looking in* (11th ed.). Belmont, CA: Wadsworth.

ALBEE, G. W., & K. D. Ryan-Finn. (1993, November/December). An overview of primary prevention. *Journal of Counseling and Development,* 115–123.

ALBRECHT, F., & Wallace, M. (1998). Detecting chronic fatigue syndrome: The role of counselors. *Journal of Counseling and Development, 76,* 183–188.

ALDWIN, C., & Gilmer, D. F. (2004). *Health, illness, and optimal aging.* Thousand Oaks, CA: Sage.

ALFORD, B. A., & Beck, A. T. (1997). Therapeutic interpersonal support in cognitive therapy. *Journal of Psychotherapy Integration, 7,* 105–117.

ALI, B. S., Rahbar, M. H., Naeem, S., & Gul, A. (2003). The effectiveness of counseling on anxiety and depression by minimally trained counselors: A randomized controlled trial. *American Journal of Psychotherapy, 57,* 324–336.

ALI, S. R., Liu, W. M., & Humedian, M. (2004). Islam 101: Understanding the religion and therapy implications. *Professional Psychology: Research and Practice, 35,* 635–642.

ALLE-CORLISS, L. A., & Alle-Corliss, R. M. (2006). *Human service agencies: An orientation to fieldwork* (2nd ed.). Belmont, CA: Wadsworth/ Thomson Learning.

ALLOY, L. B., & Riskind, J. H. (2005). *Cognitive vulnerability to emotional disorders.* Mahwah, NJ: Lawrence Erlbaum.

ALTMAIER, E. M., Russell, D. W., Kao, C. F., Lehmann, T. R., & Weinstein, J. N. (1993). Role of self-efficacy in rehabilitation outcome among chronic low back pain patients. *Journal of Counseling Psychology, 40,* 335–339.

ALVORD, M. K., & Grados, J. J. (2005). Enhancing resilience in children: A proactive approach. *Professional Psychology: Research and Practice, 36,* 238–245.

AMERICAN MEDICAL ASSOCIATION. (1992). Evidence-based medicine working group: A new approach to teaching the practice of medicine. *Journal of the American Medical Association, 268,* 2420–2425.

AMRHEIN, P. C. (2004). How does motivational interviewing work? What client talk reveals. *Journal of Cognitive Psychotherapy, 18,* 323–336.

ANDERSEN, P. A. (1999). *Nonverbal communication: Forms and functions.* Mountain View, CA: Mayfield.

ANDERSON, H., & Goolishian, H. (1992). The client is the expert: A not-knowing approach to therapy. In S. McNamee & K. J. Gergen (Eds.), *Therapy as social construction* (pp. 25–39). London: Sage.

ANDREWS, G., Creamer, M., Crino, R., Hunt, C., Lampe, L., & Page, A. (2002). *The treatment of anxiety disorders: Clinician guides and patient manuals* (2nd ed.) New York: Cambridge University Press.

ANGUS, L. E., & McLeod, J. (Eds.). (2004). *The handbook of narrative and psychotherapy: Practice, theory and research.* Thousand Oaks, CA: Sage.

ANKUTA, G. Y., & Abeles, N. (1993, February). Client satisfaction, clinical significance, and meaningful change in psychotherapy. *Professional Psychology: Research and Practice, 24,* 70–74.

ARGYRIS, C. (1999). *On organizational Learning* (2nd ed.). Cambridge, MA: Blackwell.

ARKIN, R. M., & Hermann, A. D. (2000, July). Constructing desirable identities—Self-presentation in psychotherapy and daily life. *Psychological Bulletin, 126,* 501–504

ARKOWITZ, H. (1997). The varieties of support in psychotherapy. *Journal of Psychotherapy Integration, 7,* 151–159.

ARNETT, J. J. (May, 2000). Emerging adulthood: A theory of development from the late teens through the twenties. *American Psychologist, 55,* 469–480.

ARNKOFF, D. B. (1995). Two examples of strains in the therapeutic alliance in an integrative cognitive therapy. *Psychotherapy in Practice, 1,* 33–46.

ASAY, T. P., & Lambert, M. J. (1999). The empirical case for the common factors in therapy: Quantitative findings. In M. A. Hubble, B. L. Duncan, & S. D. Miller (Eds.), *The heart and soul of change: What works in therapy* (pp. 33–56). Washington, DC: American Psychological Association.

ASPINWALL, L. G., & Staudinger, U. M. (Eds.). (2003). *A psychology of human strengths: Fundamental questions and future directions for a positive psychology.* Washington, DC: American Psychological Association.

ATKINS, D. C., & Christensen, A. (2001). Is professional training worth the bother? A review of the impact of psychotherapy training on client outcome. *Australian Psychologist, 36,* 122–131.

ATKINSON, D. R., Worthington, R. L., Dana, D. M., & Good, G. E. (1991). Etiology beliefs, preferences for counseling orientations, and counseling effectiveness. *Journal of Counseling Psychology, 38,* 258–264.

AUSTAD, S. A. (1996). *Is long term psychotherapy unethical?* San Francisco: Jossey-Bass.

AXELSON, J. A. (1999). *Counseling and development in a multicultural society* (3rd ed.). Pacific Grove, CA: Brooks/Cole/Wadsworth.

AZAR, B. (1995). Breaking through barriers to creativity. *Monitor, 26*(8), pp. 1, 20.

AZAR, B. (2000, July/August). Psychology's largest prize goes to four extraordinary scientists. *Monitor on Psychology,* 38–40.

BAILEY, K. G., Wood, H. E., & Nava, G. R. (1992). What do clients want? Role of psychological kinship in professional helping. *Journal of Psychotherapy Integration, 2,* 125–147.

BAKER, E. K. (2003). *Caring for ourselves: A therapist's guide to personal and professional well-being.* Washington, DC: American Psychological Association.

BALTES, P. B., & Staudinger, U. M. (2000, January). Wisdom: A metaheuristic (pragmatic) to orchestrate mind and virtue toward excellence. *American Psychologist, 55,* 122–135.

BANDURA, A. (1986). *Social foundations of thought and action: A social cognitive theory.* Englewood Cliffs, NJ: Prentice Hall.

BANDURA, A. (1989). Human agency in social cognitive theory. *American Psychologist, 44,* 1175–1184.

BANDURA, A. (1990). Foreword to E. A. Locke & G. P. Latham (1990), *A theory of goal setting and task performance.* Englewood Cliffs, NJ: Prentice Hall.

BANDURA, A. (1991). Human agency: The rhetoric and the reality. *American Psychologist, 46,* 157–161.

BANDURA, A. (Ed.). (1995). *Self-efficacy in changing societies.* New York: Cambridge University Press.

BANDURA, A. (1997). *Self-efficacy: The exercise of control.* New York: Freeman.

BANDURA, A. (2001). Social cognitive theory: An agentic perspective. *Annual Review of Psychology, 52,* 1–26.

BANGERT, A.W., & Baumberger, J. P. (2005). Research and statistical techniques used in the *Journal of Counseling and Development* 1990–2001. *Journal of Counseling and Development, 83,* 480–487.

BARKER, C., & Pistrang, N. (2002). Psychotherapy and social support: Integrating research on psychological helping. *Clinical Psychology Review, 22,* 361–379.

BARLOW, D. H. (2004). *Anxiety and its disorders: The nature and treatment of anxiety and panic* (2nd ed.). New York: Guilford Press.

BARLOW, D. H. (2004). Psychological treatments. *American Psychologist, 59,* 869–878.

BARLOW, D. H., & Durand, V. M. (2005). *Abnormal psychology: An integrative approach.* Belmont, CA: Wadsworth/Thomson Learning.

BARON, J. (2001). *Thinking and deciding* (3rd ed.). New York: Cambridge University Press.

BAR-ON, R., & Parker, J. D. A. (Eds.). (2000). *Handbook of emotional intelligence: Theory, development, assessment, and application at home, school, and in the workplace.* San Francisco: Jossey-Bass.

BAR-ON, R. (1997). *The emotional intelligence inventory (EQ-i): Technical manual.* Toronto: Multi-health Systems.

BARRETT-Lennard, G. T. (1981). The empathy cycle: Refinement of a nuclear concept. *Journal of Counseling Psychology, 28*, 91–100.

BASIC BEHAVIORAL SCIENCE TASK FORCE OF THE NATIONAL ADVISORY MENTAL HEALTH COUNCIL. (1996). Basic behavioral science research for mental health: Family processes and social networks. *American Psychologist, 51*, 622–630.

BAUMEISTER, R. F., Dale, K., & Sommer, K. L. (1998). Freudian defense mechanisms and empirical findings in modern social psychology: Reaction formation, projection, displacement, undoing, isolation, sublimation, and denial. *Journal of Personality, 66*, 1081–1117.

BECK, A. T. (1979). *Cognitive therapy and the emotional disorders.* New York: Penguin Books.

BECK, A. T., Freeman, A., & Davis, D. D. (2003). *Cognitive therapy of personality disorders* (2nd ed.). New York: Guilford Press.

BECKER, K. W., & Carson, D. K. (2004). When lightning strikes: Reexamining creativity in psychotherapy. *Journal of Counseling and Development, 82*, 111–115.

BEDELL, J. R., & Lennox, S. S. (1997). *Handbook for communication and problem-solving skills training: A cognitive-behavioral approach.* New York: Wiley.

BEDI, R. P. (2006). Concept mapping the client's perspective on counseling alliance formation. *Journal of Counseling Psychology, 53*, 26–35.

BENBENISHTY, R., & Schul, Y. (1987). Client-therapist congruence of expectations over the course of therapy. *British Journal of Clinical Psychology, 26*, 17–24.

BENNETT, M. I., & Bennett, M. B. (1984). The uses of hopelessness. *American Journal of Psychiatry, 141*, 559–562.

BERENSON, B. G., & Mitchell, K. M. (1974). *Confrontation: For better or worse.* Amherst, MA: Human Resource Development Press.

BERG, I. K. (1994). *Family based services: A solution-focused approach.* New York: Norton.

BERGIN, A. E. (1991). Values and religious issues in psychotherapy and mental health. *American Psychologist, 46*, 394–403.

BERGIN, A. E., & Garfield, S. L. (1994). *Handbook of psychotherapy and behavior change* (4th ed.). New York: Wiley.

BERNE, E. (1964). *Games people play.* New York: Grove Press.

BERNHEIM, K. F. (1989). Psychologists and the families of the severely mentally ill: The role of family consultation. *American Psychologist, 44*, 561–564.

BERNSTEIN, R. (1994). *Dictatorship of virtue: Multiculturalism and the battle for America's future.* New York: Knopf.

BERSOFF, D. N. (1995). *Ethical conflicts in psychology.* Hyattsville, MD: American Psychological Association.

BERTOLINO, B., & O'Hanlon, B. (2002). *Collaborative, competency-based counseling and therapy.* Boston: Allyn & Bacon.

BEUTLER, L. E. (2000). David and Goliath: When psychotherapy research meets health care delivery systems. *American Psychologist, 55*, 997–1007.

BEUTLER, L. E., & Bergan, J. (1991). Values change in counseling and psychotherapy: A search for scientific credibility. *Journal of Counseling Psychology, 38*, 16–24.

BEUTLER, L. E., Machado, P. P. P., & Neufeldt, S. A. (1994). Therapist Variables. In A. E. Bergin & S. L. Garfield (Eds.), *Handbook of psychotherapy and behavior change* (4th ed.; pp. 229–269). New York: Wiley.

BILBREY, J., & Bilbrey, P. (1995). Judging, trusting, and utilizing outcome data: A survey of behavioral healthcare payors. *Behavioral Healthcare Tomorrow, 4*, 62–65.

BINDER, J. L., & Strupp, H. H. (1997). "Negative process": A recurrently discovered and underestimated facet of therapeutic process and outcome in the individual psychotherapy of adults. *Clinical Psychology: Science and Practice, 4*, 121–139.

BISCHOFF, M. M., & Tracey, T. J. G. (1995). Client resistance as predicted by therapist behavior: A study of sequential dependence. *Journal of Counseling Psychology, 42*, 487–495.

BLACKMAN, J. S. (2004). *101 defenses: How the mind shields itself.* New York: Brunner-Routledge.

BLAIR, G. R. (2000a). *Goal-setting 101.* Tampa, FL: GoalsGuy Learning Systems.

BLAIR, G. R. (2000b). *The ten commandments of goal setting.* Tampa, FL: GoalsGuy Learning Systems.

BLAMPIED, N. M. (2000). Single-case research designs: A neglected alternative. *American Psychologist, 55*, 960.

BLOOM, B. L. (1997). *Planned short-termed psychotherapy: A clinical handbook* (2nd ed.). Boston: Allyn & Bacon.

BOGG, T., & Roberts, B. W. (2004). Conscientiousness and health-related behaviors: A meta-analysis of the leading behavioral contributors to morality. *Psychological Bulletin, 130*, 887–919.

BOHART, A. C., & Greenberg, L. S. (Eds.). (April, 1997). *Empathy reconsidered: New directions in psychotherapy.* Washington, DC: American Psychology Association.

BOISVERT, C. M., & Faust, D. (2003). Leading researchers' consensus on psychotherapy research findings: Implications for the teaching and conduct of psychotherapy. *Professional Psychology: Research and Practice, 34,* 508–513.

BONANNO, G. A. (2004). Loss, trauma, and human resilience. *American Psychologist, 59,* 20–28.

BONANNO, G. A. (2005). Clarifying and extending the construct of adult resilience. *American Psychologist, 60,* 265–267.

BORDIN, E. S. (1979). The generalizability of the psychoanalytic concept of the working alliance. *Psychotherapy: Theory, Research and Practice, 16,* 252–260.

BORNSTEIN, R. F., & Bowen, R. F. (1995). Dependency is psychotherapy: Toward an integrated treatment approach. *Psychotherapy, 32,* 520–534.

BOSWORTH, H. B., & Oddone, E. Z. (2005). *Patient treatment adherence: Concepts, interventions, and measurement.* Mahwah, NJ: Erlbaum.

BOSWORTH, H. B., Oddone, E. Z., & Weinberger, M. (Eds.). (2005). *Patient treatment adherence: Concepts, interventions, and measurement.* Mahwah, NJ: Erlbaum.

BRAMMER, L. (1973). *The helping relationship: Process and skills.* Englewood Cliffs, NJ: Prentice Hall.

BRECKLER, S. J. (2004). Legitimate psychological science. *Monitor on Psychology,* 22.

BRECKLER, S. J., & Olson, J. (2006). *Social psychology alive.* Belmont, CA: Wadsworth.

BROCK, T. C., Green, M. C., & Reich, D. A. (1998, January). New evidence of flaws in the *Consumers Report* study of psychotherapy. *American Psychologist, 53,* 62–63.

BRODER, M. S. (2000). Making optimal use of homework to enhance your therapeutic effectiveness. *Journal of Rational-Emotive & Cognitive Behavior Therapy, 18,* 3–18.

BRODSKY, S. L. (2005). Psychotherapy with reluctant and involuntary clients. In G. P. Koocher, J. C. Norcross, & S. S. Hill, III (Eds.), *Psychologists' Desk Reference* (2nd ed.; pp. 257–262). New York: Oxford University Press.

BRONFENBRENNER, U. (1977). Toward an experimental ecology of human development. *American Psychologist,* 513–531.

BROWN, R. T., Freeman, W. S., Brown, R. A., Belar, C., Hersch, L., Hornyak, L. M., Rickel, A., Rozensky, R., Sheridan, E., & Reed, G. (2002). The role of psychology in health care delivery. *Professional Psychology: Research and Practice, 33,* 536–545.

BROWNELL, K. D., Marlatt, G. A., Lichtenstein, E., & Wilson, G. T. (1986). Understanding and preventing relapse. *American Psychologist, 41,* 765–782.

BURKARD, A. W., Knox, S., Groen, M., Perez, M., & Hess, S. A. (2006). European American therapist self-disclosure in cross-cultural counseling. *Journal of Counseling Psychology, 50,* 324–332.

BURKE, B. L., Dunn, C. W., Atkins, D. C., & Phelps, J. S. (2004). The emerging evidence base for motivational interviewing: A meta-analytic and qualitative inquiry. *Journal of Cognitive Psychotherapy, 18,* 309–322.

BURKS, R. J., & Keeley, S. M. (1989). Exercise and diet therapy: Psychotherapists' beliefs and practices. *Professional Psychology: Research & Practice, 20,* 62–64.

BUSHE, G. (1995). Advances in appreciative inquiry as an organizational development intervention. *Organizational Development Journal, 1,* 14–22.

CACIOPPO, J. T., Berntson, G. G., & Semin, G. R. (2005). Scientific symbiosis: The mutual benefit of iteratively adopting the perspective of realism and instrumentalism. *American Psychologist, 60,* 347–348.

CACIOPPO, J. T., Semin, G. R., & Berntson, G. G. (2004). Realism, instrumentalism, and scientific symbiosis: Psychological theory as a search for truth and the discovery of solutions. *American Psychologist, 59,* 214–223.

CADE, B., & O'Hanlon, W. H. (1993). *A Brief Guide to Brief Therapy.* New York: W. W. Norton.

CALHOUN, L. G., & TEDESCHI, R. G. (Eds.). (2006). *The handbook of posttraumatic growth.* Mahwah, NJ: Lawrence Erlbaum.

CAMERON, J. E. (1999). Social identity and the pursuit of possible selves: Implications for the psychological well-being of university students. *Group Dynamics, 3,* 179–189.

CAMIC, P. M., Rhodes, J. E., & Yardley, L. (Eds.). (2003). *Qualitative research in psychology: Expanding perspectives in methodology and design.* Washington, DC: American Psychology Association.

CANTER, M. B., Bennett, B. E., Jones, S. E., & Nagy, T. F. (1994). *Ethics for psychologists: A commentary on the APA ethics code.* Hyattsville, MD: American Psychological Association.

CANTOR, D. W. (2005). Patients' rights in psychotherapy. In G. P. Koocher, J. C. Norcross, & S. S. Hill III (Eds.), *Psychologists' Desk Reference* (2nd ed.; pp. 181–183). New York: Oxford University Press.

CAPRARA, G. V., & Cervone, D. (2003). A conception of personality for a psychology of human strengths: Personality as an agentic, self-regulating system. In L. G. Aspinwall & U. M. Staudinger (Eds.), *A psychology of human strengths* (pp. 61–74). Washington, DC: American Psychological Association.

CAPUZZI, D., & Gross, D. R. (1999). (Eds.). *Counseling and psychotherapy: Theories and interventions* (2nd ed.). Upper Saddle River, NJ: Merrill Prentice Hall.

CARKHUFF, R. R. (1969a). *Helping and human relations: Vol. 1.* Selection and training. New York: Holt, Rinehart & Winston.

CARKHUFF, R. R. (1969b). *Helping and human relations: Vol. 2.* Practice and research. New York: Holt, Rinehart & Winston.

CARKHUFF, R. R. (1971). Training as a preferred mode of treatment. *Journal of Counseling Psychology, 18,* 123–131.

CARKHUFF, R. R. (1987). *The art of helping* (6th ed.). Amherst, MA: Human Resource Development Press.

CARKHUFF, R. R., & Anthony, W. A. (1979). *The skills of helping: An introduction to counseling.* Amherst, MA: Human Resource Development Press.

CARR, A. (2004). *Positive psychology: The science of happiness and human strengths.* New York: Brunner-Routledge.

CARSON, D. K., & Becker, K. W. (2003). *Creativity in psychotherapy: Reaching new heights with individuals, couples, and families.* Binghamton, NY: Haworth Clinical Practice Press.

CARSON, D. K., & Becker, K. W. (2004). When lightning strikes: Reexamining creativity in psychotherapy. *Journal of Counseling & Development, 82,* 111–115.

CARSON, D. K., Becker K. W., Vance K. E, & Forth N. L. (2003, March). The role of creativity in marriage and family therapy practice: A national online study. *Contemporary Family Therapy, 25,* 89–109.

CARTER, C. L. (2006, March–April). Transformers. *Alliance,* 4–7.

CARTON, J. S., Kessler, E. A., & Pape, C. L. (1999, Spring). Nonverbal decoding skills and relationship well-being in adults. *Journal of Nonverbal Behavior, 23,* 91–100.

CARTWRIGHT, B., Arredondo, P., & D'Andrea, M. (2004). Dignity, development, & diversity. *Counseling Today,* 24–26, 32.

CASHWELL, C. S., & Young, J. S. (Eds.). (2005). *Integrating spirituality and religion into counseling: A guide to competent practice.* Alexandria, VA: American Counseling Association.

CASTONGUAY, L. G. (1997). Support in psychotherapy: A common factor in need of empirical data, conceptual clarification, and clinical input. *Journal of Psychotherapy Integration, 7,* 99–103.

CERVONE, D. (2000). Thinking about self-efficacy. *Behavior modification, 24,* 30–56.

CERVONE, D., & Scott, W. D. (1995). Self-efficacy theory and behavioral change: Foundations, conceptual issues, and therapeutic implications. In W. O'Donohue & L. Krasner (Eds.), *Theories of behavior therapy: Exploring behavior change* (pp. 349–383). Washington, DC: American Psychological Association.

CHAMBLESS, D. L. (2005). Compendium of empirically supported therapies. In G. P. Koocher, J. C. Norcross, & S. S. Hill, III, *Psychologists' desk reference* (2nd ed., pp. 183–192). New York: Oxford University Press.

CHANG, E. C. (Ed.). (2001). *Optimism and pessimism: Implications for theory, research, and practice.* Washington, DC: American Psychological Association.

CHANG, E. C., D'Zurilla, T. J., & Sanna, L. J. (Eds.). (2004). *Social problem solving: Theory, research, and training.* Washington, DC: American Psychological Association.

CHARLTON, M., Fowler, T., & Ivandick, M. J. (2006). *Law and mental health professionals: Colorado.* Washington, DC: American Psychological Association.

CHRISTENSEN, A. J. (2004). *Patient adherence to medical treatment regimens: Bridging the gap between behavioral science and biomedicine.* New Haven: Yale University Press.

CHRISTENSEN, A., & Jacobson, N. S. (1994). Who (or what) can do psychotherapy?: The status and challenge of non-professional therapies. *Psychological Science,* 1–7.

CLAIBORN, C. D. (1982). Interpretation and change in counseling. *Journal of Counseling Psychology, 29,* 439–453.

CLAIBORN, C. D., Berberoglu, L. S., Nerison, R. M., & Somberg, D. R. (1994). The client's perspective: Ethical judgments and perceptions of therapist practices. *Professional Psychology: Research and Practice, 25,* 268–274.

CLARK, A. J. (1991). The identification and modification of defense mechanisms in counseling. *Journal of Counseling and Development, 69,* 231–236.

CLARK, A. J. (1998). *Defense mechanisms in the counseling process.* Thousand Oaks, CA: Sage.

COCHRAN, J. L., & Cochran, N. H. (2006). *The heart of counseling: A guide to developing therapeutic relationships.* Pacific Grove, CA: Brooks/Cole.

COCKELL, S. J., Zaitsoff, S. L., & Geller, J. (2004). Maintaining change following eating disorder treatment. *Professional Psychology: Research and Practice, 35,* 527–534.

COHEN, S., Underwood, L., & Gotlieb, B. (Eds.). (2000). *Measuring and intervening in social support.* New York: Oxford University Press.

COLAN, L. J. (2003). *Sticking to it: The art of adherence.* Dallas: Cornerstone Leadership Institute.

COLBY, S. M., Monti, P. M., Barmnett, N. P., Rohsenow, D. J., Weissman, K., Spirito, A., Woolard, R. H., & Lewander, W. J. (1998). Brief motivational interviewing in a hospital setting for adolescent smoking: A preliminary study. *Journal of Counseling and Clinical Psychology, 66,* 574–578.

COLE, H. P., & Sarnoff, D. (1980). Creativity and counseling. *Personnel and Guidance Journal, 59,* 140–146.

COLLINS, J. C., & Porras, J. I. (1994). *Built to last: Successful habits of visionary companies.* New York: Harper Business.

COMSTOCK, D. (2005). *Diversity and development: Critical contexts that shape our lives and relationships.* Monterey, CA: Brooks/Cole–Thomson.

CONSUMER REPORTS. (1994). Annual questionnaire.

CONSUMER REPORTS. (1995, November). Mental health: Does therapy help? 734–739.

CONTERIO, K., & Lader, W. (1998). *Bodily harm: The breakthrough healing program for self-injurers.* New York: Hyperion.

CONYNE, R. K., & Bemak, F. (Eds.). (2005). *Journeys to professional excellence: Lessons from leading counselor educators and practitioners.* Alexandria, VA: American Counseling Association.

CONYNE, R., & Cook, E. E. (Eds.). (2004). *Ecological counseling: An innovative approach to conceptualizing person-environment interaction.* Alexandria, VA: American Counseling Association.

COOPER, J. F. (1995). *A primer of brief psychotherapy.* New York: W. W. Norton.

COOPER, S. H. (1998). Changing notions of defense within psychoanalytic theory. *Journal of Personality, 66,* 947–964.

COREY, G. (1996). *Theory and practice of counseling and psychotherapy* (5th ed.). Pacific Grove, CA: Brooks/Cole.

COREY, G., Corey, M. S., & Callanan, P. (1997). *Issues and ethics in the helping professions* (5th ed.). Pacific Grove, CA: Brooks/Cole.

CORMIER, S., & Nurius, P. S. (2003). *Interviewing and change strategies for helpers: Fundamental skills and cognitive-behavior*

interventions (5th ed.). Belmont, CA: Wadsworth/Thomson Learning.

CORRIGAN, P. W. (1995). Wanted: Champions of psychiatric rehabilitation. *American Psychologist, 50,* 514–521.

COSIER, R. A., & Schwenk, C. R. (1990). Agreement and thinking alike: Ingredients for poor decisions. *Academy of Management Executive, 4,* 69–74.

COSTANZO, M. (1992). Training students to decode verbal and nonverbal clues: Effects on confidence and performance. *Journal of Educational Psychology, 84,* 308–313.

COTTONE, R. R., & Claus, R. E. (2000). Ethical decision-making models: A review of the literature. *Journal of Counseling & Development, 78,* 275.

COVEY, S. R. (1989). *The seven habits of highly effective people.* New York: Simon & Schuster (Fireside edition, 1990).

COYNE, J. C., & Racioppo, M. W. (2000). Never the twain shall meet?: Closing the gap between coping research and clinical intervention research. *American Psychologist, 55,* 655–664.

CRAMER, P. (1998). Coping and defense mechanisms: What's the difference? *Journal of Personality, 66,* 919–946.

CRAMER, P. (2000). Defense mechanisms in psychology today. *American Psychologist, 55,* 637–646.

CRAMER, P. (2005). *A new look at defense mechanisms.* New York: Guilford Press.

CRAMER, P. (2006). *Protecting the self: Defense mechanisms in action.* New York: Guilford Press.

CRANO, W. D. (2000, March). Milestones in the psychological analysis of social influence. *Group Dynamics, 4,* 68–80.

CROSS, S. E., & Markus, H. R. (1994). Self-schemas, possible selves, and competent performance. *Journal of Education Psychology, 86,* 423–438.

CUMMINGS, N. A. (1979). Turning bread into stones: Our modern antimiracle. *American Psychologist, 34,* 1119–1129.

CUMMINGS, N. A. (2000). *The first session with substance abusers.* San Francisco: Jossey-Bass.

CUMMINGS, N. A., Budman, S. H., & Thomas, J. L. (1998, October). Efficient psychotherapy as a viable response to scarce resources and rationing of treatment. *Professional Psychology: Research and Practice, 29,* 460–469.

CUMMINGS, N. A., & Cummings, J. L. (2000). *The essence of psychotherapy: Reinventing the art in the new era of data.* San Diego: Academic Press.

DAUSER, P. J., Hedstrom, S. M., & Croteau, J. M. (1995). Effects of disclosure of comprehensive pretherapy information on clients at a university counseling center. *Professional Psychology: Research and Practice, 26,* 190–195.

DeANGELIS, T. (2002, July/August). If you do just one thing, make it exercise. *Monitor on Psychology,* 49–51.

DeANGELIS, T. (2005, October). Putting people in their places. *Monitor on Psychology,* 34–35.

DE BONO, E. (1992). *Serious creativity: Using the power of lateral thinking to create new ideas.* New York: Harper Business.

DEEGEAR, J., & Lawson, D. M. (2003). The utility of empirically supported treatments. *Professional Psychology: Research and Practice, 34,* 271–277.

DE JONG, P., & Berg, I. K. (2002). *Interviewing for solutions* (2nd ed.). Belmont, CA: Brooks/Cole.

DENZIN, N. K., & Lincoln, Y. S. (Eds.). (2000). *Handbook of qualitative research* (2nd ed.). Thousand Oaks, CA: Sage Publications.

de SHAZER, S. (1985). *Keys to solution in brief therapy.* New York: Norton.

de SHAZER, S. (1988). *Clues: Investigating solutions in brief therapy.* New York: Norton.

DETWEILER-BEDELL, J. B., & Whisman, M. A. (2005). A lesson in assigning homework: Therapist, client, and task characteristics in cognitive therapy for depression. *Professional Psychology: Research and Practice, 36,* 219–223.

DEUTSCH, M. (1954). Field theory in social psychology. In G. Lindzey (Ed.), *The handbook of social psychology* (Vol. 1). Cambridge, MA: Addison-Wesley.

DeVITO, J. A. (2004). *The interpersonal communication book* (10th ed.). Boston: Allyn & Bacon.

DILLON, C. (2003). *Learning from mistakes in clinical practice.* Belmont, CA: Brooks/Cole.

DIMOND, R. E., Havens, R. A., & Jones, A. C. (1978). A conceptual framework for the practice of prescriptive eclecticism in psychotherapy. *American Psychologist, 33,* 239–248.

DIVISION 12 TASK FORCE. (1995). Training in and dissemination of empirically-validated psychological treatments. *The Clinical Psychologist, 49,* 3–23.

DOBMEYER, A. C., Rowan, A. B., Etherage, J. R., & Wilson, R. J. (2003). Training psychology interns in primary behavioral health care. *Professional Psychology: Research and Practice, 34,* 586–594.

DODGEN, C. E. (2005). *Nicotine dependence: Understanding and applying the most effective treatment interventions.* Washington, DC: American Psychological Association.

DONNAY, D. A. C., & Borgen, F. H. (1999). The incremental validity of vocational self-efficacy: An examination of interest, self-efficacy, and occupation. *Journal of Counseling Psychology, 46,* 432–447.

DORN, F. J. (Ed.). (1986). *The social influence process in counseling and psychotherapy.* Springfield, IL: Charles C. Thomas.

DORNER, D. (1996). *The logic of failure: Why things go wrong and what we can do to make them right.* New York: Holt.

DRAYCOTT, S., & Dabbs, A. (1998). Cognitive dissonance: An overview of the literature and its integration into theory and practice of clinical psychology. *British Journal of Clinical Psychology, 37,* 341–353.

DRISCOLL, R. (1984). *Pragmatic psychotherapy.* New York: Van Nostrand Reinhold.

DUAN, C., & Hill, C. E. (1996). The current state of empathy research. *Journal of Counseling Psychology, 43,* 261–274.

DUNCAN, B. L., Miller, S., & Sparks, J. (2004). *The heroic client: A revolutionary way to improve effectiveness.* San Francisco: Jossey-Bass.

DURAND, V. M., & Barlow, D. H. (2006). *Essentials of abnormal psychology.* Belmont, CA: Wadsworth/Thomson Learning.

DURLAK, J. A. (1979). Comparative effectiveness of paraprofessional and professional helpers. *Psychological Bulletin, 86,* 80–92.

D'ZURILLA, T. J., & Nezu, A. M. (1999). *Problem-solving therapy: A social competence approach to clinical intervention* (2nd ed.). New York: Springer.

D'ZURILLA, T., J., & Nezu, A. M. (2001). Problem-solving therapies. In K. S. Dobson (Ed.), *The handbook of cognitive-behavioral therapies* (2nd ed.; pp. 211–245). New York: Guilford.

ECHTERLING, L. G., Presbury, J. H., & McKee, J. E. (2005). *Crisis intervention: Promoting resilience and resolution in troubled times.* Upper Saddle River, NJ: Pearson-Merrill Prentice Hall.

ECONOMIST. (1995, July). Come feel the noise.

EDWARDS, C. E., & Murdock, N. L. (1994). Characteristics of therapist self-disclosure in the counseling process. *Journal of Counseling and Development, 72,* 384–389.

EDWARDS, D. J. A., Dattilio, F. M., & Bromley, D. B. (2004). Developing evidence-based practice: The role of case-based research. *Professional Psychology: Research and Practice, 35,* 589–597.

EGAN, G. (1970). *Encounter: Group processes for interpersonal growth.* Pacific Grove, CA: Brooks/Cole.

EGAN, G. (1975). *The skilled helper: A model for systematic helping and interpersonal relating.* Monterey, CA: Brooks/Cole–Thomson.

EGAN, G. (1994). *Working the shadow side.* San Francisco: Jossey-Bass.

EGAN, G. (in press). *Conversations for the 21st century: The pragmatics of dialogue.* London: BT.

EGAN, G., & Cowan, M. A. (1979). *People in systems: A model for development in the human service professions and education.* Monterey, CA: Brooks/Cole.

EKMAN, P. (1992). *Telling lies: Clues to deceit in the marketplace, politics, and marriage.* New York: Norton.

EKMAN, P. (1993). Facial expression and emotion. *American Psychologist, 48,* 384–392.

EKMAN, P., & Friesen, W. V. (1975). *Unmaking the human face: A guide to recognizing emotions from facial cues.* Englewood Cliffs, NJ: Prentice Hall.

EKMAN, P., & Rosenberg, E. L., (Eds.). (1998). *What the face reveals: Basic and applied studies of spontaneous expression using the facial action coding system (FACS).* New York: Oxford University Press.

ELIAS, M. J., & Tobias, S. E. (2002). *Social problem solving interventions in the schools.* New York: Guilford.

ELLIS, A. (1984). Must most psychotherapists remain as incompetent as they are now? In J. Hariman (Ed.), *Does psychotherapy really help people?* Springfield, IL.: Charles C. Thomas.

ELLIS, A. (1985). *Overcoming resistance: Rational-emotive therapy with difficult clients.* New York: Springer.

ELLIS, D. B. (1999). *Creating your future.* Boston, MA: Houghton Mifflin Co.

ELLIS, K. (1998). *The magic lamp: Goal setting for people who hate setting goals.* New York: Crown.

ENGELS, D. W. (2004). *The professional counselor: Portfolio, competencies, performance guidelines, and assessment* (3rd ed.). Alexandria, VA: American Counseling Association.

ENGLISH, P. W., & Sales, B. D. (2005). *More than the law: Behavioral and social facts in legal decision making.* Washington, DC: American Psychological Association.

EPSTON, D., White, M., & Murray, K. (1992). A proposal for re-authoring therapy: Rose's revisioning of her life and commentary. In S. McNamee & K. J. Gergen (Eds.), *Therapy as social construction* (pp. 96–116). London: Sage.

ETZIONI, A. (1989, July–August). Humble decision making. *Harvard Business Review,* 120–126.

EVERLY, G. S., Jr., & Lating, J. M. (2004). *Personality-guided therapy for posttraumatic stress disorder.* Washington, DC: American Psychological Association.

EYSENCK, H. J. (1952). The effects of psychotherapy: An evaluation. *Journal of Consulting and Clinical Psychology, 60,* 659–663.

FARBER, B. A. (2003a). Patient self-disclosure: A review of the research. *Journal of Clinical Psychology, 59,* 589–600.

FARBER, B. A. (2003b). Self-disclosure in psychotherapy practice and supervision: An introduction. *Journal of Clinical Psychology, 59,* 525–528.

FARBER, B. A., Berano, K. C., & Capobianco, J. A. (2004). Clients' perceptions of the process and consequences of self-disclosure in psychotherapy. *Journal of Counseling Psychology, 51*(3), 340–346.

FARBER, B. A., Manevich, I., Metzger, J., & Saypol, E. (2005). Choosing psychotherapy as a career: Why did we cross that road? *Journal of Clinical Psychology/In Session, 61,* 1009–1031.

FARMER, R. F., & Nelson-Gray, R. O. (2005). *Personality-guided behavior therapy.* Washington, DC: American Psychological Association.

FARRELLY, F., & Brandsma, J. (1974). provocative therapy. Cupertino, CA: Meta Publications.

FAUTH, J., & Williams, E. N. (2005). The in-session self-awareness of therapist-trainees: Hindering or helpful? *Journal of Counseling Psychology, 52,* 443–447.

FELLER, R. (1984). *Job-search agreements.* Monolith, Colorado State University, Fort Collins.

FERENTZ, L. R. (2002). *Understanding self-injurious behaviors.* Troy, MI: Performance Resource Press.

FERGUSON, T. (1987, January–February). Agreements with yourself. *Medical Self-Care,* 44–47.

FESTINGER, S. (1957). *A theory of cognitive dissonance.* New York: Harper & Row.

FEYMAN, R. (1974). Cargo cult science. Adapted from the Caltech commencement address given in 1974. Retrieved March 26, 2006, from http://www.lhup.edu/~dsimanek/cargocul.htm

FINN, S. E. (2005). How psychological assessment taught me compassion and firmness. *Journal of Personality Assessment, 84,* 29–32.

FISH, J. M. (1995, Spring). Does problem behavior just happen? Does it matter? *Behavior and Social Issues, 5,* 3–12.

FISHER, C. B., & Younggren, J. N. (1997). The value and unity of the 1992 ethics code. *Professional Psychology: Research and Practice, 28,* 582–592.

FLACH, F. (1997). *Resilience: How to bounce back when the going get tough.* New York: Hatherleigh Press.

FLACK, W. F., & Laird, J. D. (Eds.). (1998). *Emotions and psychopathology.* New York: Oxford University Press.

FLETCHER, A. M. (2003). Renewed hope for self-change. *American Psychologist, 58,* 822–823.

FOLKMAN, S., & Moskowitz, J. T. (2000). Positive affect and the other side of coping. *American Psychologist, 55,* 647–664.

FOXHALL, K. (2000, July/August). Research for the real world. *Monitor on Psychology,* 28–36.

FRANCE, K. (2005). Crisis intervention. In G. P. Koocher, J. C. Norcross, & S. S. Hill, III (Eds.), *Psychologists' desk reference* (2nd ed., pp. 245–249). New York: Oxford University Press.

FRANCES, A., Clarkin, J., & Perry, S. (1984). *Differential therapeutics in psychiatry.* New York: Brunner/Mazel.

FRAZIER, L. D., Hooker, K., Johnson, P. M., & Kaus, C. R. (2000). Continuity and change in possible selves in late life: A 5-year longitudinal study. *Basic & Applied Social Psychology, 22,* 237–243.

FREIRE, P. (1970). *Pedagogy of the oppressed.* New York: Seabury.

FREMONT, S. K., & Anderson, W. (1986). What client behaviors make counselors angry? An exploratory study. *Journal of Counseling and Development, 65,* 67–70.

FRIEDLANDER, M. L., & Schwartz, G. S. (1985). Toward a theory of strategic self-presentation in counseling and psychotherapy. *Journal of Counseling Psychology, 32,* 483–501.

FRIEMAN, S. (1997). *Time effective psychotherapy: Maximizing outcomes in an era of minimized resources.* Boston: Allyn & Bacon.

FRUEH, B. C., de Arellano, M. A., & Turner, S. M. (1997). Systematic desensitization as an alternative exposure strategy in the treatment of a veteran with military-related PTSD. *American Journal of Psychiatry, 154,* 287–288.

FUREDI, F. (2004). *Therapy culture: Cultivating vulnerability in an uncertain age.* New York: Routledge.

GALASSI, J. P., & Bruch, M. A. (1992). Counseling with social interaction problems: Assertion and social anxiety. In S. D. Brown &

R. W. Lent (Eds.), *Handbook of counseling psychology.* New York: Wiley.

GALOTTI, K. M. (2002). *Making decisions that matter: How people face important life choices.* Mahwah, NJ: Erlbaum.

GARBER, J., & Seligman, M. (Eds.). (1980). *Human helplessness: Theory and applications.* New York: W. H. Freeman.

GARFIELD, S. L. (1996). Some problems associated with "validated" forms of psychotherapy. *Clinical Psychology: Science and Practice, 3,* 218–229.

GASTON, L., Goldfried, M. R., Greenberg, L. S., Horvath, A. O., Raue, P. J., & Watson, J. (1995). The therapeutic alliance in psychodynamic, cognitive-behavioral, and experimental therapies. *Journal of Psychotherapy Integration, 5,* 1–26.

GATI, I., Krausz, M., & Osipow, S. H. (1996). A taxonomy of difficulties in career decision making. *Journal of Counseling Psychology, 43,* 510–526.

GELATT, H. B. (1989). Positive uncertainty: A new decision-making framework for counseling. *Journal of Counseling Psychology, 36,* 252–256.

GELATT, H. B., Varenhorst, B., & Carey, R. (1972). *Deciding: A leader's guide.* Prince-ton, NJ: College Entrance Examination Board.

GELSO, C. J., Hill, C. E., Mohr, J., Rochlen, A. B., & Zack, J. (1999). The face of transference in successful, long-term therapy: A qualitative analysis. *Journal of Counseling Psychology, 46,* 257–267.

GELSO, C. J., Kivlighan, D. M., Wise, B., Jones, A., & Friedman, S. C. (1997). Tranference, insight, and the course of time-limited therapy. *Journal of Counseling Psychology, 44,* 209–217.

GIANNETTI, E. (1997). *Lies we live by: The art of self-deception.* New York and London: Bloomsbury.

GIBB, J. R. (1968). The counselor as a role-free person. In C. A. Parker (Ed.), *Counseling theories and counselor education.* Boston: Houghton Mifflin.

GIBB, J. R. (1978). *Trust: A new view of personal and organizational development.* Los Angeles: The Guild of Tutors Press.

GIBBS, L. E. (2003). *Evidence-based practice for the helping professions: A practical guide with integrated multimedia.* Pacific Grove, CA: Thomson Learning.

GIGERENZER, G., & Selten, R. (2002). *Bounded rationality: The adaptive toolbox.* Cambridge, MA: MIT Press.

GILBERT, T. F. (1978). *Human competence: Engineering worthy performance.* New York: McGraw-Hill.

GILLILAND, B. E., & James, R. K. (1997). *Theory and strategies in counseling and psychotherapy* (4th ed.). Needham Heights, MA: Allyn & Bacon.

GILOVICH, T. (1991). *How we know what isn't so: The fallibility of human reason in everyday life*. New York: The Free Press.

GLADDING, S. T. (2002). *Becoming a counselor: The light, the bright, and the serious*. Alexandria, VA: American Counseling Association.

GLADDING, S. T. (2005). Recognizing counseling as a quiet revolution. *Counseling Today*, 5, 14.

GLASSER, W. (2000). *Reality therapy in action*. New York: HarperCollins.

GLICKEN, M. D. (2004). *Improving the effectiveness of the helping professional: An evidence-based approach to practice*. Thousand Oaks, CA: Sage.

GODWIN, G. (1985). *The finishing school*. New York: Viking.

GOLDFRIED, M. R. (2001). How therapists change: Personal and professional reflections. Washington, DC: American Psychological Association.

GOLDSMITH, D. J. (2004). *Communicating social support*. New York: Cambridge University Press.

GOLEMAN, D. (1995). *Emotional intelligence*. New York: Bantam Books.

GOLEMAN, D. (1998). *Working with emotional intelligence*. New York: Bantam Books.

GOLLWITZER, P. M. (1999). Implementation intentions strong effects of simple plans. *American Psychologist*, 54, 493–503.

GOSLIN, D. A. (1985). Decision making and the social fabric. *Society*, 22, 7–11.

GRACE, M., Kivlighan, D. M., Jr., & Kunce, J. (1995, Summer). The effect of nonverbal skills training on counselor trainee nonverbal sensitivity and responsiveness and on session impact and working alliance ratings. *Journal of Counseling and Development*, 73, 547–552.

GRAY, G. V., Brody, D. S., & Johnson, D. (2005). The evolution of behavioral primary care. *Professional Psychology: Research and Practice*, 36, 123–129.

GRAYBAR, S. R., & Leonard, M. A. (2005). In defense of listening. *American Journal of Psychotherapy*, 59, 1–18.

GREENBERG, L. S. (1986). Change process research. *Journal of Consulting and Clinical Psychology*, 54, 4–9.

GREENBERG, L. S. (2002). *Emotion-focused therapy: Coaching clients to work through their feelings*. Washington, DC: American Psychological Association.

GREENBERG, L. S., & Paivio, S. C. (1997). *Working with the emotions in psychotherapy*. New York: Guilford Press.

GREENBERG, L. S., & Watson, J. C. (2000). Alliance ruptures and repairs in experimental therapy. *Journal of Clinical Psychology*, 56, 175–186.

GREENSON, R. R. (1967). *The technique and practice of psychoanalysis*. New York: International Universities Press.

GRIFFITHS, P. E. (1997). *What emotions really are*. Chicago: University of Chicago Press.

GROOPMAN, J. (2004). *The anatomy of hope: How people prevail in the face of illness*. New York: Random House.

GUISINGER, S., & Blatt, S. J. (1994). Individuality and relatedness: Evolution of a fundamental dialectic. *American Psychologist*, 49, 104–111.

GUTERMAN, J. T. (2006). *Mastering the art of solution-focused counseling*. Alexandria, VA: American Counseling Association.

GUYATT, G., & Rennie, D. (2002). *Essentials of evidence-based clinical practice*. Chicago: AMA Press.

HAIG, B. D. (2005). Psychology needs realism, not instrumentalism. *American Psychologist*, 60, 344–345.

HALEY, J. (1976). *Problem solving therapy*. San Francisco: Jossey-Bass.

HALL, C. R., Dixon, W. A., & Mauzey, E. D. (2004). Spirituality and religion: Implications for counselors. *Journal of Counseling & Development*, 82, 504–507.

HALL, E. T. (1977). *Beyond culture*. Garden City, NJ: Anchor Press.

HALL, L. E. (2005). *Dictionary of multicultural psychology: Issues, terms, and concepts*. Thousand Oaks, CA: Sage.

HALL, P. (2005). Behind the lines. *FT Magazine*, 34–35.

HALLECK, S. L. (1988). Which patients are responsible for their illnesses? *American Journal of Psychotherapy*, 42, 338–353.

HAMMOND, J. S., Keeney, R. L., & Raiffa, H. (1998). The hidden traps in decision making. *Harvard Business Review*, 47–58.

HAMMOND, J. S., Keeney, R. L., & Raiffa, H. (1999). *Smart choices: A practical guide to making better decisions*. Boston, MA: Harvard Business School Press.

HANDELSMAN, M. M., & Galvin, M. D. (1988). Facilitating informed consent for outpatient psychotherapy: A suggested written format. *Professional Psychology: Research and Practice*, 19, 223–225.

HANNA, F. J. (1994). A dialectic of experience: A radical empiricist approach to conflicting theories in psychotherapy. *Psychotherapy, 31*, 124–136.

HANNA, F. J. (2002). *Therapy with difficult clients: Using the precursors model to awaken change.* Washington, DC: American Psychological Association.

HANNA, F. J., Hanna, C. A., & Keys, S. G. (1999). Fifty strategies for counseling defiant, aggressive adolescents: Reaching, accepting, and relating. *Journal of Counseling and Development, 77,* 395–404.

HANNA, F. J., & Ottens, A. J. (1995). The role of wisdom in psychotherapy. *Journal of Psychotherapy Integration, 5,* 195–219.

HANSEN, J. T. (2004). Thoughts on knowing: Epistemic implications of counseling practice. *Journal of Counseling Psychology, 82,* 131.

HANSEN, J. T. (2005). The devaluation of inner subjective experiences by the counseling profession: A plea to reclaim the essence of the profession. *Journal of Counseling and Development, 83,* 406–415.

HARE-MUSTIN, R., & Marecek, J. (1986). Autonomy and gender: Some questions for therapists. *Psychotherapy, 23,* 205–212.

HARKIN, J. (2005, October 14). Paralysed by choice. *Financial Times.*

HARPER, R. (2004). *Personality-guided therapy in behavioral medicine.* Washington, DC: American Psychological Association.

HARRIS, G. A. (1995). *Overcoming resistance: Success in counseling men.* Alexandria, VA: American Counseling Association.

HASTIE, R., & Dawes, R. M. (2001). *Rational choice in an uncertain world.* Thousand Oaks, CA: Sage.

HATCHER, S. L., Favorite, T. K., Hardy, E. A., Goode, R. L., Deshetler, L. A., & Thomas, R. M. (2005). An analogue study of therapist empathic process: Working with difference. *Psychotherapy: Theory, Research, Practice, Training, 42,* 198–210.

HATFIELD, D. R., & Ogles, B. M. (2004). The use of outcome measures by psychologists in clinical practice. *Professional Psychology: Research and Practice, 35,* 485–491.

HATTIE, J. A., Sharpley, C. E., & Rogers, H. J. (1984). Comparative effectiveness of professional and paraprofessional helpers. *Psychological Bulletin, 95,* 534–541.

HAVERKAMP, B. E., Morrow, S. L., & Ponterotto, J. G. (2005). A time and a place for qualitative and mixed methods in counseling research. *Journal of Counseling Psychology, 52,* 123–125.

HAYS, K. F. (1999). *Working it out: Using exercise in psychotherapy.* Washington, DC: American Psychological Association.

HAZLER, R. J., & Kottler, J. A. (2005). *The emerging professional counselor: Student dreams to professional realities* (2nd ed.). Alexandria, VA: American Counseling Association.

HEADLEE, R., & Kalogjera, I. J. (1988). The psychotherapy of choice. *American Journal of Psychotherapy, 42,* 532–542.

HECKER, L. L. (Ed.). (2002). When lightning strikes: Using creativity in psychotherapy [Special issue]. *Journal of Clinical Activities, Assignments, & Handouts in Psychotherapy Practice, 2*(2).

HECKER, L. L., & Kottler, J. A. (2002). Growing creative therapists: Introduction to the special issue. *Journal of Clinical Activities, Assignments, & Handouts in Psychotherapy Practice, 2*(2), 1–3.

HECKMAN, J. J. (2006, January 10). Investing in disadvantaged children is an economically efficient policy: Research presented at the Committee for Economic Development/The Pew Charitable Trusts/PNC financial services Group Forum on "Building the economic case for investments in preschool." Retrieved March 26, 2006, from http://www.ced.org/docs/report/report/report_2006heckman.pdf.

HEINSSEN, R. K. (1994, June). Therapeutic contracting with schizophrenic patients: A collaborative approach to cognitive-behavioral treatment. Paper presented at the 21st International Symposium for the Psychotherapy of Schizophrenia, Washington, DC.

HEINSSEN, R. K., Levendusky, P. G., & Hunter, R. H. (1995). Client as colleague: Therapeutic contracting with the seriously mentally ill. *American Psychologist, 50,* 522–532.

HELLER, K., Swindle, R., Pescosolido, B., & Kikuzawa, S. (2000). Responses to nervous breakdowns in America over a 40-year period: Mental health policy implications. *American Psychologist, 55*(7), 740–749.

HENGGELER, S. W., Schoenwald, S. K., & Pickrel, S. G. (1995, Fall). Multisystemic therapy: Bridging the gap between university and community-based treatment. *Journal of Consulting and Clinical Psychology, 63,* 709–717.

HENTSCHEL, U., Smith, G., Draguns, J. G., & Ehlers, W. (Eds.). (2004). *Defense mechanisms: Theoretical, research, and clinical perspectives.* Amsterdam: Elsevier.

HEPPNER, P. P. (1989). Identifying the complexities within clients' thinking and decision making. *Journal of Counseling Psychology, 36,* 257–259.

HEPPNER, P. P., & Claiborn, C. D. (1989). Social influence research in counseling: A review and critique (Monograph). *Journal of Counseling Psychology, 36,* 365–387.

HEPPNER, P. P., & Frazier, P. A. (1992). Social psychological processes in psychotherapy: Extrapolating basic research to counseling psychology. In S. D. Brown & R. W. Lent (Eds.), *Handbook of counseling psychology.* New York: Wiley.

HEPPNER, P. P., Witty, T. E., & Dixon, W. A. (2004). Problem-solving appraisal and human adjustment: A review of 20 years of research using the problem solving inventory. *The Counseling Psychologist, 32,* 344–428.

HERINK, R. (Ed.). (1980). *The psychotherapy handbook.* New York: Meridian.

HICKS, J. W. (2005). *50 signs of mental illness: A guide to understanding mental health.* New Haven: Yale University Press.

HICKSON, M. L., & Stacks, D. W. (1993). Nonverbal communication: Studies and applications (2nd ed.). Madison, WI: Brown & Benchmark.

HIGGINSON, J. G. (1999). Defining, excusing, and justifying deviance: Teen mothers' accounts for statutory rape. *Symbolic Interaction, 22,* 25–44.

HIGHLEN, P. S., & Hill, C. E. (1984). Factors affecting client change in counseling. In S. D. Brown & R. W. Lent (Eds.), *Handbook of counseling psychology* (pp. 334–396). New York: Wiley.

HILL, C. E. (1994). What is the therapeutic relationship? A reaction to Sexton and Whiston. *The Counseling Psychologist, 22,* 90–97.

HILL, C. E., & Corbett, M. M. (1993, January). A perspective on the history of process and outcome research in counseling psychology. *Journal of Counseling Psychology, 40,* 3–24.

HILL, C. E., Gelso, C. J., & Mohr, J. J. (2000). Client concealment and self-presentation in psychotherapy: Comment on Kelly (2000). *Psychological Bulletin, 126,* 495–500.

HILL, C. E., Nutt-Williams, E., Heaton, K. J., Thompson, B. J., & Rhodes, R. H. (1996). Therapist retrospective recall of impasses in long-term psychotherapy: A qualitative analysis. *Journal of Counseling Psychology, 43,* 207–217.

HILL, C. E., Thompson, B. J., Cogar, M. C., & Denmann, D. W., III. (1993). Beneath the surface of long-term therapy: Therapist and client report of their own and each other's covert processes. *Journal of Counseling Psychology, 40,* 278–287.

HILL, C. E., & Williams, E. N. (2000). The process of individual counseling. In R. Lent & S. Brown (Eds.), *Handbook of counseling psychology* (3rd ed., pp. 670–710). New York: Wiley.

HILLIARD, R. B. (1993, June). Single-case methodology in psychotherapy process and outcome research. *Journal of Consulting & Clinical Psychology, 61,* 373–380.

HOGAN, B. E., Linden, W., & Najarian, B. (2002). Social support interventions: Do they work? *Clinical Psychology Review, 22,* 381–440.

HOLADAY, M., & McPhearson, R. W. (1997, May/June). Resilience and severe burns. *Journal of Counseling and Development, 75,* 346–356.

HOLMES, G. R., Offen, L., & Waller, G. (1997). See no evil, hear no evil, speak no evil: Why do relatively few male victims of childhood sexual abuse receive help for abuse-related issues in adulthood? *Clinical Psychology Review, 27,* 69–88.

HOOVER, J., & DiSilvestro, R. P. (2005). *The art of constructive confrontation: How to achieve more accountability with less conflict.* New York: Wiley.

HORVATH, A. O. (2000). The therapeutic relationship: From transference to alliance. *Journal of Clinical Psychology, 56,* 163–173.

HORVATH, A. O., & Symonds, B. D. (1991). Relation between working alliance and outcome in psychotherapy: A meta-analysis. *Journal of Counseling Psychology, 38,* 139–149.

HOSKING, G., & Walsh, I. R. (2005). *The WAVE report: Violence and what to do about it.* Croydon, UK: WAVE Trust.

HOUSER, R. F., Feldman, M., Williams, K., & Fierstien, J. (1998, July). Persuasion and social influence tactics used by mental health counselors. *Journal of Mental Health Counseling, 20,* 238–249.

HOWARD, G. S. (1991). Culture tales: A narrative approach to thinking, cross-cultural psychology, and psychotherapy. *American Psychologist, 46,* 187–197.

HOWELL, W. S. (1982). *The empathic communicator.* Belmont, CA: Wadsworth.

HOYT, M. F. (1995). Brief Psychotherapies. In Gurman, A. S., & Messer, S. B. (Eds.), *Essential psychotherapies* (pp. 441–487). New York: Guilford.

HOYT, W. T. (1996). Antecedents and effects of perceived therapist credibility: A meta-analysis. *Journal of Counseling Psychology, 43,* 430–447.

HUBBLE, M. A., Duncan, B. L., & Miller, S. D. (Eds.). (1999). The heart & soul of change: What works in therapy. Washington, DC: American Psychological Association.

HUDSON, S. M., Marshall, W. L., Ward, T., Johnston, P. W., et al. (1995). Kia Marama: A cognitive-behavioral program for incarcerated child molesters. *Behavior Change, 12,* 69–80.

HUNTER, R. H. (1995). Benefits of competency-based treatment programs. *American Psychologist, 50,* 509–513.

HUTCHISON, E. D. (2003). *Dimensions of human behavior: The changing life course* (2nd ed.). Thousand Oaks, CA: Sage.

HYDE, J. S. (2005). The gender similarities hypothesis. *American Psychologist, 60,* 581–592.

ICKES, W. (1993). Empathic accuracy. *Journal of Personality, 61,* 587–610.

ICKES, W. (1997). Introduction. In W. Ickes (Ed.), *Empathic accuracy* (pp. 1–16). New York: Guilford Press.

ISHIYAMA, F. I. (1990). A Japanese perspective on client inaction: Removing attitudinal blocks through Morita therapy. *Journal of Counseling and Development, 68,* 566–570.

IVEY, A. E., & Ivey, M. B. (2007). *Intentional interviewing and counseling—Facilitating client development in a multicultural society* (6th ed.). Pacific Grove, CA: Brooks/Cole.

JACKSON, P. Z., & McKergow, M. (2002). *The solution focus.* London: Nicholas Brealey.

JACOBSON, N. S., & Christensen, A. (1996, October). Studying the effectiveness of psychotherapy: How well can clinical trials do the job? *American Psychologist, 51,* 1031–1039.

JANIS, I. L., & Mann, L. (1977). *Decision making: A psychological analysis of conflict, choice, and commitment.* New York: Free Press.

JANOSIK, E. H. (Ed.). (1984). *Crisis counseling: A contemporary approach.* Belmont, CA: Wadsworth.

JEFFREY, N. A. (2004, September 3). "Very interesting—Me, that is." *Wall Street Journal,* pp. W1, W7.

JENSEN, J. P., Bergin, A. E., & Greaves, D. W. (1990). The meaning of eclecticism: New survey and analysis of components. *Professional Psychology: Research and Practice, 21,* 124–130.

JONES, B. F., Rasmussen, C. M., & Moffitt, M. C. (1997). *Real-life problem solving: A collaborative approach to interdisciplinary learning.* Washington, DC: American Psychology Association.

JOSEPH, S., & Linley, P. A. (2005). Positive adjustment to threatening events: An organismic valuing theory of growth through adversity. *Review of General Psychology, 9,* 262–280.

KAGAN, J. (1996). Three pleasing ideas. *American Psychologist, 51,* 901–908.

KAGAN, N. (1973). Can technology help us toward reliability in influencing human interaction? *Educational Technology, 13,* 44–51.

KAHN, M. (1990). *Between therapist and client.* New York: Freeman.

KANFER, F. H., & Schefft, B. K. (1988). *Guiding therapeutic change.* Champaign, IL: Research Press.

KARASU, T. B. (1986, July). The psychotherapies: Benefits and limitations. *American Journal of Psychotherapy, 40,* 324–342.

KAROLY, P. (1995). Self-control theory. In W. O'Donohue & L. Krasner (Eds.), *Theories of behavior therapy: Exploring behavior change* (pp. 259–285). Washington, DC: American Psychological Association.

KAROLY, P. (1999). A goal systems-self-regulatory perspective on personality, psychopathology, and change. *Review of General Psychology, 3,* 264–291.

KASHDAN, T. B., Rose, P., & Fincham, F. D. (2004). Curiosity and exploration: Facilitating positive subjective experiences and personal growth opportunities. *Journal of Personality Assessment, 82,* 291–305.

KAUFMAN, G. (1989). *The psychology of shame.* New York: Springer.

KAYE, H. (1992). *Decision power.* Upper Saddle River, NJ: Prentice Hall.

KAZANTIS, N. (2000). Power to detect homework effects in psychotherapy outcome research. *Journal of Consulting & Clinical Psychology, 68,* 166–170.

KAZANTIS, N., Lampropoulos, G. L., & Deane, F. P. (2005). A national survey of practicing psychologists' use and attitudes towards homework in psychotherapy. *Journal of Consulting and Clinical Psychology, 73,* 742–748.

KAZDIN, A. E. (1996). Validated treatments: Multiple perspectives and issues: Introduction to the series. *Clinical Psychology: Science & Practice, 3,* 216–217.

KAZDIN, A. E. (2006, January). Arbitrary metrics: Implications for identifying evidence-based treatments. *American Psychologist, 61,* 42–49.

KEITH-SPIEGEL, P. (1994, November). The 1992 ethics code: Boon or bane? *Professional Psychology: Research and Practice, 25,* 315–316.

KELLEY, T. M. (2005). Natural resilience and innate mental health. *American Psychologist, 60,* 265.

KELLY, A. E. (2002). *The psychology of secrets.* New York: Plenum.

KELLY, A. E., Kahn, J. H., & Coulter, R. G. (1996). Client self-presentations at intake. *Journal of Counseling Psychology, 43,* 300–309.

KELLY, E. W., Jr. (1994). *Relationship-centered counseling: An integration of art and science.* New York: Springer.

KELLY, E. W., Jr. (1997, May/June). Relationship-centered counseling: A humanistic model of integration. *Journal of Counseling and Development, 75*, 337–345.

KENDALL, P. C. (1998). Empirically supported psychological therapies. *Journal of Consulting & Clinical Psychology, 66*, 3–6.

KENNEDY, A. (2004). Creativity in counseling group earns divisional status at ACA. *Counseling Today,* 1, 36.

KERR, B., & Erb, C. (1991). Career counseling with academically talented students: Effects of a value-based intervention. *Journal of Counseling Psychology, 38*, 309–314.

KERSTING, K. (2003). Lessons in resilience. *Monitor on Psychology,* 30–31.

KERSTING, K. (2005). Resilience: The mental muscle everyone has. *Monitor on Psychology,* 32–33.

KIERULFF, S. (1988). Sheep in the midst of wolves: Person-responsibility therapy with criminals. *Professional Psychology: Research and Practice, 19*, 436–440.

KIM, B. S. K., Hill, C. E., Gelso, C. J., Goates, M. K., Asay, P. A., & Harbin, J. M. (2003). Counselor self-disclosure, East Asian American client adherence to Asian cultural values, and counseling process. *Journal of Counseling Psychology, 53*, 15–25.

KIRSCH, I. (Ed.). (1999). *How expectancies shape experience.* Washington, DC: American Psychology Association.

KIRSCHENBAUM, D. S. (1985). Proximity and specificity of planning: A position paper. *Cognitive Therapy and Research, 9*, 489–506.

KIRSCHENBAUM, D. S. (1987). Self-regulatory failure: A review with clinical implications. *Clinical Psychological Review, 7*, 77–104.

KIVLIGHAN, D. M., Jr., & Shaughnessy, P. (2000). Patterns of working alliance development: A typology of client's working alliance ratings. *Journal of Counseling Psychology, 47*, 362–371.

KLUCKHOHN, C., & Murray, H. A. (1953). Personality formation: The determinants. In C. Kluckhohn & H. Murray (Eds.), *Personality in nature, society, and culture* (2nd ed., pp. 53–67). New York: Knopf.

KNAPP, M. L., & Hall, J. A. (1997). *Nonverbal communication in human interaction* (4th edition). San Diego: Harcourt, Brace, Jovanovich.

KNIGHT, B. G. (2004) *Psychotherapy with older adults.* Thousand Oaks, CA: Sage.

KNOX, S., Hess, S. A., Petersen, D. A., & Hill, C. E. (1997). A qualitative analysis of client perceptions of the effects of helpful therapist self-disclosure in long-term therapy. *Journal of Counseling Psychology, 44*, 274–283.

KOHUT, H. (1978). The psychoanalyst in the community of scholars. In P. H. Ornstein (Ed.), *The search for self: Selected writings of H. Kohut.* New York: International Universities Press.

KOOCHER, G. P., Norcross, J. C., & Hill, S. S., III. (Eds.). (2005). *Psychologists' desk reference.* New York: Oxford University Press.

KOTTLER, J. A. (1992). *Compassionate therapy: Working with difficult clients.* San Francisco: Jossey-Bass.

KOTTLER, J. A. (2000). *Doing good: Passion and commitment for helping others.* Philadelphia: Brunner/Routledge.

KOTTLER, J. A. (Ed.). (2002). *Counselors finding their way.* Alexandria, VA: American Counseling Association.

KUSHNER, M. G., & Sher, K. J. (1989). Fear of psychological treatment and its relation to mental health service avoidance. *Professional Psychology: Research and Practice, 20*, 251–257.

KUTHER, T. L. (2006). *Your career in psychology: Clinical and counseling psychology.* Belmont, CA: Wadsworth/Thomson Learning.

LAMBERT, M. J. (1992). Implications of outcome research for psychotherapy integration. In J. C. Norcross & M. R. Goldstein (Eds.), *Handbook of psychotherapy integration* (pp. 94–129). New York: Basic Books.

LAMBERT, M. J., Bybee, T., Houston, R., Bishop, M., Sanders, A. D., Wilkinson, R., & Rice, S. (2005). Compendium of psychotherapy treatment manuals. In G. P. Koocher, J. C. Norcross, & S. S. Hill, III, *Psychologists' desk reference* (2nd ed., 192–202). New York: Oxford University Press.

LAMBERT, M. J., & Cattani-Thompson, K. (1996, Summer). Current findings regarding the effectiveness of counseling: Implications for practice. *Journal of Counseling and Development, 74*, 601–608.

LAMBERT, M. J., & Hawkins, E. J. (2004). Measuring outcome in professional practice: Considerations in selecting and using brief outcome instruments. *Professional Psychology: Research and Practice, 35*, 492–499.

LAMBERT, M. J., Jasper, B. W., & White, J. (2005). Key principles in the assessment of psychotherapy outcome. In G. P. Koocher, J. C. Norcross, & S. S. Hill III (Eds.), *Psychologists' Desk Reference* (2nd ed.; pp. 236–239). New York: Oxford University Press.

LAMBERT, M. J., & Ogles, B. M. (2004). The efficacy and effectiveness of psychotherapy. In M. J. Lambert (Ed.), *Bergin & Garfield's handbook of psychotherapy and behavior changes* (5th ed.; pp. 93–139). New York: Wiley.

LAMBERT, W., Salzer, M. S., & Bickman, L. (1998, April). Clinical outcome, consumer satisfaction, and ad hoc ratings of improvement in children's mental health. *Journal of Consulting and Clinical Psychology, 66,* 270–279.

LAMBERT, M. J., Whipple, J. L., Smart, D. W., Vermeersch, D. A., Nielsen, S. L., & Hawkins, E. J. (2001). The effects of providing therapists with feedback on patient progress during psychotherapy: Are outcomes enhanced? *Psychotherapy Research, 11,* 49–68.

LANDRETH, G. L. (1984). Encountering Carl Rogers: His views on facilitating groups. *Personnel and Guidance Journal, 62,* 323–326.

LANDRO, L. (2005a, November 16). Bringing surgeons down to earth. *Wall Street Journal,* pp. D1, D4.

LANDRO, L. (2005b). Teaching doctors to be nicer: New accreditation rules spur medical schools to beef up interpersonal-skills training. *Wall Street Journal,* D1, D4.

LANG, P. J. (1995, May). The emotion probe: Studies of motivation and attention. *American Psychologist, 50,* 372–385.

LAU, M. Y. (2005). Ontological and epistemic claims of realism and instrumentalism. *American Psychologist, 60,* 345–346.

LAZARUS, A. A. (1993, Fall). Tailoring the therapeutic relationship, or being an authentic chameleon. *Psychotherapy, 30,* 404–407.

LAZARUS, A. A. (1997). *Brief but comprehensive psychotherapy: The multimodal way.* New York: Springer.

LAZARUS, A. A., Beutler, L. E., & Norcross, J. C. (1992). The future of technical eclecticism. *Psychotherapy, 29,* 11–20.

LAZARUS, R. S. (1999). Hope: An emotion and a vital coping resource against despair. *Social Research, 66,* 653–660.

LAZARUS, R. S. (2000). Toward better research on stressing and coping. *American Psychologist, 55,* 665–673.

LEAHEY, M., & Wallace, E. (1988). Strategic groups: One perspective on integrating strategic and group therapies. *Journal for Specialists in Group Work, 13,* 209–217.

LEAHY, R. L. (1999). Strategic self-limitation. *Journal of Cognitive Psychotherapy, 13,* 275–293.

LEBOW, J. (2002, September/October). Learning to love assessment. *Psychotherapy Networker,* 63–65.

LEBOW, J. (2006, January/February). The verdict is clear; ESTs have an incontestable track record with anxiety. *Psychotherapy Networker,* 77–79.

LEE, C. C. (Ed.). (2006). *Counseling for social justice* (2nd ed.). Alexandria, VA: ACA Foundation.

LEE, D. L., & Axelrod, S. (2005). *Behavior modification.* Austin, TX: Pro-Ed.

LEVANT, R. F. (2005). Evidence-based practice in psychology. *Monitor on Psychology, 5.*

LEVENKRON, S. (1999). *Cutting: Understanding and overcoming self-mutilation.* New York: Norton.

LEVIN, L. S., & Shepherd, I. L. (1974). The role of the therapist in Gestalt therapy. *The Counseling Psychologist, 4,* 27–30.

LEVITT, H. M., & Rennie, D. L. (2004). The act of narrating: Narrating activities and the intentions that guide them. In L. Angus & J. McLeod (Eds.), *The handbook of narrative and psychotherapy: Practice, theory and research* (pp. 299–314). Thousand Oaks, CA: Sage.

LEWIN, K. (1969). Quasi-stationary social equilibria and the problem of permanent change. In W. G. Bennis, K. D. Benne, & R. Chin (Eds.), *The planning of change.* New York: Holt, Rinehart & Winston.

LIGHTSEY, O. R., Jr. (1996, October). What leads to wellness? The role of psychological resources in well-being. *The Counseling Psychologist, 24,* 589–735.

LIN, Y. (2002). Taiwanese university students' perspectives on helping. *Counseling Psychology Quarterly, 15,* 47–58.

LINLEY, P. A., & Joseph, S. (2005). The human capacity for growth through adversity. *American Psychologist, 60,* 262–264.

LITTLE, B. R. (2001). Personality psychology: Havings, doings, and beings in context. Retrieved March 26, 2006, from http://www.brianrlittle.com/articles/

LITZ, B. T. (2005). Has resilience to severe trauma been underestimated? *American Psychologist, 60,* 262.

LLEWELYN, S. P. (1988). Psychological therapy as viewed by clients and therapists. *British Journal of Clinical Psychology, 27,* 223–237.

LOCKE, E. A., & Latham, G. P. (1984). *Goal setting: A motivational technique that works.* Englewood Cliffs, NJ: Prentice Hall.

LOCKE, E. A., & Latham, G. P. (1990). *A theory of goal setting and task performance.* Englewood Cliffs, NJ: Prentice Hall.

LOCKE, E. A., & Latham, G. P. (2002). Building a practically useful theory of goal setting and task motivation: A 35-year odyssey. *American Psychologist, 57,* 705–717.

LOCKWOOD, P., & Kunda, Z. (1999). Increasing the salience of one's best selves can undermine inspiration by outstanding role models. *Journal of Personality and Social Psychology, 76,* 214–228.

LOPEZ, F. G., Lent, R. W., Brown, S. D., & Gore, P. A. (1997). Role of social-cognitive expectations in high school students' mathematics-related interest and performance. *Journal of Counseling Psychology, 44,* 44–52.

LOPEZ, S. J., & Snyder, C. R. (2003). *Positive psychological assessment: A handbook of models and measures.* Washington, DC: American Psychological Association.

LOWE, M. R. (2003). Dieting: False hope or falsely accused? *American Psychologist, 58,* 819–820.

LOWENSTEIN, L. (1993). Treatment through traumatic confrontation approaches: The story of S. *Education Today, 43,* 198–201.

LOWMAN, R. L. (Ed.). (1998). *The ethical practice of psychology in organizations.* Washington, DC: American Psychology Association.

LUBORSKY, L. (1993, Fall). The promise of new psychosocial treatments or the inevitability of nonsignificant differences—A poll of the experts. *Psychotherapy and Rehabilitation Research Bulletin,* 6–8.

LUBORSKY, L., Crits-Christoph, P., McLellan, A. T., Woody, G., Piper, W., Liberman, B., Imber, S., & Pilkonis, P. (1986). The nonspecific hypothesis of therapeutic effectiveness: A current assessment. *American Journal of Orthopsychiatry, 56,* 501–512.

LUNDERVOLD, D. A., & Belwood, D. A. (2000, Winter). The best kept secret in counseling: Single-case experimental designs. *Journal of Counseling and Development, 78,* 92–102.

LYND, H. M. (1958). *On shame and the search for identity.* New York: Science Editions.

MACHADO, P. P. P., Beutler, L. E., & Greenberg, L. S. (1999). Emotion recognition in psychotherapy: Impact of therapist level of experience and emotional awareness. *Journal of Clinical Psychology, 55,* 39–57.

MADDI, S. R. (2005). On hardiness and other pathways to resilience. *American Psychologist, 60,* 261–262.

MADDUX, J. E. (Ed.). (1995). *Self-efficacy, adaptation, and adjustment: Theory, research, and application.* New York: Plenum.

MAGER, R. F. (1992, April). No self-efficacy, no performance. *Training,* 32–36.

MAGNAVITA, J. J. (2005). *Personality-guided relational psychotherapy: A unified approach.* Washington, DC: American Psychological Association.

MAHALIK, J. R. (1994). Development of the client resistance scale. *Journal of Counseling Psychology, 41,* 58–68.

MAHRER, A. R. (1993, Fall). The experiential relationship: Is it all-purpose or is it tailored to the individual client? *Psychotherapy, 30,* 413–416.

MAHRER, A. R., Gagnon, R., Fairweather, D. R., Boulet, D. B., & Herring, C. B. (1994). Client commitment and resolve to carry out postsession behaviors. *Journal of Counseling Psychology, 41,* 407–414.

MALLINCKRODT, B. (1996). Change in working alliance, social support, and psychological symptoms in brief therapy. *Journal of Counseling Psychology, 43,* 448–455.

MANCILLAS, A. (2005, October). Empathic invalidations. *Counseling Today,* pp. 9, 19.

MANTHEI, R. (1997). *Counseling: The skills of finding solutions to problems.* New York: Routledge.

MARCH, J. G. (1994). *A primer on decision making: How decisions happen.* New York: The Free Press.

MARKUS, H., & Nurius, P. (1986). Possible selves. *American Psychologist, 41,* 954–969.

MARSH, D., & Johnson, D. (1997). The family experience of mental illness: Implications for intervention. *Professional Psychology: Research and Practice, 28,* 229–237.

MARTIN, J. (1994). *The construction and understanding of psychotherapeutic change.* New York: Teachers College Press.

MARTIN, S. (2002, July–August). Don't think thin, think realistic. *Monitor on Psychology,* 52–54.

MASH, E. J., & Hunsley, J. (1993). Assessment considerations in the identification of failing psychotherapy: Bringing the negatives out of the darkroom. *Psychological Assessment, 5,* 292–301.

MASLACH, C., & Jackson, S. E. (1981). *Maslach Burnout Inventory* (2nd ed.). Palo Alto, CA: Consulting Psychologists Press.

MASLOW, A. H. (1968). *Toward a psychology of being* (2nd ed.). New York: Van Nostrand Reinhold.

MASSIMINI, F., & Delle Fave, A. (2000, January). Individual development in a biocultural perspective. *American Psychologist, 55,* 24–33.

MASSON, J. F. (1988). *Against therapy: Emotional tyranny and the myth of psychological healing.* New York: Atheneum.

MATHEWS, B. (1988). The role of therapist self-disclosure in psychotherapy: A survey of therapists. *American Journal of Psychotherapy, 42,* 521–531.

MATT, G. E., & Navarro, A. M. (1997). What meta-analyses have and have not taught us about psychotherapy effects: A review and future directions. *Clinical Psychology Review, 17,* 1–32.

MAYER, J. D., & Salovey, P. (1997). What is emotional intelligence? In P. Salovey & D. Sluyter (Eds.), *Emotional development and emotional intelligence: Educational implications* (pp. 3–31). New York: Basic Books.

McAULEY, E., Mihalko, S. L., & Bane, S. M. (1997). Exercise and self-esteem in middle-aged adults: Multidimensional relationships and physical-fitness and self-efficacy influences. *Journal of Behavioral Medicine, 20,* 67–83.

McAULIFFE, G. J., & Eriksen, K. P. (1999). Toward a consructivist and developmental identity for the counseling profession: The context-phase-stage-style model. *Journal of Counseling and Development, 77,* 267–279.

McCARTHY, W. C., & Frieze, I. H. (1999, Spring). *Journal of Social Issues, 55,* 33–50.

McCRAE, R. R., & Costa, P. T., Jr. (1997, May). Personality trait structure as a human universal. *American Psychologist, 52,* 509–516.

McCROSKEY, J. C. (1993). *An introduction to rhetorical communication* (5th ed.). Englewood Cliffs, NJ: Prentice Hall.

McDERMOTT, D., & Snyder, C. R. (1999). *Making hope happen.* Oakland/San Francisco: New Harbinger Press.

McKAY, G. D., & Dinkmeyer, D. (1994). *How you feel is up to you: The power of emotional choice.* San Luis Obispo, CA: Impact.

McKAY, M., Davis, M., & Fanning, P. (1997). *Thoughts and feeling: Taking control of your moods and your life.* Oakland, CA: New Harbinger Publications.

McMILLEN, C., Zuravin, S., & Rideout, G. (1995). Perceived benefit from child sexual abuse. *Journal of Consulting and Clinical Psychology, 63,* 1037–1043.

McNEILL, B., & Stolenberg, C. D. (1989). Reconceptualizing social influence in counseling: The elaboration likelihood model. *Journal of Counseling Psychology, 36,* 24–33.

McWHIRTER, E. H. (1996). *Counseling for empowerment.* Alexandria, VA: American Counseling Association.

McWILLIAMS, N. (2005a). Preserving our humanity as therapists. *Psychotherapy: Therapy, Research, Practice, Training, 42,* 139–151.

McWILLIAMS, N. (2005b). Response to Norcross. *Psychotherapy: Therapy, Research, Practice, Training, 42,* 156–159.

MEHRABIAN, A. (1972). *Nonverbal communication.* Chicago: Aldine-Atherton.

MEHRABIAN, A. (1981). *Silent messages: Implicit communication of emotions and attitudes* (2nd ed.). Belmont, CA: Wadsworth.

MEHRABIAN, A., & Reed, H. (1969). Factors influencing judgments of psychopathology. *Psychological Reports, 24,* 323–330.

MEICHENBAUM, D. (1994). *A clinical handbook/practical therapist manual for assessing and treating adults with post-traumatic stress disorder (PTSD).* Waterloo, Ontario: Institute Press.

MEICHENBAUM, D., & Turk, D. C. (1987). *Facilitating treatment adherence: A practitioner's guidebook.* New York: Plenum Press.

MEIER, S. T., & Letsch, E. A. (2000). What is necessary and sufficient information for outcome assessment? *Professional Psychology: Research and Practice, 31,* 409–411.

MESSER, S. B. (2004). Evidence-based practice: Beyond empirically supported treatments. *Professional Psychology: Research and Practice, 35,* 580–588.

METCALF, L. (1998). *Solution focused group therapy: Ideas for groups in private practice, schools, agencies, and treatment programs.* New York: The Free Press.

MILAN, M. A., Montgomery, R. W., & Rogers, E. C. (1994). Theoretical orientation revolution in clinical psychology: Fact or fiction? *Professional Psychology: Research and Practice, 4,* 398–402.

MILBANK, D. (1996, October 31). Hiring welfare people, hotel chain finds, is tough but rewarding. *Wall Street Journal,* pp. A1, A14.

MILLER, G. A., Galanter, E., & Pribram, K. H. (1960). *Plans and the structure of behavior.* New York: Holt, Rinehart & Winston.

MILLER, L. K. (2006). *Principles of everyday behavior analysis.* Belmont, CA: Wadsworth/Thomson Learning.

MILLER, L. M. (1984). *American spirit: Visions of a new corporate culture.* New York: Morrow.

MILLER, S. D., Hubble, M. A., & Duncan, B. L. (Eds.). (1996). *Handbook of solution-focused brief therapy.* San Francisco: Jossey-Bass.

MILLER, W. C. (1986). *The creative edge: Fostering innovation where you work.* Reading, MA: Addison-Wesley.

MILLER, W. R. (1999). *Integrating spirituality into treatment: Resources for practitioners.* Washington, DC: American Psychological Association.

MILLER, W. R., & Rollnick, S. (1995). What is motivational interviewing? *Behavior and cognitive Psychotherapy, 23,* 325–334.

MILLER, W. R., & Rollnick, S. (Eds.). (2002). *Motivational interviewing: Preparing people for change.* New York: Guilford Press.

MILLER, W. R., & Rollnick, S. (2004). Talking oneself into change: Motivational interviewing, stages of change, and therapeutic process. *Journal of Cognitive Psychotherapy, 18,* 299–308.

MILLER, W. R., & Thoresen, C. E. (2003). Spirituality, religion, and health: An emerging research field. *American Psychologist, 58,* 24–35.

MILTENBERGER, R. G. (2001). *Behavior modification: Principles and procedures* (2nd ed.). Belmont, CA: Wadsworth/Thomson Learning.

MILTON, J. (2002). *The Road to Malpsychia.* New York: Encounter.

MOHR, D. C. (1995, Spring). Negative outcome in psychotherapy: A critical review. *Clinical Psychology: Science and Practice, 2,* 1–27.

MOLINARI, V., Karel, M., Jones, S., Zeiss, A., Cooley, S. G., Wray, L., Brown, E., & Gallagher-Thompson, D. (2003). Recommendations about the knowledge and skills required of psychologists working with other adults. *Professional Psychology: Research and Practice, 34,* 435–443.

MORRIS, T. (1994). *True success: A new philosophy of excellence.* New York: Grosset/Putman.

MULTON, K. D., Brown, S. D., & Lent, R. W. (1991). Relation of self-efficacy beliefs to academic outcomes: A meta-analytic investigation. *Journal of Counseling Psychology, 38,* 30–38.

MURPHY, J. J. (1997). *Solution-focused counseling in middle and high schools.* Alexandria, VA: American Counseling Association.

MYER, R. A., & Moore, H. B. (2006). Crisis in context theory: An ecological model. *Journal of Counseling and Development, 84,* 139–147.

NAJAVITZ, L. M., Ghinassi, F., Van Horn, A., Weiss, R. D., Siqueland, L., Frank, A., Thase, M. E., Luborsky, L., & Simon-Onken, L. (2004). Therapist satisfaction with four manual-based treatments on a national multisite trial: An exploratory study. *Psychotherapy: Theory, Research, Practice, Training, 41,* 26–37.

NAJAVITS, L. M., Weiss, R. D., Shaw, S. R., & Dierberger, A. E. (2000). Psychotherapists views of treatment manuals. *Professional Psychology: Research and Practice, 31,* 404–408.

NATHAN, P. E. (1998, March). Practice guidelines: Not yet ideal. *American Psychologist, 53,* 290–299.

NATHAN, P. E., & Gorman, J. M. (2002). *A guide to treatments that work.* New York: Oxford University Press.

NELSON-JONES, R. (2005). *Practical counseling & helping skills* (5th ed.). Thousand Oaks, CA: Sage Publications.

NEUBAUER, A. C., & Freudenthaler, H. H. (2005). Models of emotional intelligence. In R. Schulze & R. D. Roberts (Eds.), *Emotional intelligence: An international handbook.* (pp. 32–50). Cambridge, MA: Hogrefe.

NEWMAN, B. M., & Newman, P. R. (2006). *Development through life: A psychosocial approach* (9th ed.). Belmont, CA: Wadsworth.

NEWMAN, C. F., & Strauss, J. L. (2003). When clients are untruthful: Implications for the therapeutic alliance, case conceptualization, and intervention. *Journal of Cognitive Psychotherapy, 17,* 241–252.

NEWMAN, R. (2005). APA's resilience initiative. *Professional Psychology: Research and Practice, 36,* 227–229.

NEZU, A. M., Nezu, C. M., Friedman, S. H., Faddis, S., & Houts, P. S. (1998). *Helping cancer patients cope. A problem-solving approach.* Washington, DC: American Psychological Association.

NICHOLSON, B., Anderson, M., Fox, R., & Brenner, V. (2002). One family at a time: A prevention program for at-risk parents. *Journal of Counseling and Development, 80,* 362–371.

NIELSEN, S. L., Smart, D. W., Isakson, R. L., Worthen, V. E., Gregersen, A. T., & Lambert, M. J. (2004). The Consumer Reports effectiveness score: What did consumers report? *Journal of Counseling Psychology, 51,* 25–37.

NORCROSS, J. C. (2002). *Psychotherapy relationships that work: Therapists contributions and responsiveness to patients.* Washington, DC: American Psychological Association.

NORCROSS, J. C., Beutler, L. E., & Levant, R. E. (2005). *Evidence-based practices in mental health: Debate and dialogue on the fundamental questions.* Washington, DC: American Psychological Association.

NORCROSS, J. C., & Farber, B. A. (2005). Choosing psychotherapy as a career: Beyond "I want to help people." *Journal of Clinical Psychology/In Session, 61,* 939–943.

NORCROSS, J. C., & Hill, C. E. (2005). Compendium of empirically supported therapy relationships. In G. P. Koocher, J. C. Norcross, & S. S. Hill III (Eds.), *Psychologists' Desk Reference* (2nd ed.; pp. 202–208). New York: Oxford University Press.

NORCROSS, J. C., & Kobayashi, M. (1999). Treating anger in psychotherapy: Introduction and cases. *Journal of Clinical Psychology: In Session, 55,* 275–282.

NORCROSS, J. C., Santrock, J. W., Zuckerman, E. L., Campbell, L. F., Smith, T. F., & Sommer, R. (2003). *Authoritative guide to self-help resources in mental health*. New York: Guilford Press.

NORCROSS, J. C., & Wogan, M. (1987). Values in psychotherapy: A survey of practitioners' beliefs. *Professional Psychology: Research and Practice, 18*, 5–7.

NORENZAYAN, A., & Heine, S. J. (2005). Psychological universals: What are they and how can we know? *Psychological Bulletin, 131*, 763–784.

NORTON, R. (1983). *Communicator style: Implicit communication of emotions and attitudes*. Beverly Hills, CA: Sage.

O'CONNELL, B. (2005). *Solution-focused therapy* (2nd ed.). Thousand Oaks, CA: Sage.

O'CONNELL, B., & Palmer, S. (2003). *Handbook of solution-focused therapy*. Thousand Oaks, CA: Sage.

O'DONOHUE, W., & Levensky, E. (Eds.). (2004). *Handbook of forensic psychology*. San Diego: Academic Press.

O'DONOVAN, A., Bain, J. D., & Dyck, M. J. (2005). Does clinical psychology education enhance the clinical competence of practitioners? *Professional Psychology: Research and Practice, 36*, 104–111.

O'HANLON, W. H., & Weiner-Davis, M. (1989). *In search of solutions: A new direction in psychotherapy*. New York: Norton.

O'LEARY, A. (1985). Self-efficacy and health. *Behavior Research and Therapy, 23*, 437–451.

OMER, H. (2000). Troubles in the therapeutic relationship: A pluralistic perspective. *Journal of clinical psychology, 56*, 201–210.

O'NEIL, J. M. (2004). Response to Heppner, Witty, and Dixon: Inspiring and high-level scholarship that can change people's lives. *The Counseling Psychologist, 32*, 439–449.

ORLINSKY, D. E., & Howard, K. I. (1987, Spring). A generic model of psychotherapy. *Journal of Integrative and Eclectic Psychotherapy, 6*, 6–27.

ORLINSKY, D. E., & Ronnestad, M. H. (2005). *How psychotherapists develop: A study of therapeutic work and professional growth*. Washington, DC: American Psychological Association.

OTANI, A. (1989). Client resistance in counseling: Its theoretical rationale and taxonomic classification. *Journal of Counseling and Development, 67*, 458–461.

PADGETT, D. K. (2004). *The qualitative research experience*. Pacific Grove, CA: Thomson Learning.

PARÉ, D., & Lysack, M. (2004). The willow and the oak: From monologue to dialogue in the scaffolding of therapeutic conversations. *Journal of Systemic Therapies, 23*, 6–20.

PATTERSON, C. H. (1985). *The therapeutic relationship: Foundations for an eclectic psychotherapy*. Pacific Grove, CA: Brooks/Cole.

PAUL, G. L. (1967). Strategy of outcome research in psychotherapy. *Journal of Consulting Psychology, 31*, 109–118.

PAULHUS, D. L., Fridhandler, B., & Hayes, S. (1997). Psychological defense: Contemporary theory and research. In J. Johnson, R. Hogan, & S. R. Briggs (Eds.), *Handbook of personality* (pp. 544–580). New York: Academic Press.

PAYNE, E. C., Robbins, S. B., & Dougherty, L. (1991). Goal directedness and older-adult adjustment. *Journal of Counseling Psychology, 38*, 302–308.

PEKARIK, G., & Guidry, L. L. (1999). Relationship of satisfaction to symptom change, follow-up adjustment, and clinical significance in private practice. *Professional Psychology: Research and Practice, 5*, 474–478.

PEKARIK, G., & Wolff, C. B. (1996). Relationship of satisfaction to symptom change, follow-up adjustment, and clinical significance. *Professional Psychology: Research and Practice, 27*, 202–208.

PENNEBAKER, J. W. (1995). Emotion, disclosure, and health: An overview. In J. W. Pennebaker (Ed.), *Emotion, disclosure, and health* (pp. 3–10). Washington, DC: American Psychological Association.

PERSONS, J. B. (1991, February). Psychotherapy outcome studies do not accurately represent current models of psychotherapy: A proposed remedy. *American Psychologist, 46*, 99–106.

PERVIN, L. A. (1996). *The science of personality*. New York: Wiley.

PETERSON, C., Maier, S. F., & Seligman, M. E. P. (1995). *Learned helplessness: A theory for the age of personal control*. New York: Oxford University Press.

PETERSON, C., & Seligman, M. E. P. (2004). *Character strengths and virtues: A classification handbook*. New York: Oxford University Press/ Washington, DC: American Psychological Association.

PETERSON, C., Seligman, M. E. P., & Vaillant, G. E. (1988). Pessimistic explanatory style as a risk factor for physical illness: A thirty-five-year longitudinal study. *Journal of Personality and Social Psychology, 55*, 23–27.

PETERSON, D. R. (2003). Unintended consequences: Ventures and misadventure in the education of professional psychologists. *American Psychologist, 58,* 791–800.

PETERSON, Z. D. (2002). More than a mirror: The ethics of therapist self-disclosure. *Psychotherapy: Theory, Research, Practice, Training, 39,* 21–31.

PFEFFER, J., & Sutton, R. I. (2006, January). Evidence-based management. *Harvard Business Review,* 63–74.

PFEIFFER, A. M., Whelan, J. P., & Martin, J. M. (2000). Decision-making bias in psychotherapy: Effects of hypothesis source and accountability. *Journal of Counseling Psychology, 47,* 429–436.

PHILLIPS, A., & Daniluk, J. C. (2004). Beyond "survivor": How childhood sexual abuse informs the identity of adult women at the end of the therapeutic process. *Journal of Counseling & Development, 82,* 177–184.

PHILLIPS, E. L. (1987). The ubiquitous decay curve: Service delivery similarities in psychotherapy, medicine, and addiction. *Professional Psychology: Research and Practice, 18,* 650–652.

PHILLIPS, M. M. (2005, December 8). Two recent wars give widows' group a new call to duty. *Wall Street Journal,* A1, A6.

PINSOF, W. M. (1995). *Integrative problem-centered therapy. A synthesis of family, individual, and biological therapies.* New York: Basic Books.

PLOUS, S. (1993). *The psychology of judgment and decision making.* New York: McGraw-Hill.

PLUTCHIK, R. (2001). *Emotions in the practice of psychotherapy: Clinical implications affect theories.* Washington, DC: American Psychological Association.

PLUTCHIK, R. (2003). *Emotions and life: Perspectives from psychology, biology, and evolution.* Washington, DC: American Psychological Association.

POLIVY, J., & Herman, C. P. (2002). If at first you don't succeed: False hopes of self-change. *American Psychologist, 57,* 677–689.

POLLARD, C. A., & Zuercher-White, E. (2003). *The Agoraphobia Workbook: A comprehensive program to end your fear of symptom attacks.* Oakland, CA: New Harbinger.

PONTEROTTO, J. G., Fuertes, J. N., & Chen, E. C. (2000). Models of multicultural counseling. In S. D. Brown & R. W. Lent (Eds.), *Handbook of counseling psychology* (3rd ed.; pp. 639–669). New York: Wiley.

POPE, K. S., Sonne, J. L., & Greene, B. (2006). *What therapists don't talk about and why: Understanding taboos that hurt us and our clients.* Washington, DC: American Psychological Association.

POPE, K. S., & Vasquez, M. J. T. (2005). *How to survive and thrive as a therapist: Information, ideas, and resources for psychologists in practice.* Washington, DC: American Psychological Association.

POPE, M. (2006). *Professional counseling 101: Building a strong professional identity.* Alexandria, VA: American Counseling Association.

PRESTON, J. (1998). *Integrative brief therapy: Cognitive, psychodynamic, humanistic & neurobehavioral approaches.* San Luis Obispo, CA: Impact Publishers.

PROCHASKA, J. O., & Norcross, J. C. (1994). *Systems of psychotherapy: A transtheoretical analysis* (3rd ed.). Pacific Grove, CA: Brooks/Cole.

PROCHASKA, J. O., & Norcross, J. C. (2002). *Systems of psychotherapy: A transtheoretical analysis.* (5th ed.). Pacific Grove, CA: Brooks/Cole.

PROCHASKA, J. O., Norcross, J. C., & DiClemente, C. C. (2005). Stages of change: Prescriptive guidelines. In G. P. Koocher, J. C. Norcross, & S. S. Hill III (Eds.), *Psychologists' desk reference* (2nd ed.; pp. 226–231). New York: Oxford University Press.

PROCTOR, E. K., & Rosen, A. (1983). Structure in therapy: A conceptual analysis. *Psychotherapy: Theory, Research, and Practice, 20,* 202–207.

PUTNAM, R. D. (2000). *Bowling alone.* New York: Simon & Schuster.

PYSZCZYNSKI, T., & Greenberg, J. (1987). Self-regulatory preservation and the depressive self-focusing style: A self-awareness theory of depression. *Psychological Bulletin, 102,* 122–138.

QUALLS, S. H., & Abeles, N. (2000). *Psychology and the aging revolution: How we adapt to long life.* Washington, DC: American Psychological Association.

RAMEY, C. H., & Chrysikou, E. (2005). The scientific denial of the real and the dialectic of scientism and humanism. *American Psychologist, 60,* 346–347.

RASMUSSEN, P. R. (2005). *Personality-guided cognitive-behavioral therapy.* Washington, DC: American Psychological Association.

RATNER, H. (1998). Solution-focused brief therapy: From hierarchy to collaboration. In Bayne, R., Nicholson, P., & Horton, I. (Eds.), *Counseling and Communication Skills for Medical and Health Practitioners.* London: BPS Books.

RAUP, J. L., & Myers, J. E. (1989). The empty nest syndrome: Myth or reality? *Journal of Counseling & Development, 68,* 180–183.

READ, J. P., & Brown, R. A. (2003). The role of physical exercise in alcoholism treatment and recovery. *Professional Psychology: Research and Practice, 34,* 49–56.

REANDEAU, S. G., & Wampold, B. E. (1991). Relationship of power and involvement to working alliance: A multiple-case sequential analysis of brief therapy. *Journal of Counseling Psychology, 38,* 107–114.

RENNIE, D. L. (1994). Clients' deference in psychotherapy. *Journal of Counseling Psychology, 41,* 427–437.

RICHARDS, P. S., & Bergin, A. E. (2005). *A spiritual strategy for counseling and psychotherapy* (2nd ed.). Washington, DC: American Psychological Association.

RICHMOND, V. P., & McCroskey, J C. (2000). *Nonverbal Behavior in Interpersonal Relations.* (4th ed.). Needham Heights, MA: Allyn & Bacon.

RIGGIO, R. E., & Feldman, R. S. (2005). *Applications of nonverbal communication.* Mahwah, NJ: Erlbaum.

ROBERTS, B. W., Walton, K. E., & Bogg, T. (2005). Conscientiousness and health across the life course. *Review of General Psychology, 9,* 156–168.

ROBERTSHAW, J. E., Mecca, S. J., & Rerick, M. N. (1978). *Problem-solving: A systems approach.* New York: Petrocelli Books.

ROBINS, R. W., Gosling, S. D., & Craik, K. H. (1999, February). An empirical analysis of trends in psychology. *American Psychologist, 54,* 117–128.

ROBINSON, B. E., & Bacon, J. G. (1996). The "If only I were thin . . ." treatment program: Decreasing the stigmatizing effects of fatness. *Professional Psychology: Research and Practice, 27,* 175–183.

ROBINSON, B. S., Davis, K. L., & Meara, N. M. (2003). Motivational attributes of occupational possible selves for low-income rural women. *Journal of Counseling Psychology, 50,* 156–164.

ROFFMAN, A. E. (2004, Summer). Is anger a thing-to-be-managed? *Psychotherapy: Theory, Research, Practice, Training, 41,* 161–171.

ROGERS, C. R. (1951). *Client-centered therapy.* Boston: Houghton Mifflin.

ROGERS, C. R. (1957). The necessary and sufficient conditions of therapeutic personality change. *Journal of Consulting Psychology, 21,* 95–103.

ROGERS, C. R. (1965). *Client-centered therapy: Its current practice, implications and theory.* Boston: Houghton, Mifflin.

ROGERS, C. R. (1975). Empathic—An unappreciated way of being. *The Counseling Psychologist 5*(2), 2–10.

ROGERS, C. R. (1980). *A way of being.* Boston: Houghton Mifflin.

ROGERS, C. R., Perls, F., & Ellis, A. (1965). *Three approaches to psychotherapy 1* [Film]. Orange, CA: Psychological Films, Inc.

ROGERS, C. R., Shostrom, E., & Lazarus, A. (1977). *Three approaches to psychotherapy 2* [Film]. Orange, CA: Psychological Films, Inc.

ROISMAN, G. I. (2005). Conceptual clarifications in the study of resilience. *American Psychologist, 60,* 264–265.

ROLLINS, J. (2005). A campaign for counselor wellness. *Counseling Today,* 22–23, 42.

ROLLNICK S., & Miller, W. R. (1995). What is motivational interviewing? *Behavioural and Cognitive Psychotherapy, 23,* 325–334.

ROSEN, S., & Tesser, A. (1970). On the reluctance to communicate undesirable information: The MUM effect. *Sociometry, 33,* 253–263.

ROSEN, S., & Tesser, A. (1971). Fear of negative evaluation and the reluctance to transmit bad news. Proceedings of the 79th Annual Convention of the American Psychological Association, 6, 301–302.

ROSENTHAL, R. (2005). *The era of choice: The ability to choose and its transformation of contemporary life.* Cambridge, MA: MIT Press.

ROTH, A., & Fonagy, P. (2004). *What works for whom? A critical review of psychotherapy research* (2nd ed.). New York: Guilford Publications.

ROTHBAUM, B. O. (2006). *Pathological anxiety: Emotional processing in etiology and treatment.* New York: Guilford.

RUPERT, P. A., & Morgan, D. J. (2005). Work setting an burnout among professional psychologists. *Professional Psychology: Research and Practice, 36,* 544–550.

RUSCIO, J. (2006). *Critical thinking in psychology: Separating sense from nonsense.* Belmont, CA: Wadsworth/Thomson Learning.

RUSSELL, J. A. (1995). Facial expressions of emotion: What lies beyond minimal universality? *Psychological Bulletin, 118,* 379–391.

RUSSELL, J. A., Fernandez-Dols, J.-M., & Mandler, G. (Eds.). (1997). *The psychology of facial expression.* New York: Cambridge University Press.

SALES, B. D., Miller, M. O., & Hall, S. R. (2005). *Laws affecting clinical practice.* Washington, DC: American Psychological Association.

SALOVEY, P., Brackett, M. A., & Mayer, J. D. (2004). *Emotional intelligence: Key readings on the Mayer and Salovey model.* Port Chester, NY: Dude Publishing.

SALOVEY, P., & Mayer, J. D. (1990). Emotional intelligence. *Imagination, Cognition, and Personality, 9,* 185–211.

SALOVEY, P., Rothman, A. J., Detweiler, J. B., & Steward, W. T. (2000, January). Emotional states and physical health. *American Psychologist, 55,* 110–121.

SATCHER, D. (2000). Mental health: A report of the surgeon general—executive summary. *Professional Psychology: Research and Practice, 31,* 5–13.

SCHEEL, M., Hanson, W., & Razzhavaikina, T. (2004). The process of recommending homework in psychotherapy: A review of therapist delivery methods, client acceptability, and factors that affect compliance. *Psychotherapy: Theory, Research, Practice, Training, 41,* 38–55.

SCHIFF, J. L. (1975). *Cathexis reader: Transactional analysis treatment of psychosis.* New York: Harper & Row.

SCHMIDT, F. L., & Hunter, J. E. (1993). Tacit knowledge, practical intelligence, general mental ability, and job knowledge. *Current Directions in Psychological Science, 1,* 8–9.

SCHMIDT, F., & Taylor, T. K. (2002). Putting empirically supported treatments into practice. *Professional Psychology: Research and Practice, 33,* 483–489.

SCHOEMAKER, P. J. H., & Russo, J. E. (1990). *Decision traps.* New York: Doubleday.

SCHULZE, R., & Roberts, R. D. (2005). *Emotional intelligence: An international handbook.* Cambridge, MA: Hogreff.

SCHWARTZ, B. (2004). *The paradox of choice: Why more is less.* New York: HarperCollins.

SCHWARTZ, R. S. (1993). Managing closeness in psychotherapy. *Psychotherapy, 30,* 601–607.

SCHWARZER, R. (Ed.). (1992). *Self-efficacy: Thought control of action.* Bristol, PA: Taylor & Francis.

SCHWARZER, R., & Fuchs, R. (1996). Self-efficacy and health behaviors. In M. Conner & P. Norman (Eds.), *Predicting health behavior: Research and practice with social cognition models* (pp. 163–196). Buckingham, UK: Open University Press.

SCHWARZER, R., & Renner, B. (2000). Social-cognitive predictors of health behavior: Action self-efficacy and coping self-efficacy. *Health Psychology, 19,* 487–495.

SCOTT, B. (2000). *Being real: An ongoing decision.* Berkeley, CA: Frog.

SCOTT, N. E., & Borodovsky, L. G. (1990). Effective use of cultural role taking. *Professional Psychology: Research and Practice, 21,* 167–170.

SECUNDA, A. (1999). *The 15 second principle: Short, simple steps to achieving long-term goals.* New York: Berkley.

SEIKKULA, J., & Trimble, D. (2005). Healing elements of therapeutic conversation: Dialogue as an embodiment of love. *Family Process, 44,* 461–473.

SELIGMAN, D. (2003, December 8). It's all the rage. *Forbes,* p. 89.

SELIGMAN, M. (1975). *Helplessness: On depression, development, and death.* San Francisco: Freeman.

SELIGMAN, M. (1991). *Learned optimism.* New York: Knopf.

SELIGMAN, M. (1995). The effectiveness of psychotherapy: The Consumer Reports study. *American Psychologist, 50,* 965–974.

SELIGMAN, M., & Csikszentmihalyi, M. (2000). Positive psychology: An introduction. *American Psychologist, 55,* 5–14.

SELIGMAN, M. P. (1998). Positive psychology network concept paper. Retrieved November 20, 2005 from http://www.psyche.upenn.edu/seligman/ppgrant.htm.

SELIGMAN, M. P. (2004). *Authentic happiness: Using the new positive psychology to realize your potential for lasting fulfillment.* New York: Free Press.

SELIGMAN, M. P., Steen, T. A., Park, N., & Peterson, C. (2005). Positive psychology progress: Empirical validation of interventions. *American Psychologist, 60,* 410–421.

SERVAN-SCHREIBER, D. (2004, July/August). Run for your life. *Psychotherapy Networker,* 47–51, 67.

SEXTON, T. L., & Whiston, S. C. (1994). The status of the counseling relationship: An empirical review, theoretical implications, and research directions. *The Counseling Psychologist, 22,* 6–78.

SHARF, R. S. (1996). *Theories of psychotherapy and counseling: Concepts and cases.* Belmont, CA: Wadsworth.

SHARF, R. S. (2003). *Theories of psychotherapy and counseling* (3rd ed.). Belmont, CA: Wadsworth.

SHEFET, O. M., & Curtis, R. C. (2005). Guidelines for terminating psychotherapy. In G. P. Koocher, J. C. Norcross, & S. S. Hill, III, *Psychologists' desk reference* (2nd ed., pp. 354–359). New York: Oxford University Press.

SHELDON, K. M. (2004). *Optimal human being: An integrated multi-level perspective.* Mahwah, NJ: Erlbaum.

SHORT, D. (2006, January-February). Erickson's legacy. *Psychotherapy Networker,* 71–75.

SIGELMAN, C. K., & Rider, E. A. (2006). *Life-span human development* (5th ed.). Belmont, CA: Wadsworth.

SILVERMAN, D. K. (2005). What works in psychotherapy and how do we know? What evidence-based practice has to offer. *Psychoanalytic Psychology, 22,* 306–312.

SIMON, H. A., Dantzig, G. B., Hogarth, R., Piott, C. R., Raiffa, H., Schelling, T. C., Shepsie, K. A., Thaier, R., Tversky, A., & Winter, S. (1986). *Report of the research briefing panel on decision making and problem solving.* Washington, DC: National Academy Press.

SIMON, J. C. (1988). Criteria for therapist self-disclosure. *American Journal of Psychotherapy, 42,* 404–415.

SIMONTON, D. K., & Baumeister, R. F. (2005). Positive psychology at the summit. *Review of General Psychology, 9,* 99–102.

SKOVHOLT, T. M. (2001). *The resilient practitioner: Burnout prevention and self-care strategies for counselors, therapists, teachers, and health professionals.* Boston: Allyn & Bacon.

SLOAN, D. M., & Marx, B. P. (2004a). A closer examination of the structured written disclosure paradigm. *Journal of Consulting and Clinical Psychology, 72,* 365–175.

SLOAN, D. M., & Marx, B. P. (2004b). Taking pen to hand: Evaluating theories underlying the written disclosure paradigm. *Clinical Psychology: Science and Practice, 11,* 121–137.

SLOAN, D. M., Marx, B. P., & Epstein, E. M. (2005). Further examination of the exposure model underlying the efficacy of written emotional disclosure. *Journal of Consulting and Clinical Psychology, 73,* 549–554.

SMABY, M., & Tamminen, A. W. (1979). Can we help belligerent counselees? *Personnel and Guidance Journal, 57,* 506–512.

SMITH, E. R., & Mackie, D. M. (2000). *Social psychology.* 2nd edition. Philadelphia: Psychology Press.

SMITH, P. L., & Fouad, N. A. (1999). Subject-matter specificity of self-efficacy, outcome expectancies, interests, and goals: Implications for the social-cognitive model. *Journal of Counseling Psychology, 4,* 461–471.

SNYDER, C. R. (1989). Reality negotiation: From excuses to hope and beyond. *Journal of Social and Clinical Psychology, 8,* 130–157.

SNYDER, C. R. (1994). *The psychology of hope: You can get there from here.* New York: The Free Press.

SNYDER, C. R. (1995). Conceptualizing, measuring, and nurturing hope. *Journal of Counseling and Development, 73,* 355–360.

SNYDER, C. R. (1998). A case for hope in pain, loss and suffering. In J. H. Harvey, J. Omarzy, & E. Miller (Eds.), *Perspectives on loss: A sourcebook.* Washington, DC: Taylor & Francis.

SNYDER, C. R. (Ed.). (1999). *Coping: The psychology of what works.* New York: Oxford University Press.

SNYDER, C. R., & Elliott, T. R. (2005, September). Twenty-first century graduate education in clinical psychology: A four level matrix model. *Journal of Clinical Psychology, 61,* 1033–1054.

SNYDER, C. R., Harris, C., Anderson, J. R., Holleran, S. A., Irving, L. M., Sigmon, S. T., Yoshinobu, L., Gibb, J., Langelle, C., & Harney, P. (1991). The will and the ways: Development and validation of an individual differences measure of hope. *Journal of Personality and Social Psychology, 60,* 570–585.

SNYDER, C. R., & Higgins, R. L. (1988). Excuses: Their effective role in the negotiation of reality. *Psychological Bulletin, 104,* 23–35.

SNYDER, C. R., Higgins, R. L., & Stucky, R. J. (1983). *Excuses: Masquerades in search of grace.* New York: Wiley.

SNYDER, C. R., & Lopez, S. J. (Eds.). (2005). *Handbook of positive psychology* (2nd ed.). New York: Oxford University Press.

SNYDER, C. R., & Lopez, S. J. (2006). *Positive psychology: The scientific and practical explorations of human strengths.* Thousand Oaks, CA: Sage.

SNYDER, C. R., McDermott, D., Cook, W., & Rapoff, M. (1997). *Hope for the journey: Helping children through the good times and the bad.* New York: Basic Books.

SNYDER, C. R., Michael, S. T., & Cheavens, J. (1999). Hope as a psychotherapeutic foundation for nonspecific factors, placebos, and expectancies. In Huble, M. A., Duncan, B., & Miller, S. (Eds.), *Heart and soul of change.* Washington, DC: American Psychological Association.

SNYDER, C. R., & Rand, K. L. (2003). The case against false hope. *American Psychologist, 58,* 820–821.

SNYDER, C. R., Sympson, S. C., Ybasco, F. C., Borders, T. F., Babyak, M. A., & Higgins, R. L. (1996). Development and validation of the State Hope Scale. *Journal of Personality and Social Psychology, 2,* 321–335.

SOMBERG, D. R., Stone, G. L., & Claiborn, C. D. (1993). Informed consent: Therapists' beliefs and practices. *Professional Psychology: Research and Practice, 24,* 153–159.

SOMMERS, C. H., & Satel, S. (2005). *One nation under therapy.* New York: St. Martin's.

SOMMERS-FLANAGAN, J., & Sommers-Flanagan, R. (1995). Psychotherapeutic techniques with treatment-resistant adolescents. *Psychotherapy, 32,* 131–140.

SPERRY, L., & Shafranske, E. P. (Eds.). (2005). *Spiritually oriented psychotherapy.* Washington, DC: American Psychological Association.

SREBALUS, D. J., & Brown, D. (2001). *A guide to the helping professions.* Boston: Allyn & Bacon.

STEIN, D. M., & Lambert, M. J. (1995). Graduate training in psychotherapy: Are therapy outcomes enhanced? *Journal of Consulting and Clinical Psychology, 63,* 182–196.

STERNBERG, R. J. (1990). Wisdom and its relations to intelligence and creativity. In R. J. Sternberg (Ed.), *Wisdom: Its nature, origins, and development* (pp. 124–159). New York: Cambridge University Press.

STERNBERG, R. J. (1998). Abilities are forms of developing expertise. *Educational Researcher, 27*(3), 11–20.

STERNBERG, R. J. (Ed.). (2002). *Why smart people can be so stupid.* New Haven: Yale University Press.

STERNBERG, R. J. (2003, March). Responsibility: One of the other three Rs. *Monitor on Psychology,* p. 5.

STERNBERG, R. J., Grigorenko, E. L., & Singer J. L. (Eds.). (2004). *Creativity: From potential to realization.* Washington, DC: American Psychological Association.

STERNBERG R. J., & Lubart, T. I. (1996). Investing in creativity. *American Psychologist, 51,* 677–688.

STERNBERG R. J., Wagner, R. K., Williams, W. M., & Horvath, J. A. (1995, Winter). Testing common sense. *American Psychologist, 50,* 912–927.

STICKER, G., & Fisher, M. (Eds.). (1990). *Self-disclosure in therapeutic relationships.* London: Plenum.

STILES, W. B., Glick, M. J., Osatuke, K., Hardy, G. E., Shapiro, D. A., Agnew-Davies, R., Rees, A., & Barkham, M. (2004). Patterns of alliance development and the rupture-repair hypothesis: Are productive relationships U-shaped or V-shaped? *Journal of Counseling Psychology, 51,* 81–92.

STOTLAND, E. (1969). *The psychology of hope.* San Francisco: Jossey-Bass.

STOUT, C. E., & Hayes, R. A. (Eds.). (2004). *The evidence-based practice: Methods, models, and tools for mental health professionals.* New York: Wiley.

STRAUSS, R., & Goldberg, W. A. (1999). Self and possible selves during the transition to fatherhood. *Journal of Family Psychology, 13,* 244–259.

STRICKER, G., & Fisher, A. (Eds.). (1990). *Self-disclosure in the therapeutic relationship.* New York: Plenum.

STROH, P., & Miller, W. W. (1993, May). HR professionals should thrive on paradox. *Personal Journal,* 132–135.

STRONG, S. R. (1968). Counseling: An interpersonal influence process. *Journal of Counseling Psychology, 15,* 215–224.

STRONG, S. R. (1991). Social influence and change in therapeutic relationships. In C. R. Snyder & D. R. Forsyth (Eds.), *Handbook of social and clinical psychology: The health perspective* (pp. 540–562). New York: Pergamon Press.

STRONG, S. R., & Claiborn, C. D. (1982). *Change through interaction: Social psychological processes of counseling and psychotherapy.* New York: Wiley.

STRONG, S., Yoder, B., & Corcoran, J. (1995). Counseling: A social process for encouraging personal powers. *The Counseling Psychologist, 23,* 374–384.

STRUPP, H. H., Hadley, S. W., & Gomes-Schwartz, B. (1977). *Psychotherapy for better or worse: The problem of negative effects.* New York: Jason Aronson.

SUE, D. W., & Sue, D. (1990). Counseling the culturally different: Theory and practice (2nd ed.). New York: Wiley.

SULLIVAN, T., Martin, W., Jr., & Handelsman, M. (1993). Practical benefits of an informed-consent procedure: An empirical investigation. *Professional Psychology: Research and Practice, 24,* 160–163.

SUNSTEIN, C. (2005). *Laws of fear: Beyond the cautionary principle.* Cambridge, MA: Cambridge University Press.

SYKES, C. J. (1992). *A nation of victims.* New York: St. Martin's Press.

TAN, S. (1997). The role of psychologist in paraprofessional helping. *Professional Psychology: Research and Practice, 28,* 368–372.

TASHAKKORI, A., & Teddlie, C. (Eds.). (2003). *Handbook of the mixed methods in the behavioral and social sciences.* Thousand Oaks, CA: Sage.

TAUSSIG, I. M. (1987). Comparative responses of Mexican Americans and Anglo-Americans to early goal setting in a public mental health clinic. *Journal of Counseling Psychology, 34,* 214–217.

TAYLOR, S. E., Kemeny, M. E., Reed, G. M., Bower, J. E., & Gruenewald, T. L. (2000, January). Psychological resources: Positive illusions and health. *American Psychologist, 55.* 99–109.

TAYLOR, S. E., Pham, L. B., Rivkin, I. D., & Armor, D. A. (1998). Harnessing the imagination: Mental stimulation, self-regulation, and coping. *American Psychologist, 53,* 429–439.

TEDESCHI, R. G., & Kilmer, R. P. (2005). Assessing strengths, resilience, and growth to guide clinical interventions. *Professional Psychology: Research and Practice, 36,* 230–237.

TESSER, A., & Rosen, S. (1972). Similarity of objective fate as a determinant of the reluctance to transmit unpleasant information: The MUM effect. *Journal of Personality and Social Psychology, 23,* 46–53.

TESSER, A., Rosen, S., & Batchelor, T. (1972). On the reluctance to communicate bad news (the MUM effect): A role play extension. *Journal of Personality, 40,* 88–103.

TESSER, A., Rosen, S., & Tesser, M. (1971). On the reluctance to communicate undesirable messages (the MUM effect): A field study. *Psychological Reports, 29,* 651–654.

TEYBER, E. (2005). *Interpersonal process in therapy: An integrative model* (5th ed.) Belmont, CA: Wadsworth/Thomson Learning.

TICKELL, C. (2005). A review of books in human evolution and where the human race seems to be headed. *Financial Times,* p. W5.

TIEDENS, L. Z., & Leach, C. W. (2004). *The social life of emotions.* New York: Cambridge University Press.

TKACHUK, G. A., & Martin, G. L. (1999, June). Exercise therapy for patients with psychiatric disorders: Research and clinical implications. *Professional Psychology: Research and Practice, 30,* 275–282.

TOMPKINS, M. A. (2005). Six steps to improve psychotherapy homework compliance. In G. P. Koocher, J. C. Norcross, & S.S. Hill, III (Eds.), *Psychologists' desk reference* (2nd ed., pp. 319–324). New York: Oxford University Press.

TRACEY, T. J. (1991). The structure of control and influence in counseling and psychotherapy: A comparison of several definitions and measures. *Journal of Counseling Psychology, 38,* 265–278.

TRACY, B. (2003). *Goals! How to get everything you want, faster than you ever thought possible.* San Francisco, CA: Berrett-Koehler.

TREVINO, J. G. (1996, April). Worldview and change in cross-cultural counseling. *Counseling Psychologist, 24,* 198–215.

TUGADE, M. M., & Fredrickson, B. L. (2004). Resilient individuals use positive emotions to bounce back from negative emotional experiences. *Journal of Personality and Social Psychology, 86,* 320–333.

TURNER, V. J. (2002). *Secret scars: Uncovering and understanding the addiction of self-injury.* Center City, MN: Hazelden.

TYLER, F. B., Pargament, K. I., & Gatz, M. (1983). The resource collaborator role: A model for interactions involving psychologists. *American Psychologist, 38,* 388–398.

VACHON, D. O., & Agresti, A. A. (1992). A training proposal to help mental health professionals clarify and manage implicit values in the counseling process. *Professional Psychology: Research and Practice, 23,* 509–514.

VAILLANT, G. E. (1994). Ego mechanisms of defense and personality psychopathology. *Journal of Abnormal Psychology, 103,* 44–50.

VAILLANT, G. E. (2000, January). Adaptive mental mechanisms: Their role in a positive psychology. *American Psychologist, 55,* 89–98.

VALSINER, J. (1986). Where is the individual subject in scientific psychology. In J. Valsiner (Ed.), *The individual subject and scientific psychology* (pp. 1–16). New York: Plenum.

VOGEL, L., & Wester, S. R. (2003). To seek help or not to seek help: The risks of self-disclosure. *Journal of Counseling Psychology, 50,* 351–361.

WACHTEL, P. L. (1989, August 6). Isn't insight everything? [Book review]. *New York Times,* p. 18.

WACHTEL, P. L., & Messer, S. B. (Eds.). (1997). Theories of psychotherapy: Origins and evolution. Washington, DC: American Psychology Association.

WAEHLER, C. A., Kalodner, C. R., Wampold, B. E., & Lichtenberg, J. W. (2000, September). Empirically supported treatments (ESTs) in perspective: Implications for counseling training. *The Counseling Psychologist, 28,* 657–671.

WAHLSTEIN, D. (1991). Nonverbal behavior and self-presentation. *Psychological Bulletin, 110,* 587–595.

WALKER, L. E., & Shapiro, D. (2004). *Introduction to forensic psychology.* New York: Springer.

WALTER, J. L., & Peller, J. E. (1992). *Becoming solution-focused in brief therapy.* New York: Brunner/Mazel.

WAMPOLD, B. E. (2003). Bashing positivism and revering a medical model under the guise of evidence. *The Counseling Psychologist, 5,* 539–545.

WAMPOLD, B. E., & Bhati, K. S. (2004). Attending to the omissions: A historical examination of evidence-based practice movements. *Professional Psychology: Research and Practice, 35,* 563–570.

WANG, Y.-W., Davidson, M. M., Yakushko, O. F., Savoy, H. B., Tan, J. A., & Bleier, J. K.

(2003). The scale of ethnocultural empathy: Development, validation, & reliability. *Journal of Counseling Psychology, 50,* 221–234.

WATKINS, C. E., Jr. (1990). The effects of counselor self-disclosure: A research review. *The Counseling Psychologist, 18,* 477–500.

WATSON, D. L., & Tharp, R. G. (1993). *Self-directed behavior* (6th ed.). Pacific Grove, CA: Brooks/Cole.

WATSON, D. L., & Tharp, R. G. (2007). *Self-directed behavior.* Belmont, CA: Wadsworth.

WEICK, K. E. (1979). *The social psychology of organizing* (2nd ed.). Reading, MA: Addison-Wesley.

WEINBERGER, J. (1995, Spring). Common factors aren't so common: The common factors dilemma. *Clinical Psychology: Science and Practice, 2,* 45–69.

WEINER, I. B., & Hess, A. K. (2005). *The handbook of forensic psychology* (3rd ed.) Hoboken, NJ: Wiley.

WEINER, M. F. (1983). *Therapist disclosure: The use of self in psychotherapy* (2nd ed.). Baltimore: University Park Press.

WEINRACH, S. G. (1989). Guidelines for clients of private practitioners: Committing the structure to print. *Journal of Counseling and Development, 67,* 299–300.

WEISZ, J. R., Donenberg, G. R., Han, S. S., & Weiss, B. (1995, Fall). Bridging the gap between laboratory and clinic in child and adolescent psychotherapy. *Journal of Consulting and Clinical Psychology, 63,* 688–701.

WEISZ, J. R., Rothbaum, F. M., & Blackburn, T. C. (1984). Standing out and standing in: The psychology of control in America and Japan. *American Psychologist, 39,* 955–969.

WEISZ, J. R., Sandler, I. N., Durlak, J. A., & Anton, B. S. (2005). Promoting and protecting youth mental health through evidence-based prevention and treatment. *American Psychologist, 60,* 628–648.

WEISZ, J. R., Weersing, V. R., & Henggeler, S. W. (2005). Jousting with straw men: Comment on Westen, Novotny, & Thompson-Brenner. *Psychological Bulletin, 131,* 418–426.

WELLENKAMP, J. (1995). Cultural similarities and differences regarding emotional disclosure: Some examples from Indonesia and the Pacific. In J. W. Pennebaker (Ed.), *Emotion, Disclosure, and Health,* 293–311.

WERTH, J. L., Jr., Kopera-Frye, K., Blevins, D., & Bossick, B. (2003). Older adult representation in the counseling psychology literature. *Counseling Psychologist, 31,* 789–814.

WESSLER, R., Hankin, S., & Stern, J. (2001). *Succeeding with difficult clients: Applications of cognitive appraisal therapy.* New York: Academic Press.

WESTEN, D., Novotny, C. M., & Thompson-Brenner, H. (2004). The empirical status of empirically supported psychotherapies: Assumptions, findings, and reporting in controlled clinical trials. *Psychological Bulletin, 130,* 631–663.

WESTEN, D., Novotny, C. M., & Thompson-Brenner, H. (2005). EBP ≠ EST: Reply to Crits-Christoph et al. (2005) and Weisz et al. (2005). *Psychological Bulletin, 131,* 427–433.

WHEELER, D. D., & Janis, I. L. (1980). *A practical guide for making decisions.* New York: Free Press.

WHISTON, S. C., & Sexton, T. L. (1993, February). An overview of psychotherapy outcome research: Implications for practice. *Professional Psychology: Research and Practice, 24,* 43–51.

WHYTE, G. (1991). Decision failures: Why they occur and how to prevent them. *Academy of Management Executive, 5,* 23–31.

WILLIAMS, R. (1989, January–February). The trusting heart. *Psychology Today,* 36–42.

WILSON, G. T. (1998). Manual-based treatment and clinical practice. *Clinical Psychology: Science & Practice, 5,* 363–375.

WINBORN, B. (1977). Honest labeling and other procedures for the protection of consumers of counseling. *Personnel and Guidance Journal, 56,* 206–209.

WOODSIDE, M. R., & McClam, T. (2006). *An introduction to human services* (5th ed.). Belmont, CA: Wadsworth/Thomson Learning.

WOODY, R. H. (1991). *Quality care in mental health.* San Francisco: Jossey-Bass.

WOSKET, V. (2006). *Egan's skilled helper model: Developments and applications in counselling.* London: Brunner-Routledge.

WRIGHT, R. H., & Cummings, N. A. (Eds.). (2005). *Destructive trends in mental health: The well-intentioned path to harm.* New York: Routledge.

YALOM, V., & Bugental, F. T. (1997). Support in existential-humanistic psychotherapy. *Journal of Psychotherapy Integration, 7,* 119–128.

YANKELOVICH, D. (1992, October 5). How public opinion really works. *Fortune,* pp. 102–108.

YUN, K. A. (1998). Relational closeness and production of excuses in events of failure. *Psychological Reports, 83,* 1059–1066.

ZASTROW, C. (2004). *Introduction to social work and social welfare* (8th ed.). Belmont, CA: Brooks/Cole.

ZIEGLER, P., & Hiller, T. (2001). *Recreating partnership: A solution-oriented, collaborative approach to couples therapy.* New York: Norton.

ZIMMERMAN, B. J. (1995). Self-efficacy and educational development. In A. Bandura (Ed.), *Self-efficacy in changing societies* (pp. 202–231). New York: Cambridge University Press.

ZIMMERMAN, B. J. (1996, April). *Measuring and mismeasuring academic self-efficacy: Dimensions, problems, and misconceptions.* Symposium presented at the meeting of the American Educational Association, New York.

ZIMMERMAN, T. S., Prest, L. A., & Wetzel, B. E. (1997). Solution-focused couples therapy groups: An empirical study. *Journal of Family Therapy, 19,* 125–144.

ZIMRIN, H. (1986). A profile of survival. *Child Abuse and Neglect, 10,* 339–349.

ZUERCHER-WHITE, E. (1998). *An end to panic: Breakthrough techniques for overcoming panic disorder* (2nd ed.). Oakland, CA: New Harbinger.

TO THE OWNER OF THIS BOOK:

I hope that you have found *The Skilled Helper: A Problem-Management and Opportunity-Development Approach to Helping,* Eighth Edition, useful. So that this book can be improved in a future edition, would you take the time to complete this sheet and return it? Thank you.

School and address: _____

_____

Department: _____

Instructor's name: _____

1. What I like most about this book is:_____

_____

_____

2. What I like least about this book is:_____

_____

_____

3. My general reaction to this book is:_____

_____

_____

4. The name of the course in which I used this book is:_____

_____

5. Were all of the chapters of the book assigned for you to read? _____

   If not, which ones weren't?_____

6. In the space below, or on a separate sheet of paper, please write specific suggestions for improving this book and anything else you'd care to share about your experience in using this book.

_____

_____

FOLD HERE

THOMSON

BROOKS/COLE ™

BUSINESS REPLY MAIL
FIRST-CLASS MAIL   PERMIT NO. 34   BELMONT CA

POSTAGE WILL BE PAID BY ADDRESSEE

Attn: Marquita Flemming
          Senior Acquisitions Editor, Counseling

BrooksCole/Thomson Learning
10 Davis Drive
Belmont CA 94002-9801

FOLD HERE

OPTIONAL:

Your name:_____ Date: _____

May we quote you, either in promotion for *The Skilled Helper,* Eighth Edition, or in future publishing ventures?

Yes: _____ No: _____

Sincerely yours,

Gerard Egan